CompTIA Convergence+™
Certification Study Guide

(Exam CTO-101)

Tom Carpenter

New York   Chicago   San Francisco   Lisbon   London   Madrid
Mexico City   Milan   New Delhi   San Juan   Seoul   Singapore   Sydney   Toronto

The McGraw·Hill Companies

Cataloging-in-Publication Data is on file with the Library of Congress

McGraw-Hill books are available at special quantity discounts to use as premiums and sales promotions, or for use in corporate training programs. To contact a special sales representative, please visit the Contact Us page at www.mhprofessional.com.

CompTIA Convergence+™ Certification Study Guide (Exam CTO-101)

1 2 3 4 5 6 7 8 9 0   FGR FGR   0 1 9 8

ISBN-13: Book P/N 978-0-07-159682-4 and CD P/N 978-0-07-159683-1
of set 978-0-07-159680-0

ISBN-10: Book P/N 0-07-159682-8 and CD P/N 0-07-159683-6
of set 0-07-159680-1

Sponsoring Editor Tim Green	**Technical Editor** Shan Nichols	**Composition** International Typesetting and Composition
Editorial Supervisor Jody McKenzie	**Copy Editor** Robert Campbell	**Illustration** International Typesetting and Composition
Project Manager Vibha Bhatt, International Typesetting and Composition	**Proofreader** Carol Shields	**Art Director, Cover** Jeff Weeks
Acquisitions Coordinator Jennifer Housh	**Indexer** Kevin Broccoli	
	Production Supervisor Jim Kussow	

CompTIA Authorized Quality Curriculum

The logo of the CompTIA Authorized Quality Curriculum (CAQC) program and the status of this or other training material as "Authorized" under the CompTIA Authorized Quality Curriculum program signifies that, in CompTIA's opinion, such training material covers the content of CompTIA's related certification exam.

The contents of this training material were created for the CompTIA Convergence+ exam covering CompTIA certification objectives that were current as of 2008.

CompTIA has not reviewed or approved the accuracy of the contents of this training material and specifically disclaims any warranties of merchantability or fitness for a particular purpose. CompTIA makes no guarantee concerning the success of persons using any such "Authorized" or other training material in order to prepare for any CompTIA certification exam.

How to Become CompTIA Certified

This training material can help you prepare for and pass a related CompTIA certification exam or exams. In order to achieve CompTIA certification, you must register for and pass a CompTIA certification exam or exams.

In order to become CompTIA certified, you must:

1. Select a certification exam provider. For more information please visit http://www.comptia.org/certification/general_information/exam_locations.aspx

2. Register for and schedule a time to take the CompTIA certification exam(s) at a convenient location

3. Read and sign the Candidate Agreement, which will be presented at the time of the exam(s). The text of the Candidate Agreement can be found at http://www.comptia.org/certification/general_information/candidate_agreement .aspx

4. Take and pass the CompTIA certification exam(s).

For more information about CompTIA's certifications, such as its industry acceptance, benefits, or program news, please visit www.comptia.org/certification.

CompTIA is a not-for-profit information technology (IT) trade association. CompTIA's certifications are designed by subject matter experts from across the IT industry. Each CompTIA certification is vendor-neutral, covers multiple technologies, and requires demonstration of skills and knowledge widely sought after by the IT industry. To contact CompTIA with any questions or comments, please call (630) 678-8300 or email questions@comptia.org.

I dedicate this book to my family and friends. You have all helped me become who I am. To my family, thank you for your patience and love. To my church family, thank you for your support. To my friends around the world, thank you for your e-mails and kind words.

ABOUT THE AUTHOR

Tom Carpenter is a technical experts' expert. He teaches in-depth courses on Microsoft technologies, wireless networking, and security, and professional development skills such as project management, team leadership, and communication skills for technology professionals. Tom holds Convergence+, CWNA, CSWP, and Wireless# certifications with the CWNP program and is also a Microsoft Certified Partner. The Wireless Networking, Windows Administration, and IT Project Management Bootcamps that Tom offers annually provide the in-depth knowledge IT professionals need to succeed. He lives with his lovely wife, Tracy, and their four children, Faith, Rachel, Thomas, and Sarah, in Ohio. His company, SYSEDCO, provides training and consulting services throughout the United States. For more information about Tom and the services offered by his company, visit www.SYSEDCO.com.

About the Technical Editor

Shan Nichols is a Senior Consultant with AT&T Consulting Solutions. He is involved with the planning, design, and implementation of large customer deployments throughout the world. His main technology focus is unified communications and contact center technologies. He currently holds Cisco partner certifications for Unified Contact Center Enterprise and is a Cisco Qualified Specialist (CQS) for Unified Contact Center Express. He is a graduate of the University of Florida and is currently pursuing his MBA degree through Arizona State University. He lives in Marietta, Georgia, with his wonderful wife Antoinette, their son William, and their three dogs. He can be reached at shannich@gmail.com.

About LearnKey

LearnKey provides self-paced learning content and multimedia delivery solutions to enhance personal skills and business productivity. LearnKey claims the largest library of rich streaming-media training content that engages learners in dynamic media-rich instruction complete with video clips, audio, full motion graphics, and animated illustrations. LearnKey can be found on the Web at www.LearnKey.com.

CONTENTS

ix

ACKNOWLEDGMENTS

I would like to acknowledge all of the readers who have provided me with such valuable feedback on my previous books. From the critiques to the laudations, they have all added to my knowledge of the writing and teaching process. Additionally, I want to thank the excellent staff at McGraw-Hill and Shan Nichols, who was a very patient technical editor. Shan, you definitely increased the quality of the technical material in this book. Finally, I want to acknowledge yet another person from history who has influenced my life: Phineas Taylor Barnum. That's right. P.T. Barnum taught me how to get up when you're down and how to think creatively when solutions seem few. He taught me through books. For that, and the many great books by other great authors in my library, I am forever grateful.

T he objective of this study guide is to prepare you for the Convergence+ exam by familiarizing you with the technology or body of knowledge tested on the exam. Because the primary focus of the book is to help you pass the test, I don't always cover every aspect of the related technology. Some aspects of the technology are only covered to the extent necessary to help you understand what you need to know to pass the exam, but I hope this book will serve you as a valuable professional resource after your exam. In particular, you will find many on-the-job facts and tips scattered throughout the book. These tips will prove valuable as you begin working with VoIP and video over IP technologies.

In This Book

This book is organized in such a way as to serve as an in-depth review for the Convergence+ exam for both experienced networking professionals and newcomers to networking voice and video technologies. Each chapter covers a major aspect of the exam, with an emphasis on the "why" as well as the "how to" of working with and supporting converged networks.

On the CD

For more information on the CD-ROM, please see the Appendix, "About the CD-ROM," at the back of the book.

Exam Readiness Checklist

At the end of the Introduction you will find an Exam Readiness Checklist. This table has been constructed to allow you to cross-reference the official exam objectives with the objectives as they are presented and covered in this book. The checklist also allows you to gauge your level of expertise on each objective at the outset of your studies. This should allow you to check your progress and make sure you spend the time you need on more difficult or unfamiliar sections. References have been provided for the objective exactly as the vendor presents it, the section of the study guide that covers that objective, and a chapter and page reference.

In Every Chapter

We've created a set of chapter components that call your attention to important items, reinforce important points, and provide helpful exam-taking hints. Take a look at what you'll find in every chapter:

- Every chapter begins with **Certification Objectives**—what you need to know in order to pass the section on the exam dealing with the chapter topic. The Objective headings identify the objectives within the chapter, so you'll always know an objective when you see it!

- **Exam Watch** notes call attention to information about, and potential pitfalls in, the exam. These helpful hints are written by authors who have taken the exams and received their certification—who better to tell you what to worry about? They know what you're about to go through!

It is important that you understand the basic operations that take place at each layer of the OSI model.

- **Step-by-Step Exercises** are interspersed throughout the chapters. These are typically designed as hands-on exercises that allow you to get a feel for the real-world experience you need in order to pass the exams. They help you master skills that are likely to be an area of focus on the exam. Don't just read through the exercises; they are hands-on practice that you should be comfortable completing. Learning by doing is an effective way to increase your competence with a product.

- **On the Job** notes describe the issues that come up most often in real-world settings. They provide a valuable perspective on certification- and product-related topics. They point out common mistakes and address questions that have arisen from on-the-job discussions and experience.

- **Inside the Exam** sidebars highlight some of the most common and confusing problems that students encounter when taking a live exam. Designed to anticipate what the exam will emphasize, they will help ensure you know what you need to know to pass the exam. You can get a leg up on how to respond to those difficult-to-understand questions by focusing extra attention on these sidebars.

- The **Certification Summary** is a succinct review of the chapter and a restatement of salient points regarding the exam.

■ The **Two-Minute Drill** at the end of every chapter is a checklist of the main points of the chapter. It can be used for last-minute review.

Q&A ■ The **Self Test** offers questions similar to those found on the certification exams. The answers to these questions, as well as explanations of the answers, can be found at the end of each chapter. By taking the Self Test after completing each chapter, you'll reinforce what you've learned from that chapter while becoming familiar with the structure of the exam questions.

■ The **Lab Question** at the end of the Self Test section offers a unique and challenging question format that requires the reader to understand multiple chapter concepts to answer correctly. These questions are more complex and more comprehensive than the other questions, as they test your ability to take all the knowledge you have gained from reading the chapter and apply it to complicated, real-world situations. These questions are aimed to be more difficult than what you will find on the exam. If you can answer these questions, you have proved that you know the subject!

Some Pointers

Once you've finished reading this book, set aside some time to do a thorough review. You might want to return to the book several times and make use of all the methods it offers for reviewing the material:

1. *Re-read all the Two-Minute Drills,* or have someone quiz you. You also can use the drills as a way to do a quick cram before the exam. You might want to make some flash cards out of 3 × 5 index cards that have the Two-Minute Drill material on them.

2. *Re-read all the Exam Watch notes and Inside the Exam elements.* Remember that these notes are written by authors who have taken the exam and passed. They know what you should expect—and what you should be on the lookout for.

3. *Re-take the Self Tests.* Taking the tests right after you've read the chapter is a good idea, because the questions help reinforce what you've just learned. However, it's an even better idea to go back later and do all the questions in the book in one sitting. Pretend that you're taking the live exam. When you go through the questions the first time, you should mark your answers on a separate piece of paper. That way, you can run through the questions as many times as you need to until you feel comfortable with the material.

4. *Complete the Exercises.* Did you do the exercises when you read through each chapter? If not, do them! These exercises are designed to cover exam topics, and there's no better way to get to know this material than by practicing. Be sure you understand why you are performing each step in each exercise. If there is something you are not clear on, re-read that section in the chapter.

I'll provide more recommendations that will help improve your chances of success on exam day in the Introduction.

INTRODUCTION

The *Convergence+ Study Guide* will be helpful to two categories of students: those who simply wish to learn more about converged networks and those who are preparing for the Convergence+ certification by CompTIA. If you are in the latter category, you can rest assured that this book covers the objectives sufficiently to help you pass the exam. If you are reading this book for the knowledge alone, you hold in your hands an excellent overview of the technologies and processes used in voice and video over IP networks.

This book will help you understand the following concepts:

- Networking hardware fundamentals, including cabling and devices
- TCP/IP communications, including addressing and protocols
- VoIP protocols, including H.323 and SIP
- Video over IP solutions, including broadcasting, multicasting, and unicasting
- Troubleshooting a converged network
- Securing a converged network

You are not required to posses any previous knowledge of the topics; however, the knowledge acquired through studying for the Network+ and Security+ exams would be beneficial. If you have the ability to implement and test the TCP/IP protocol suite on a small network, that would be helpful as well, though this action is also not required.

How This Book Is Organized

The *Convergence+ Study Guide* is organized into four major sections. Chapters 1 through 4 cover data networking fundamentals. Topics include infrastructure design, network protocols, infrastructure hardware, and client devices. Chapters 5 and 6 cover the basics of telephony. Here, the topics include the telephony networks, such as the PSTN, and telephony hardware. Chapters 7 through 11 address the convergence of data and voice networks. In these chapters, you will learn about VoIP protocols, streaming media solutions, management of converged networks, and converged network troubleshooting. Finally, Chapters 12 and 13 cover security as it relates to converged networks. You first learn about security threats in Chapter 12 and then learn about security solutions in Chapter 13.

As you can see, the book provides you with complete coverage of the knowledge needed to pass the Convergence+ exam, and it also helps you understand how these technologies fit into current data networks.

Exam Objectives

Subject Area	Approximate Percent of Exam
Telephony	22%
Network Engineering	20%
Applications	16%
Hardware & Architecture	17%
Management	12%
Security	13%

Tips for Succeeding on the Convergence+ Exam

The following tips will prove beneficial as you study for the Convergence+ exam:

- Read every chapter of this book and take notes.
- Rewrite the Exam Watches in a way that is memorable to you.
- As you're reading the book, be sure to reread sections that are new to you. Take note of facts such as protocol names and features, as well as hardware specifications.
- Write your own definitions for key terms and test those definitions against the glossary, trusted web sites, and vendor literature.
- As you take the practice exams on the CD or at the end of each chapter, take special note of the questions you miss. Review the sections related to those questions.
- Download and read vendor literature from at least two or three VoIP solution providers. Study the specifications for their systems and the recommended implementation procedures.
- Be sure to get a good night's sleep the night before the exam, eat a healthy breakfast, and drink a glass of water 30–45 minutes before testing.

- Review the Glossary before traveling to the testing center or just before entering the facility.
- Review the Two-Minute Drills at the end of each chapter just before going into the testing center.
- During the exam, answer every question. Give each one your best shot.

In addition to these tips, feel free to e-mail me if you have any questions along the journey. I'm always glad to hear from my readers and will usually respond within 48 hours. My e-mail address is carpenter@sysedco.com.

CompTIA Convergence+ Exam CTO-101

Exam Readiness Checklist

Exam Objective	Chapter #	Page #	Beginner	Intermediate	Advanced
Telephony					
Demonstrate application of traffic engineering concepts	9	304			
Describe fundamentals of voice systems	5	204			
Describe the components of number and dialing plans	5	223			
Identify the various endpoints used in a converged environment	4, 7	166, 254			
Network Engineering					
Define QoS, describe implementation techniques, and show the importance of QoS	9	310			
Analyze network performance	9	294			
Describe networking technologies used in a converged network	1, 4	24, 186			
Identify methods of encoding, decoding, and compression	7, 8	258, 277			
Applications					
Identify different types of messaging applications	7	261			
Identify different types of collaboration applications	7	262			
Identify different components of a contact center	10	326			

Exam Readiness Checklist

Exam Objective	Chapter #	Page #	Beginner	Intermediate	Advanced
Identify components of mobility	7	264			
Identify methods used for rich media transmission	8	273			
Identify benefits of using different standards and the impact on the performance of the network	8	280			
Hardware and Architecture					
Identify the layers of the OSI model and know its relevance to converged networks	1	5			
Recognize network models and how they affect the converged network	3	100			
Identify the functions of hardware components as used on a converged network	1, 3, 7	42, 108, 250			
Management					
Identify and execute problem solving and analysis processes	11	348			
Identify common symptoms and problems on a converged network	11	363			
Describe and use tools and commands to monitor network performance in a converged environment	10	330			
Identify and describe proper administration tasks and procedures	10	334			
Security					
Explain concepts and components of security design and how they affect the converged network	12, 13	382, 415			

1

Networking Infrastructure and Design

> *The very first step toward success in any occupation is to become interested in it.*
>
> —*Sir William Osler*

Y ou are fortunate in that you've shown interest in convergence technologies. I am fortunate in that I have the attention of someone who has shown enough interest to make the initial investment of time and resources needed to acquire this book. You have what it takes to succeed on this journey: interest in the topic and a desire to excel. For this enthusiasm, I applaud you and welcome you to the world of converged communications.

Convergence is defined as "the process or state of converging" (The *Oxford American College Dictionary*, Copyright 2002, Oxford University Press, Inc.). The same resource defines *converge* as to "come together from different directions so as to eventually meet." Ultimately, convergence is the coming together of previously disconnected or separated concepts, things, or people. For our purpose here, I will define convergence as "the process of bringing *voice, multimedia,* and *data* communications *together* on a *shared network.*" The key is that these three separate elements actually become one element: data. Of course, data is computerized data already. Voice or sound becomes computerized data. Video becomes computerized data. Since all three are computerized data, they can all traverse our existing data networks; however, they do not all have the same priority, and so we must implement solutions that allow us to favor the highest-priority data at any moment.

This is the ultimate goal of this book: I want you to be confident in your ability to implement and troubleshoot a converged network that allows for the transfer of voice, multimedia, and other data. In order to do this, you'll need to start with the right foundation. You'll need to begin by understanding the general concept of a network and then move on to the specific components and technologies used in a computer network. This chapter will begin by explaining networks in general and will introduce topics such as computer network topologies, computer network types, and the hardware used on these networks. You will also learn about a network communications model that will play an important role throughout the remaining chapters of the book. We'll begin by defining a network.

Networks Defined

A *network* is a "group of connected or interconnected people or things." When those of us working as technology professionals hear the term *network*, we tend to immediately think of a computer network, but the reality is that this defines only one type of network. The currently popular web site known as LinkedIn (www.linkedin.com) is a networking site that allows people to form connections with one another based on shared likes, dislikes, experiences, or simply the desire to connect. Network marketing is a phrase that has been used for years to reference a form of marketing that takes advantage of an individual's network of connections with other people that is very similar to the networks that have been built on the LinkedIn web site. The point is that networks are groups of connected entities and are not limited to modern computer networks.

In fact, the telephone system is a perfect example of a network different from the typical computer networks we implement as local area networks in our organizations. Sometimes called the *public switched telephone network (PSTN)*, it is a network that consists of the telephone endpoints and the cables and devices used between these endpoints. This network allows me to use a traditional land line telephone to place a call to my friend across town or my extended family several states away.

When two networks combine or connect in some way, these networks are said to be converged. Many years ago I modulated TCP/IP communications (data communications) over an analog phone line using the PSTN that was designed to carry the human voice and not data. The purpose of this modulation was to implement a connection that allowed communications on the Internet using a standard phone line. Today, we digitize the human voice over an Internet Protocol (IP) network that was designed to carry computer data and not the human voice. Ultimately, when implementing Voice over IP (VoIP), the IP network eventually connects to the PSTN (in many cases) to route calls outside of the IP network and connect to others' telephones that are simply connected to a standard land line. This connection point between the IP network and the traditional telephone network is one way in which we say the networks are converged.

Additionally, the fact that voice data is now traveling on a network that was designed for computer data also suggests convergence. We could say that the two data types have been converged. The computer data may consist of a word processor document or an e-mail; voice data is a digitized version of the human-generated sound waves. In the end, both data types become IP packets, which become Ethernet frames (on Ethernet networks), which become 1's and 0's on the communications medium that is utilized.

In the end, there are at least two points of convergence in the areas of voice and data. The first point is where our IP networks connect to traditional telephone networks. The second point is the unification of upper-level data such as voice and computer data into shared lower-level communications such as IP packets. These convergence points will all become very clear as you read through this book. For now, let's explore the historical development of information networks. A brief overview of the major developments along the way will help you understand why things work the way they do today and how to best take advantage of the available technologies.

The Evolution of Information Networks

The electric telegraph was the precursor to modern electronic or digital information networks. While other networks, like the pony express or Claude Chappe's non-electric telegraph, existed before the electric telegraph, they required complete involvement of humans on an ongoing basis either from end to end or at each end. The electric telegraph was a revolutionary leap forward in that it eventually allowed messages to be sent to a remote location and recorded onto paper tape as raised dots and dashes matching up with Morse Code.

It is interesting to note that tests were performed over many years in order to validate the potential of electricity in communications. In 1746 Jean-Antoine Nollet asked about two hundred monks to form a long snaking line. He had each one hold the end of a twenty-five-foot iron wire connecting him to the next monk in the line. Then, without warning, he injected an electrical current into the line. The fact that monks in a line nearly one mile long all exclaimed their literal shock at the same time showed that electricity travels long distances very rapidly. Of course, we have much more humane methods of testing today, but these early tests did indeed reveal the knowledge that eventually led to the implementation of high-speed electric telegraphs.

By 1861, the United States was implementing the transcontinental telegraph, which allowed nearly instant communications across the country. Much as we are implementing IPv6 alongside our IPv4 networks, the transcontinental telegraph wire was placed alongside the existing Pony Express route. Once the telegraph network was in place, the Pony Express became obsolete and was dissolved. Some of us look forward to the day when IPv6 does the same to IPv4.

In a quick jump to the peak of the era of the telegraph, within 30 years of its beginning, over 650,000 miles of wire was in place and some 20,000 towns were connected to the network. You could send messages from London to Bombay or from New York to Sacramento in just a few minutes. Previously, such communications would have taken weeks or months. The telegraph showed that electronic communications were indeed the way to send information over long distances.

Believe it or not, the use of electricity in sending telegraph messages eventually led to the use of electronic voice communications. The telephone, invented in the late 1800s, converted sound waves into electronic signals on the sending end and then converted electronic signals into sound waves on the receiving end. Sound travels at an average speed of about 1130 feet (344 meters) per second, while electromagnetic waves travel at about 186,400 miles (300,000 kilometers) per second. This speed means that a message sent electronically could travel around the Earth more than seven times in a second. By converting sound waves to electromagnetic waves and back, we can transmit the human voice very rapidly, which is why you can have a conversation with someone on the other side of the globe with very little delay.

The telephone system continued to evolve into the PSTN that we utilize today. In the same way, new forms of information delivery were also being developed, and nearing the end of the twentieth century two distinct networks had evolved: voice networks and data networks. Voice networks allowed the transfer of human conversations, and data networks allowed the transfer of all other kinds of information. At times, our data networks connected to one another using the voice networks as the infrastructure, and at times, our voice networks crossed over our data networks in the form of packet delivery; however, with the new millennium also came much greater interest in the convergence of voice and data networks.

on the
()o b

If you would like to learn more about the history of the telegraph and telephone, I suggest the book **The Victorian Internet** *by Tom Standage (Walker & Company, 2007) as a starting point. Studying these historical developments can help you better understand the technologies we use today and why convergence is important and beneficial to the future of data networks.*

CERTIFICATION OBJECTIVE 1.01

Identify the Layers of the OSI Model and Know Its Relevance to Converged Networks

In order to help you understand how the various networking components work together to form a converged network, I will first explain the Open System Interconnection (OSI) model. While this model is not directly implemented in the

TCP/IP networks that are most common today, it is a valuable conceptual model that helps you to relate different technologies to one another and implement the right technology in the right way.

According to document ISO/IEC 7498-1, which is the OSI Basic Reference Model standard document, the OSI model provides a "common basis for the coordination of standards development for the purpose of systems interconnection, while allowing existing standards to be placed into perspective within the overall reference model." In other words, the model is useful for new standards as they are developed and for thinking about existing standards. In Chapter 2, I will show you this reality when I relate the TCP/IP protocol suite to the OSI model. Even though TCP/IP was developed before the OSI model, it can be *placed in perspective* in relation to the model.

The OSI model allows us to think about our network in chunks or layers. You can focus on securing each layer, optimizing each layer, and troubleshooting each layer. This allows you to take a very complex communications process apart and evaluate its components. In order to understand this, you'll need to know that the OSI model is broken into seven layers. The seven layers are (from top to bottom):

- Application
- Presentation
- Session
- Transport
- Network
- Data Link
- Physical

Each layer is defined as providing services and receiving services. For example, the Data Link layer provides a service to the Physical layer and receives a service from the Physical layer. How is this? In a simplified explanation, the Data Link layer converts packets into frames for the Physical layer, and the Physical layer transmits these frames as bits on the chosen medium. The Physical layer reads bits off of the chosen medium and converts these into frames for the Data Link layer.

The layered model allows for abstraction. In other words, the higher layers do not necessarily have to know how the lower layers are doing their jobs. In addition, the lower layers do not necessarily have to know what the upper layers are actually doing with the results of the lower layers' labors. This abstraction means that you have the ability to use the same web browser and HTTP protocol to communicate on the Internet whether the lower-layer connection is a dial-up modem, a high-speed Internet connection, or somewhere in between. The resulting speed or performance will certainly vary, but the functionality will remain the same.

Figure 1-1 illustrates the concept of the OSI model. As you can see, data moves down through the layers, across the medium, and then back up through the layers on the receiving machine. Remember, most networking standards allow for the substitution of nearly any Data Link and Physical layer. While this example shows a wired Ethernet connection between the two machines, it could have just as easily been a wireless connection using the IEEE 802.11 and IEEE 802.2 standards for the descriptions of the Data Link and Physical layers. This example uses the IEEE 802.3 Ethernet standard and the IEEE 802.2 LLC standard (a layer within the Data Link layer) for the lower layers. The point is that the most popular upper-layer protocol suite, TCP/IP, can work across most lower-layer standards such as IEEE 802.2 (Logical Link Control), 802.3 (Ethernet), 802.5 (Token Ring), 802.11 (Wireless LANs), and 802.16 (WiMAX).

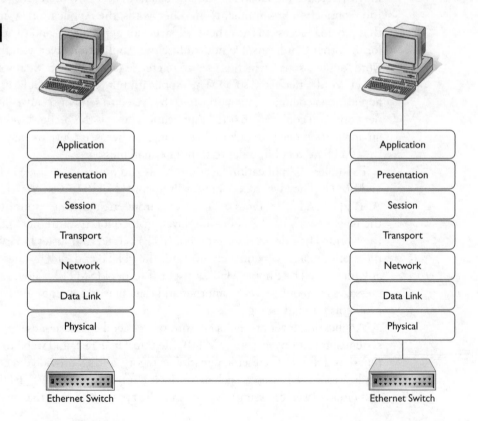

FIGURE 1-1

OSI model depiction

Application	Application
Presentation	Presentation
Session	Session
Transport	Transport
Network	Network
Data Link	Data Link
Physical	Physical

Ethernet Switch   Ethernet Switch

In order to fully understand the OSI model and be able to relate to it throughout the rest of this book, it is important that we evaluate each layer. You will need to understand the basic description of each layer and the services it provides to the networking process. I will define each layer and then give examples of its use, starting with the topmost layer, which is the Application layer, since this is the order in which they are documented in the standard.

Application Layer

The seven layers of the OSI model are defined in Clause 7 of the document ISO/IEC 7498-1. The Application layer is defined in Subclause 7.1 as the highest layer in the reference model and as the sole means of access to the OSIE (Open System Interconnection Environment). In other words, the Application layer is the layer that provides access to the other OSI layers for applications and to applications for the other OSI layers. Do not confuse the Application layer with the general word "applications," which is used to reference programs like Microsoft Excel, Corel WordPerfect, and so on. The Application layer is the OSI layer that these applications communicate with when they need to send or receive data across the network. You could say that the Application layer is the higher-level protocols that an application needs to talk to. For example, Microsoft Outlook may need to talk to the SMTP protocol in order to transfer e-mail messages.

Examples of Application layer protocols and functions include HTTP, FTP, and SMTP. The Hypertext Transfer Protocol (HTTP) is used to transfer HTML, ASP, PHP, and other types of documents from one machine to another. HTTP is the most heavily used Application layer protocol on the Internet and, possibly, in the world. The File Transfer Protocol (FTP) is used to transfer binary and ASCII files between a server and a client. Both the HTTP and FTP protocols can transfer any file type. The Simple Mail Transport Protocol (SMTP) is used to move e-mail messages from one server to another and usually works in conjunction with other protocols for mail storage.

Application layer processes fall into two general categories: user applications and system applications. E-mail (SMTP), file transfer (FTP), and web browsing (HTTP) functions fall into the user application category, as they provide direct results to applications used by users such as Outlook Express (e-mail), WS_FTP (file transfer), and Firefox (web browsing). Notice that the applications or programs used by the user actually take advantage of the application services in the Application layer, or Layer 7. In other words, Outlook Express takes advantage of SMTP. Outlook Express does not reside in Layer 7, but SMTP does. For examples of system applications, consider DHCP and DNS. The Dynamic Host Configuration Protocol (DHCP) provides for dynamic TCP/IP configuration, and the Domain Name Service (DNS) protocol provides for name to IP address resolution. Both of these are considered system-level applications because they are not usually directly accessed by the user (this is open for debate, since administrators are users too and they use command-line tools or programs to directly access these services quite frequently).

The processes operating in the Application layer are known as *application-entities*. An application-entity is defined in the standard as "an active element embodying a set of capabilities which is pertinent to OSI and which is defined for the Application Layer." In other words, application-entities are the services that run in Layer 7 and communicate with lower layers while exposing entry points to the OSI model for applications running on the local computing device. SMTP is an application-entity, as are HTTP and other Layer 7 protocols.

Imagine that you are sending an e-mail using the Simple Mail Transport Protocol (SMTP), which is the most popular method of sending an e-mail message. Your e-mail application will connect to an SMTP server in order to send the e-mail message. Interestingly, from the e-mail application's perspective, it is connecting directly to the SMTP server and is completely unaware of all the other layers of operation that allow this connection to occur. Figure 1-2 shows the e-mail as it exists at Layer 7.

Presentation Layer

The Presentation layer is defined in Subclause 7.2 of the standard as the sixth layer of the OSI model, and it provides services to the Application layer above it and the Session layer below it. The Presentation layer, or Layer 6, provides for the representation of the information communicated by or referenced by application-entities. The Presentation layer is not used in all network communications, and it, as well as the Application and Session layers, is similar to the single Application layer of the TCP/IP model. The Presentation layer provides for syntax management and conversion as well as encryption services. Syntax management refers to the process of ensuring that the sending and receiving hosts communicate with a shared syntax or language. When you understand this concept, you will realize why encryption is often handled at this layer. After all, encryption is really a modification of the data in such a way that must be reversed on the receiving end. Therefore, both the sender and receiver must understand the encryption algorithm in order to provide the proper data to the program that is sending or receiving on the network.

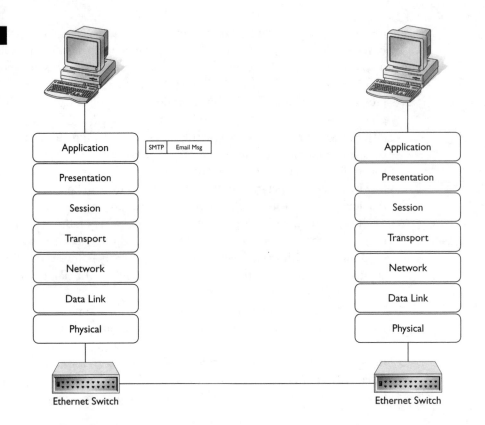

FIGURE 1-2

SMTP e-mail
being sent
through Layer 7

Don't be alarmed to discover that the TCP/IP model has its own Application layer that differs from the OSI model's Application layer. The TCP/IP protocol existed before the OSI model was released. For this reason, we relate the TCP/IP protocol suite to the OSI model, but we cannot say that it complies with the model directly. It's also useful to keep in mind the reality that the TCP/IP protocol is an implemented model and the OSI model is a "reference" model. This definition simply means that we use it as a reference to understand our networks and network communications.

Examples of Presentation layer protocols and functions include any number of data representation and encryption protocols. For example, if you choose to use HTTPS instead of HTTP, you are indicating that you want to use Secure Sockets Layer (SSL) encryption. SSL encryption is related to the Presentation layer, or Layer 6 of the OSI model.

Ultimately Layer 6 is responsible, at least in part, for three major processes: data representation, data security, and data compression. *Data representation* is the process of ensuring that data is presented to Layer 7 in a useful way and that it is passed to Layer 5 in a way that can be processed by the lower layers. *Data security* usually includes authentication, authorization, and encryption. Authentication is used to verify the identity of the sender and the receiver. With solid authentication, we gain a benefit known as *non-repudiation*. Non-repudiation simply means that the sender cannot deny the sending of data. This differentiation is often used for auditing and incident handling purposes. *Authorization* ensures that only valid users can access the data being accessed, and *encryption* ensures the privacy and integrity of the data as it is being transferred.

The processes running at Layer 6 are known as presentation-entities in the OSI model documentation. Therefore, an application-entity is said to depend on the services of a presentation-entity, and the presentation-entity is said to serve the application-entity.

As your e-mail message moves down to the Presentation layer, and since it uses SMTP, it is sent as clear text by default. This transfer is accomplished today using the Layer 6 Multipurpose Internet Mail Extensions (MIME) representation protocol that allows for binary attachments to SMTP messages. This means that the Presentation layer is converting your e-mail message, whatever its origination, into the standard MIME format or syntax. If you wanted to secure the message, the Secure/MIME (S/MIME) protocol could also be used. The S/MIME protocol, still operating at Layer 6, uses encryption to secure the data as it traverses the network. This encrypted data is sometimes said to be enveloped data. You can see the e-mail now as it exists at Layer 6 in Figure 1-3.

Session Layer

The Session layer is defined in Subclause 7.3 of the standard as "providing the means necessary for cooperating presentation-entities to organize and to synchronize their dialog and to manage their data exchange." This exchange is accomplished by establishing a connection between two communicating presentation-entities. The result is simple mechanisms for orderly data exchange and session termination.

A session includes the agreement to communicate and the rules by which the communications will transpire. Sessions are created, communications occur, and sessions are destroyed or ended. Layer 5 is responsible for establishing the session, managing the dialogs between the endpoints, and conducting the proper closing of the session.

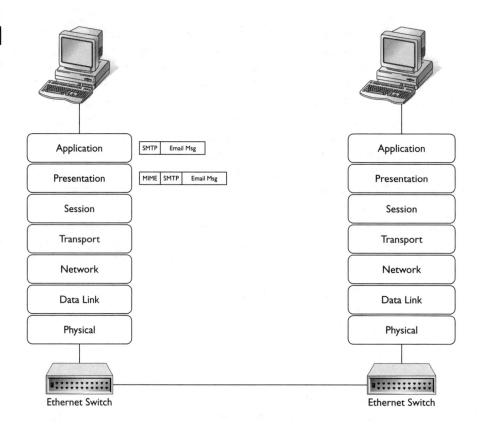

FIGURE I-3

SMTP e-mail after
reaching Layer 6

Examples of Session layer protocols and functions include the iSCSI protocol, RPC, and NFS. The iSCSI protocol provides access to SCSI devices on remote computers or servers. The protocol allows a SCSI command to be sent to the remote device. The Remote Procedure Call (RPC) protocol allows subroutines to be executed on remote computers. A programmer can develop an application that calls the subroutine in the same way as a local subroutine. RPC abstracts the Network layer and allows the application running above Layer 7 to execute the subroutine without knowledge of the fact that it is running on a remote computer. The Network File System (NFS) protocol is used to provide access to files on remote computers as if they were on the local computer. NFS actually functions using an implementation of RPC known as Open Network Computing RPC (ONC RPC) that was developed by Sun Microsystems for use with NFS; however, ONC RPC has also been used by other systems since that time. Remember that these protocols are provided only as examples of the protocols available at Layer 5 (as were the other protocols mentioned for Layers 6 and 7). By learning the functionality of protocols that operate at each layer, you can better understand the intention of each layer.

The services and processes running in Layer 5 are known as session-entities. Therefore, RPC and NFS would be session-entities. These session-entities will be served by the Transport layer.

At the Session layer, your e-mail message can begin to be transmitted to the receiving mail server. The reality is that SMTP e-mail uses the TCP protocol from the TCP/IP suite to send e-mails and the analogy is not perfect at this point. This imperfection in the comparison of models is because the TCP/IP protocol does not map directly to the OSI model, as you will learn in the next chapter. For now, know that Layer 5 is used to establish sessions between these presentation-entities. In Windows, the Winsock API provides access to the TCP/IP protocol suite. We could, therefore, say that your e-mail is passed through to the TCP/IP suite using Winsock here at Layer 5. Figure 1-4 shows the e-mail as it is passed through the Winsock API at Layer 5.

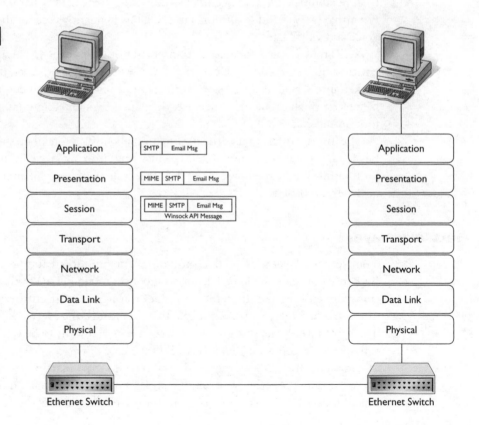

FIGURE 1-4

SMTP e-mail at Layer 5

Transport Layer

Layer 4, the Transport layer, is defined as providing "transparent transfer of data between session entities and relieving them from any concern with the detailed way in which reliable and cost effective transfer of data is achieved." This definition simply means that the Transport layer, as its name implies, is the layer where the data is segmented for effective transport in compliance with Quality of Service (QoS) requirements and shared medium access.

Examples of Transport layer protocols and functions include TCP and UDP. The Transmission Control Protocol (TCP) is the primary protocol used for the transmission of connection-oriented data in the TCP/IP suite. HTTP, SMTP, FTP, and other important Layer 7 protocols depend on TCP for reliable delivery and receipt of data. The User Datagram Protocol (UDP) is used for connectionless data communications. For example, when speed of communications is more important than reliability, UDP is frequently used. Because voice data either has to arrive or not arrive (as opposed to arriving late), UDP is frequently used for the transfer of voice and video data.

TCP and UDP are examples of transport-entities at Layer 4. These transport-entities will be served by the Network layer. At the Transport layer, the data is broken into segments if necessary. If the data will fit in one segment, then the data becomes a single segment. Otherwise, the data is segmented into multiple segments for transmission.

The Transport layer takes the information about your e-mail message from the Session layer and begins dividing it (segmenting) into manageable chunks (packets) for transmission by the lower layers. Figure 1-5 shows the e-mail after the processing at the Transport layer.

Network Layer

The Network layer is defined as providing "the functional and procedural means for connectionless-mode (UDP) or connection-mode (TCP) transmission among transport-entities and, therefore, provides to the transport-entities independence of routing and relay considerations." In other words, the Network layer says to the Transport layer, "You just give me the segments you want to be transferred and tell me where you want them to go. I'll take care of the rest." This segregation of communication is why routers do not have to expand data beyond Layer 3 to route the data properly. For example, an IP router does not care if it's routing an e-mail message or voice conversation. It only needs to know the IP address for

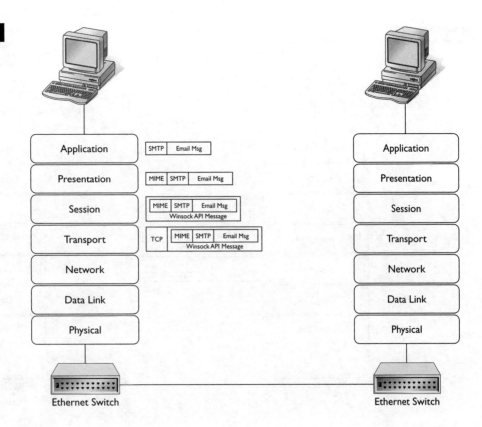

FIGURE 1-5

E-mail after
Layer 4
processing

which the packet is destined and any relevant QoS parameters in order to move the packet along.

Examples of Network layer protocols and functions include IP, ICMP, and IPsec. The Internet Protocol (IP) is used for addressing and routing of data packets in order to allow them to reach their destination. That destination can be on the local network or a remote network. The local machine is never concerned with this, with the exception of the required knowledge of an exit point, or default gateway, from the local machine's network. The Internet Control Message Protocol (ICMP) is used for testing the TCP/IP communications and for error message handling within Layer 3. Finally, IP Security (IPsec) is a solution for securing IP communications using authentication and/or encryption for each IP packet. While security protocols such as SSL, TLS, and SSH operate at Layers 4 through 7 of the OSI model, IPsec sits solidly at Layer 3. The benefit is that, since IPsec sits below Layer 4, any protocols running at or above Layer 4 can take advantage of this secure foundation. For this reason, IPsec has become more and more popular since it was first defined in 1995.

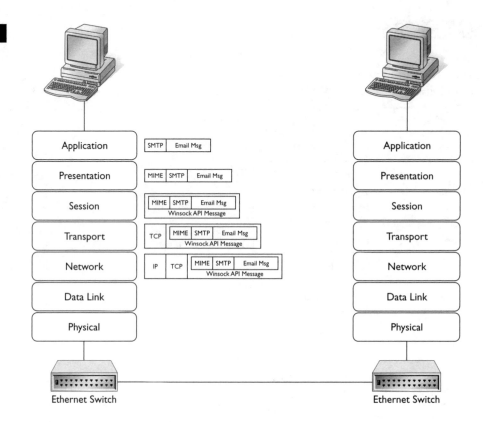

FIGURE 1-6

Layer 6 SMTP
e-mail

The services and processing operating in the Network layer are known as network-entities. These network-entities depend on the services provided by the Data Link layer. At the Network layer, Transport layer segments become packets. These packets will be processed by the Data Link layer.

At the Network layer, your e-mail message that was broken into segments at Layer 4 is now appended with appropriate destination and source addressing information in order to ensure that it arrives at the destination. The results of Layer 3 processing are shown in Figure 1-6.

Data Link Layer

The Data Link layer is defined as providing communications between connectionless-mode or connection-mode network entities. This method may include the establishment, maintenance, and release of connections for connection-mode network entities. The Data Link layer is also responsible for detecting errors that may occur in the Physical layer. Therefore, the Data Link layer provides services to Layer 3 and Layer 1. The Data Link layer, or Layer 2, may also correct errors detected in the Physical layer automatically.

Examples of Data Link layer protocols and functions include Ethernet, PPP, and HDLC. Ethernet is the most widely used protocol for local area networks (LANs) and will be the type of LAN you deal with when using most modern LAN technologies. Ethernet comes in many different implementations from 10 Mbps (megabits per second or million bits per second) to 1000 Mbps in common implementations. Faster Ethernet technologies are being developed and implemented on a small scale today. The Point to Point Protocol (PPP) is commonly used for wide area network (WAN) links across analog lines and other tunneling purposes across digital lines. The High-Level Data Link Control (HDLC) protocol is a solution created by the ISO for bit-oriented synchronous communications. It is a very popular protocol used for WAN links and is the default WAN link protocol for many Cisco routers.

The IEEE has divided the Data Link layer into two sublayers: the Logical Link Control (LLC) sublayer and the Medium Access Control (MAC) sublayer. The LLC sublayer is not actually used by many transport protocols, such as TCP. The varied IEEE standards identify the behavior of the MAC sublayer within the Data Link layer and the PHY layer as well.

The results of the processing in Layer 2 are that the packet becomes a frame that is ready to be transmitted by the Physical layer, or Layer 1. So the segments became packets in Layer 3 and now the packets have become frames. Remember, this is just the collection of terms that we use; the data is a collection of ones and zeros all the way down through the OSI layers. Each layer is simply manipulating or adding to these ones and zeros in order to perform that layer's service. As in the other layers before it, the services and processes within the Data Link layer are named after the layer and are called data-link-entities.

The Data Link layer adds the necessary header to the e-mail packets received from Layer 3, and your e-mail message, in its one or many parts, is now a frame or set of frames. The frames are ready to be transmitted by the Physical layer. In Figure 1-7 we see the e-mail message after the Data Link layer processing is complete.

Physical Layer

The Physical layer, sometimes called the PHY, is responsible for providing the mechanical, electrical, functional, or procedural means for establishing physical connections between data-link-entities. The connections between all other layers are really logical connections, as the only real physical connection that results in true transfer of data is at Layer 1—the Physical layer. For example, we say that the

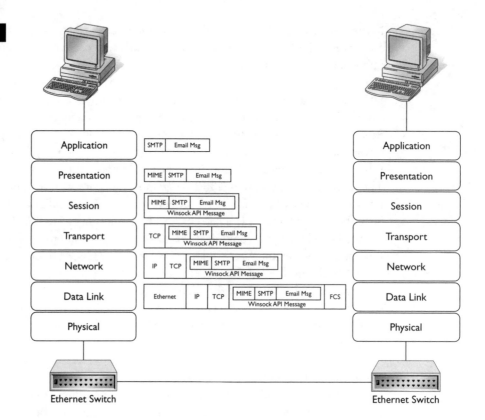

FIGURE 1-7

Layer 2 e-mail
results

Layer 7 HTTP protocol on a client creates a connection with the Layer 7 HTTP
protocol on a web server when a user browses an Internet web site; however, the
reality is that this connection is logical and the real connections happen at the
Physical layer.

It is really amazing to think that my computer—the one I'm using to type these
words—is connected to a wireless access point (AP) in my office, which is connected
to my local network that is in turn connected to the Internet. Through connections—
possibly both wired and wireless—I can send signals (that's what happens at Layer 1)
to a device on the other side of the globe. To think that there is a potential electrical
connection path between these devices and millions of others is really quite amazing.

It is Layer 1 that is responsible for taking the data frames from Layer 2 and
transmitting them on the communications medium as binary bits (ones and zeros).
This medium may be wired or wireless. It may use electrical signals or light pulses
(both actually being electromagnetic in nature). Whatever you've chosen to use
at Layer 1, the upper layers can communicate across it as long as the hardware and

drivers abstract that layer so that it provides the services demanded of the upper-layer protocols.

Examples of Physical layer protocols and functions include Ethernet, Wi-Fi, and DSL. You probably noticed that Ethernet was mentioned as an example of a Data Link layer protocol. This is because Ethernet defines both the MAC sublayer functionality within Layer 2 and the PHY for Layer 1. Wi-Fi technologies (IEEE 802.11) are similar in that both the MAC and PHY are specified in the standard. Therefore the Data Link and Physical layers are often defined in standards together. You could say that Layer 2 acts as an intermediary between Layers 3 through 7 so that you can run IPX/SPX (though hardly anyone uses this protocol today) or TCP/IP across a multitude of network types (network types being understood as different MAC and PHY specifications).

Your e-mail is finally being transmitted across the network. First a one and then a zero, then maybe another one or zero and on and on until the entire e-mail message is transmitted. Figure 1-8 shows the final results with the e-mail, now broken into frames, being transmitted on the medium.

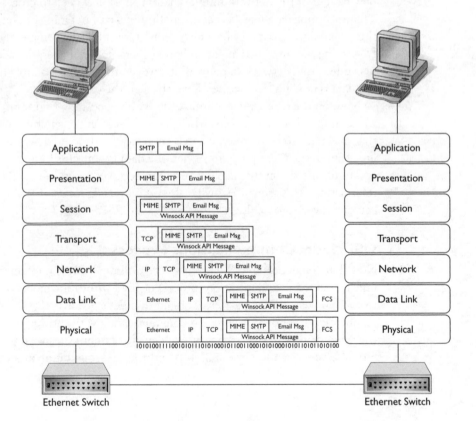

FIGURE 1-8

Layer 1 e-mail transmission

SCENARIO & SOLUTION

You are sending a file to an FTP server. The data that is to be transmitted needs to be encrypted. Which layer of the OSI model is the likely area where this will happen?	The Presentation layer. This is because encryption, compression, and syntax are frequently applied at this layer. It is important to keep in mind the possibility that encryption may also occur at other layers. For example, IPsec encrypts data at Layer 3.
Information about the source and destination MAC addresses is being added to a packet. Which layer of the OSI model is performing this operation?	The Data Link layer. Packets are created at the Network layer and are sent down to the Data Link layer, where MAC addresses are added in the frame's header for both the source and the destination.

The example of the e-mail transmission has been simplified in comparison to what really takes place. For example, each packet (from Layer 3) will be transmitted by Layer 1 (after being converted to frames by Layer 2), and then the next packet may be sent or the network interface card (NIC) may need to process incoming data. That incoming data may be a confirmation of a past outgoing packet that was part of the e-mail message, it may be a retry request, or it may be completely unrelated data. Due to the nature of varying underlying Layer 1 technologies, the actual transfer may differ from network to network. However, this example simply illustrates how the data is modified as it passes down through the OSI model.

Now, on the receiving machine, exactly the opposite would transpire. In other words, frames become packets, which become segments, which become the data that may need to be represented, decompressed, or decrypted before being forwarded upstream to the user's program. When the data is sent, it is formatted, chunked, and transmitted. On the receiving end it is received, aggregated, and possibly reformatted. This sequence is what the OSI layers do for us. It is also what many actual network protocols do for us, such as TCP/IP.

OSI Model Communications Process

Now that you understand the layers of the OSI model, it is important for you to understand the communications process utilized within the model. Each layer is said to communicate with a peer layer on another device. This process means that the Application layer on one device is communicating with the Application layer on the other device. In the same way, each layer communicates with its peer layer. This virtual communication is accomplished through segmentation and encapsulation.

INSIDE THE EXAM

Why Is the OSI Model Important?

The OSI model is more than a set of facts that you memorize for certification exams. It has become the most common method for referencing all things networking. Many resources assume that you understand this model and reference it without explanation. You may read statements like the following:

"Web authentication is a *Layer 3* security feature that causes the controller to not allow IP traffic (except DHCP-related packets) from a particular client until that client has correctly supplied a valid username and password. When you use web authentication to authenticate clients, you must define a username and password for each client. When the clients attempt to join the wireless LAN (WLAN), their users must enter the username and password when prompted by a login window."

This statement is quoted from an article at Cisco's web site. Within the article there is no explanation of what is meant by *Layer 3*. It is simply assumed that you know what this means. The OSI model, therefore, has become required foundational knowledge for

anyone seeking to work in the computer or data networking industry. Many certification exams will not test you on the OSI model directly but will phrase questions in such a way so that you will have to understand the OSI model—as well as some other set of facts—in order to answer the question correctly.

For example, it is not uncommon to see questions like this: You are a network administrator working for a manufacturing company. You want to enable secure Voice over IP communications at Layer 3. What technologies can you use to implement this security?

The possible answers will, of course, be a list of protocols. You'll have to know which of these protocols both provide security and operate at Layer 3 of the OSI model. While you will not see an exact question such as this on the Convergence+ examination, you will be greatly benefited by learning the OSI model for both your certification examination and everyday workload. Not to mention the fact that you'll actually be able to understand all those articles, whitepapers, and books that refer to various layers of the OSI model.

Segmentation is the process of segmenting or separating the data into manageable or allowable sizes for transfer. As an example, the standard Ethernet frame can include a payload (the actual data to be transferred) of no more than 1500 octets. An *octet* is eight bits and is usually called a *byte*. Therefore, data that is larger than

1500 bytes will need to be segmented into chunks that are 1500 bytes or smaller before they can be transmitted. This segmentation begins at Layer 4, where TCP segments are created, and may continue at Layer 3, where IP fragmentation can occur in order to reduce packet sizes so that they can be processed by Layer 2 as Ethernet frames.

Encapsulation is the process of enveloping information within headers so that the information can be passed across varied networks. For example, IP packets (also called *datagrams*) are encapsulated inside of Ethernet frames to be passed on an Ethernet network. This encapsulation means that the IP packet is surround by header and possibly footer information that allows the data to be transmitted. Ethernet frames consist of a header that includes the destination and source MAC addresses and the type of frame in the header. The frames also have a footer that consists of a frame check sequence (FCS) used for error correction. Figures 1-2–1-8 depict the way the data changes as it travels down through the OSI model; notice how encapsulation begins to occur at Layers 5–7 in an almost vague way (because there is no direct mapping of TCP/IP to the OSI model) and then becomes very clear as we approach Layers 1–4.

The most important thing to remember about all of this is that in actuality the Application layer on one device never talks directly to the Application layer on another device even though they are said to be peers. Instead, the communications travel through many intermediaries (OSI layers) on the way to the final destination. This layered effect is really no different than human communications. Layering is seen in human interactions as well.

Notice, in Figure 1-9, that we have two humans communicating. Behind the communications is an initial thought that needs to be transferred from the Fred to Barney. This thought may or may not already be in a language that Fred and Barney know. In this case, we assume that Fred's native speaking language is French and Barney's is English. The result is that Fred's thought is in French and he must

FIGURE 1-9

Layering in human communications

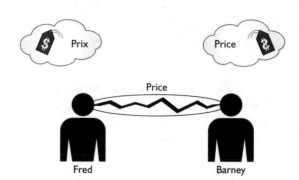

translate it into English before he speaks it. After the thought is translated into English, his brain must send signals to the vocal chords and mouth to transmit the signals of sound that result in English enunciation. Now the signals (sound waves) travel through the environment in which they are spoken until they reach Barney's ears. The eardrums receive these signals and send the received information to the brain. Here the information is interpreted and may or may not have been received correctly. Barney can send back a signal (verbal, visual, or kinesthetic) to let Fred know his understanding so that Fred can be sure Barney received the communication properly.

Do you see the similarities? Much as the Session layer represents data in a way that the remote machine can understand it, Fred's brain had to translate the original French thought into a shared language. Much as the Physical layer has to transmit electrical signals on a wired network, the vocal cords and mouth had to transmit signals as sound waves to Barney's ears. The point is that we could break human communications into layers that are similar to that which is defined in the OSI model. Also, the goal here is to provide peer communications from the "thought area" of the brain to another person's "thought area."

The most important thing for you to remember is that the OSI model is a reference and not an actual implementation. It is also useful to remember that data travels down through the OSI model on the sending machine and up through the OSI model on the receiving machine. Finally, remember that every device on a network will not need to extract everything within the encapsulated data in order to do its job. For example, a Layer 3 router can extract only to the point of the Layer 3 data and still route the data packets just fine.

exam
Watch

You will hear of many different techniques for memorizing the layers of the OSI model. While I can sympathize with these techniques for exam preparation, I encourage you to fully understand the communications process that occurs within the OSI reference model. When you remember what each layer does, it is almost automatic that you'll remember the layers in the proper order. This correlation is because communications should occur in the order in which the layers define them. It's really easy to remember a story, so think of the story of an e-mail message traveling down the stack and across the network to its destination.

Describe Networking Technologies Used in a Converged Network

If you've been working with data networks for any amount of time, you know that there are many different types of technologies and devices that are used to implement such a network. The good news is that converged networks use the same basic technologies and devices. They simply use them in a way that provides support for prioritization of voice and multimedia packets when necessary. In this section, we'll explore the various technologies used to make a network work. These include:

- Transmission media
- Signaling
- Switching
- Routing
- Network topologies

Transmission Media

Once data reaches the Physical layer, or Layer 1, it must be transmitted in some way. This data will be transmitted across a selected medium. The available media include both wired and wireless. For example, IEEE 802.11 wireless networks use the wireless medium, and IEEE 802.3 Ethernet networks use a wired medium. In this section, I will review the most common wired and wireless media available for utilization on our modern networks.

Wired Media

All wired media use cables of some sort for the transmission of data. The data will be transmitted either as electrical signals or as pulses of light across the cables. Coaxial and twisted pair cables transmit electrical signals, and fiber optic cables transmit pulsating light signals. In this section, I'll review the various wired cabling types, including coaxial, twisted pair, and fiber optic. I'll also cover different Internet connection types, such as dial-up, DSL, and cable.

Coaxial   *Coaxial* cable (also known as *coax* for short) is implemented with a center conductor made of copper surrounded by a sheathing made of some type of plastic.

Coaxial cabling

This sheathing is surrounded by a mesh of thinner copper wires that acts as a shield against electromagnetic interference (EMI) and helps contain the electrical signals within the cable to prevent leakage. Finally, this mesh is covered by a plastic coating often called a *plenum-rated coating*. Plenum-rated coating is a special plastic that is rated for use in air plenums that contain ventilation duct work. Figure 1-10 shows an example of coax cabling.

In computer networks, coax is usually used with BNC connectors, terminators, and T-connectors in a bus topology. Figure 1-11 shows each of these three items. The use of a bus topology means that it is usually used in smaller networks today if it is used at all. It is much easier to work with twisted pair cabling, and this ease of use means that it is the popular technology in all new installations and large implementations. As you will learn later in this chapter, the bus topology reaches a maximum device threshold very quickly due to its method of signal transmission. You may encounter coax cabling in small networks that have existed for more than seven to ten years; however, even that use is rare today. Certainly, you are unlikely to see coax used in any environment that is implementing a Voice over IP solution as is a major focus in this book.

The connectors are important to understand if you need to implement a coax-based network. The T-connector connects to the back of the network interface card (NIC) in your computer, and coax cables with BNC connectors will be connected to either side of the T-connector. The device at each end of the bus will have a terminator connected to the "end" side of the T-connector. This configuration is depicted in the diagram in Figure 1-12. RG-58 cable is most commonly used in computer networks.

FIGURE 1-11

Coax connectors

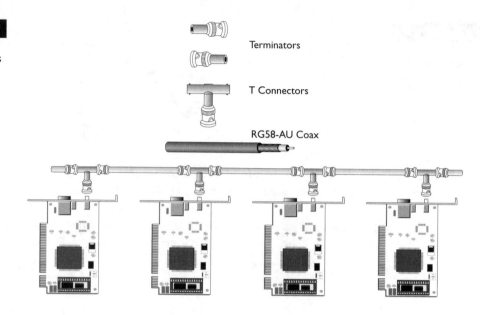

FIGURE 1-12

BNC connectors
and coax cable

Terminators

T Connectors

RG58-AU Coax

Coax cable, implemented as Thinnet or RG-58 cable, has a maximum distance per segment of 185 meters. Most implementations of coax cabling today are Thinnet implementations. Thicknet, another coax standard, allows segments with lengths up to 500 meters (due to a thicker cabling called RG-8). This is even less common than the Thinnet cabling I've chosen to focus on here. There is an additional coax cabling type, RG-62, that was used in ARCNet networks. This one is definitely only in the history books for modern network engineers. The only other coax cable type that you may encounter is RG-6, which is a cable even thicker than RG-8. RG-6 cables are used for cable television (CATV) connections and, therefore, also cable modems. It is the implementation of a cable modem for high-speed Internet connectivity where you may encounter RG-6 cabling in your network engineering work.

e x a m

ⓦatch   *RG-8 coax is used in 10BASE-5 Thicknet installations, and RG-58 is used in 10BASE-2 Thinnet installations. Both installations have a maximum data rate of 10 Mbps. The 5 in 10BASE-5 indicates 500 meters, which is the maximum length of a Thicknet segment,*   *and the 2 in 10BASE-2 indicates 200 meters (though it's actually 185 meters) as the maximum cable length in a segment for Thinnet. This limitation of 10 Mbps is one of the major reasons coax is rarely used in modern networks.*

FIGURE 1-13

UTP cable
example

Twisted Pair   Without question, the most common network medium used in modern networks is the twisted pair cable. It is thinner and easier to work with than coax and works by implementing multiple conductor wires instead of just one center wire. These wires are twisted in pairs, hence the name *twisted pair*. Two kinds of twisted pair cable types exist: unshielded twisted pair and shielded twisted pair. Most IEEE 802.3 Ethernet networks are implemented using unshielded twisted pair cabling as the medium.

Unshielded twisted pair (UTP) is implemented as an even number of wires twisted together in pairs and enclosed in an insulating sheath. Shielded twisted pair (STP) is implemented in the same way except the individual pairs are also insulated by a foil shield. This foil shield helps insulate the twisted pairs from each other within the STP cable. Due to the lack of standards, STP cables are rarely used. Figure 1-13 shows an example of a UTP cable, as this is the most common type implemented in IEEE 802.3 Ethernet networks.

UTP cables are classified in different categories and use wire *pinouts* (or connection patterns) that are defined in the Electronic Industries Alliance (EIA)/ Telecommunications Industries Association (TIA) 568 Commercial Building Wiring Standard. Table 1-1 provides a listing of the categories that are defined in the EIA/TIA 568 standard. Note that these categories are often read or written as, for instance, CAT4 for category 4 or CAT5 for category 5.

While telephone cables usually use an RJ(registered jack)-11 connector, network cables use an RJ-45 connector and jack as seen in Figure 1-14. This connector plugs into RJ-45 ports in network cards, switches, routers, firewalls, wall mounts, hubs, and many other networking devices.

Fiber Optic   *Fiber optic* cable is a high-speed cabling technology that transmits light across glass fibers instead of electricity across copper wires. In the center of the fiber optic cable is a flexible plastic or glass that is clear. This core is surrounded by a reflective cladding material that is in turn surrounded by a protective sheath.

TABLE 1-1	UTP EIA/TIA Cable Classifications

Category	Application
Category 1	Traditional telephone connections. This is considered voice grade cabling and is not recommended for data.
Category 2	Provides bandwidth of up to 4 Mbps and includes four pairs of wire (eight total wires). This category is rarely used due to its limited bandwidth.
Category 3	Provide bandwidth of up to 10 Mbps and includes four pairs of wire, as do all UTP cables. This category implements signaling rates up to 16 MHz and may still be seen in some 10BASE-T Ethernet implementations, though it should be considered obsolete at this time.
Category 4	This is the first category listed as data grade by the EIA/TIA and can provide up to 16 Mbps. Because it cannot provide 100 Mbps, it is not much more useful than CAT3 and is not commonly used even though it will support 10BASE-T Ethernet at 10 Mbps. This cable provides a signaling rate of up to 20 MHz.
Category 5	This is the most common UTP cable used in the first decade of the new millennium. It provides up to 100 Mbps and a signaling rate of up to 100 MHz. 100BASE-TX utilizes either CAT5 or CAT6 cabling. There is also a CAT5e cable that is useful for 1000BASE-TX connections running at 1000 Mbps or 1 Gbps, depending on the syntax you prefer.
Category 6	CAT6 is the most commonly recommended medium for 1 Gbps connections. The same jack is used for CAT5 and CAT6 cables (an RJ-45 jack), so the CAT6 cables are backward compatible. CAT6 is rated for signaling up to 200 MHz.

The light is reflected off the cladding material and passes down the glass fiber so that practically all the transmitted light reaches the receiving end. Figure 1-15 shows an example of fiber optic cable in diagram form.

Signals are transmitted by injecting light into/onto the fibers with pulsations. Think of the childhood game of flashlight signals. I remember using a flashlight to

FIGURE 1-14

RJ-45 jack

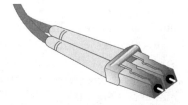

FIGURE 1-15

Fiber optic cable

send signals to my cousin. We used a long light for vowels and short flashes of light for various consonants. This system allowed us to have one signal for all vowels and to use our human intelligence to figure out which vowels fit with the provided consonants. The end result was a signaling system based on pulsating lights.

Computers can "read" these pulsating lights and process them much faster than us mere mortals, and so they can transfer data at much higher rates. Where my cousin and I would require more than three or four minutes to transmit a single sentence, fiber optic cabling can be used to transmit a twenty-volume encyclopedia set in the same amount of time. Theoretically, fiber optic cabling can transmit data at speeds up to 50 Gbps and can do this over very long distances of up to 30 kilometers.

The speed and distance will depend on the mode of fiber implemented. Single-mode fiber will allow for longer distances (up to 30 km) and higher bandwidth, but it is more expensive than the alternative: multimode fiber (only 2 km distance). Single-mode fiber uses a single path through the fiber. You might call it a straight-through fiber implementation. Multimode fiber uses multiple paths through the fiber, and you might call it reflective fiber, since it uses reflection algorithms to process multiple paths. Table 1-2 provides an overview of the three main wired cabling types.

Dial-Up Versus High-speed   In today's networks, Internet connectivity has become essential. Very few organizations have no need for the Internet. At the very least, e-mail is utilized. When connecting your network to the Internet, you have two fundamental choices: very slow or faster. This choice may seem like an oversimplification, but with the current technology, it is also the reality. In the 90s, I remember dialing up to the Internet and then downloading one to ten megabyte files and not feeling frustrated by the fact that it took between ten minutes and ten hours, depending on my connection quality. When the 56k modems came along, I thought I was in heaven and had reached the pinnacle.

TABLE 1-2

Cabling Types, Speeds, and Distances

Cabling	Speed	Distance
Coax (RG-58)	10 Mbps	185 meters
UTP	4 Mbps to 1 Gbps	100 meters
Fiber	100 Mbps to 10 Gbps	2 km (multimode) or 30 km (single-mode)

Of course, these faster modems were quickly followed by ISDN and then DSL and even cable and satellite technologies that were much faster. The reality is that a dial-up line to the Internet is not likely to provide a fast enough connection for shared Internet in any installation. Even if only two or three users are browsing web pages, it will be unacceptably slow. Today, you will need DSL or business-class cable at a minimum. Larger companies will need dedicated or partial T1s and faster connections.

For this reason, your decision is not really between dial-up and high-speed Internet connections. The decision is to be made among the various high-speed technologies. If your organization has a few dozen people or more that will need access to the Internet (e-mail, web browsing, etc.) concurrently, you will likely need to acquire a high-speed T1 or fractional T1 line from a local provider. If your organization requires that only a few individuals access the Internet concurrently, you may be able to make your decision between the less expensive cable or DSL options.

DSL Versus Cable The Digital Subscriber Line (DSL) service has been a phenomenal option for small businesses and home users for a few years now. It provides speeds up to 52 Mbps on Very High Bit Rate DSL (VDSL), though this standard greatly limits the distance between the service provider and the subscriber unless fiber cabling is used. Asymmetrical DSL (ADSL) is the more common implementation in consumer and small business installations. ADSL provides up to 6 Mbps (6000 Kbps) down speeds and up to 640 Kbps up speeds. This means that you can download faster than you can upload. This is an important point of decision if you are implementing any locally hosted services that must be accessed from the Internet side across the DSL connection. Table 1-3 provides a breakdown of the common DSL types and their features. It is important to remember that the actual speed of your DSL connections will depend on line quality, distance from the provider, and the speed of the provider's core network.

Cable Internet service has been in existence for more than seven years now and is very popular in many countries in larger metropolitan areas. This service is sometimes called *broadband cable* or *high-speed cable Internet*. Business-class cable Internet service can provide data rates of greater than 50 Mbps, while consumer grade service is usually less than or equal to 10 Mbps.

DSL is a dedicated technology. This means that your connections should be very stable. What you get one day is very likely to be what you'll get the next. Cable Internet is a shared technology comparable to Ethernet. Your bandwidth will vary depending on the utilization of the network by other subscribers. Business-class cable subscriptions can provide guaranteed bandwidth, but consumer-class

TABLE I-3	DSL Types and Speeds

DSL Type	Speeds	Distance from Provider
ADSL	6 Mbps down and 640 Kbps up	Usually less than 3 km
ADSL Lite (also called G.lite ADSL or Universal DSL)	1.5 Mbps down and 512 Kbps up	Usually less than 3 km
Rate Adaptive DSL (RADSL)	Variable line speeds adjusted based on current conditions with maximums equal to ADSL	Usually less than 3 km
VDSL	52 Mbps down and 16 Mbps up	About 300 meters for maximum bit rate; bit rate degrades as the signal attenuates (or the bit rate decreases over longer distances)
Symmetric or Single Line DSL (SDSL)	2 Mbps down and up	Usually less than 3 km

connections usually provide up to a certain bandwidth with either minimum bandwidth guarantees or no guarantees. Pay close attention to the contract when signing up for cable or DSL connections. You want to make sure you have the bandwidth you need for your intended use.

Wireless Media

Wired networking is not the only game in town. Wireless networking has become extremely popular in the last decade. There are many different wireless networking standards and technologies and they vary in their implementation, but one thing is consistent: electromagnetic waves. They all use electromagnetic waves in order to transmit and receive data. In this section, I will briefly introduce the wireless media that are available. Some are used in wireless LANs (WLANs), and others are used in wireless MANs or even wireless WANs.

Line of Sight   Technically, any wireless technology can be made a line of sight technology by using highly directional antennas; however, some technologies are implemented with the intention of utilizing line of sight for communications. When two wireless endpoints communicate with each other in a highly directional, point-to-point fashion, they are said to be *line of sight* connections. Line of sight connections are used for bridge connections that connect two otherwise disconnected networks and for high-speed connections to data centers for distant areas in a facility. Figure 1-16 depicts a line of sight wireless link.

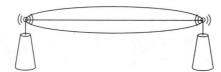

Non–Line of Sight The technologies used in WLANs are non–line of sight technologies. This is because they use semidirectional or omnidirectional antennas. Semidirectional antennas transmit the radio frequency (electromagnetic waves) signals in a wide path in one direction from the antenna, whereas omnidirectional antennas transmit the signal in all directions fairly evenly around the antenna. The result is that semidirectional antennas provide coverage over a greater distance in a specified direction, and omnidirectional antennas provide coverage over a lesser distance from the antenna in all directions. Figure 1-17 depicts a non–line of sight wireless configuration.

Satellite Satellite technology has come into common use for both television service and high-speed Internet service, particularly in rural areas where DSL runs would be too long or cable service is not provided. Additionally, satellite is an excellent technology—assuming the bandwidth is sufficient—for both land and sea mobile stations. Satellite Internet connections have traditionally used *one-way with terrestrial return*. In other words, the data received from the Internet is transmitted from the satellite to the receiver and the date transmitted to the Internet is sent through a terrestrial connection (either a land line or a mobile phone). *Two-way* satellite Internet is much more expensive and requires line of sight. Therefore, two-way satellite installations require a well-trained engineer to install and align the equipment. One-way with terrestrial return is great for mobile and stationary connection requiring mostly downloads and little in the way of uploads, and two-way is suitable for stationary applications such as rural areas without other service provision.

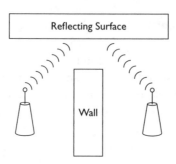

Wi-Fi   The most popular wireless LAN technology is Wi-Fi, which is based on the IEEE 802.11 standards. The last five or six years have seen an explosion of interest and implementations in the Wi-Fi marketplace. I'll discuss Wi-Fi in more detail in Chapters 3 and 4 as we investigate infrastructure and client devices.

EVDO   *Evolution Data Optimized,* or *EVDO,* is a broadband mobile wireless standard that has been implemented by several carriers around the world. In the United States, both Verizon Wireless and Sprint Nextel provide EVDO service. EVDO service typically provides down speeds of between 512 Kbps and 1.4 Mbps. At my home in Ohio I can download at close to 1 Mbps in the right area (I have to be in the corner of the laundry room for best reception—it might not be ideal, but it works). Thankfully, I have 6 Mbps DSL that really works at 6 Mbps piped into my house.

EVDO is not the only game in town when it comes to cell-phone company high-speed Internet provision. Cingular (now AT&T), for example, provides its High-Speed Downlink Packet Access (HSDPA) broadband service. The HSDPA averages between 400 and 700 Kbps down speeds and acceptable latency (the minimum time required to move data from one point to another).

Using Virtual Private Networks

Whether your connection to the Internet is wired or wireless, you can create a virtual WAN by tunneling your organization's data through a *virtual private network (VPN)*. A VPN is really nothing more than a session between two endpoints that encrypts all data transmitted across that session. It is not uncommon for small businesses to create WAN links by purchasing high-speed Internet connections at each physical location wherein the business operates and then establishing VPN connections between these locations. As long as secure tunneling technologies are used, the data is protected against eavesdropping and other attacks and the WAN connection can be very inexpensive.

VPN connections can be created across DSL, cable, satellite, EVDO, and any other Internet connection type that provides an always-on connection or does not "timeout" the connection. In the past, many dial-up service providers have disconnected a user after a period of inactivity to free the phone line for other subscribers. DSL, cable, and other types of Internet connections discussed here do not use these timeout mechanisms and are considered always on.

Most VPN connections start at either Layer 2 or Layers 4–7. For example, an L2TP (Layer 2 Tunneling Protocol) VPN creates a Layer 2 connection and then uses IP Security (IPsec) to encrypt the data that passes through the Layer 2 tunnel. This encryption allows IP traffic to be encapsulated in the tunnel in a secure fashion. You will learn more about VPNs in Chapter 13 as you learn about methods used to secure your converged network.

Signaling

Depending on context, signaling can mean many different things in the telecommunications industry. When talking about the Physical layer of a data network, signaling references the method used to transmit data. When talking about the public switched telephone network (PSTN), signaling may refer to the conversion of sound waves to analog electrical signals or it may refer to the alerts detected related to line conditions and other factors that often fall into the categories of supervisory, address, alerting, or control signals. In this chapter, since we are laying the foundation of our data networks on which we will later transfer our voice traffic, I will focus on signaling from the perspective of transmission methods. Throughout the book, I'll come back to signaling again and again as the topics require.

It's All Binary

One fact that is important to remember is that computers only work with binary data. If the computer is going to manipulate an image, a sound file, or a spreadsheet, it must be able to read and work with it in a binary fashion. This is because computers think in 1's and 0's, unlike those of us who think in base 10. This difference is important to keep in mind because analog information (information with wide and smooth variations), such as voice signals, will ultimately have to become digital information (information represented by strict contrasts—aka 1's and 0's). With that said, let's look at analog and digital signals.

Analog Versus Digital

Figure 1-18 shows a typical representation of an analog signal. Notice how the signal can be represented in graphical form as having both wide and smooth (or gradual) variations. Your voice is this way. It can smoothly transition up and down from

FIGURE 1-18

Analog signal

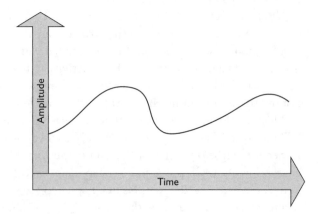

FIGURE 1-19

Digital signal

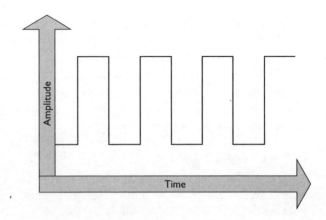

one frequency to another in a melodic fashion. The human voice also can sharply change from one frequency to another such as when you mother was yelling at you and then the phone rang, which resulted in the sudden change to a soft and serene voice as she spoke to the person on the other end of the connection and sweetly said, "Hello?" The human voice is an analog transmitter, and the sound waves can be said to be analog signals.

If this human voice is to travel a data network, the sound waves must be converted to digital signals. Figure 1-19 shows a typical representation of a digital signal. Notice that a digital signal uses only sharp contrasts to represent 1's and 0's. You will learn in Chapter 7 all the details of how the analog human sound wave is converted into digital computer data. For now, just remember that there are two kinds of signals: digital and analog. Digital signals are binary in nature, and analog signals are variant in nature. We'll come back to this on several occasions through the book.

Switching

Telecommunications networks can implement two primary kinds of switching: circuit switching and packet switching. *Circuit switching* is used to reserve a route or path between the two endpoints that need to communicate. Because a circuit is reserved, the entire communication is sent in sequence and there is no rebuilding of the data at the receiver, as it is certain to arrive in order. Of course, this reservation means the bandwidth cannot be utilized by any other devices that may need it and can make circuit switching rather costly in today's packet-switched world. The benefit of circuit switching is that the connection is always there and the bandwidth is guaranteed as long as the connection exists.

Packet switching (also called *datagram switching*) is used to segment a message into small parts and then send those parts across a shared network. The first part may actually travel a different route than the second part and could in fact arrive at the destination after the second part. Voice over IP implementations, which are a large focus of the later chapters in this book, rely on packet switching as opposed to circuit switching. This design does introduce concerns because the voice data must arrive quickly at the destination or calls can be dropped and sound quality can suffer. You will learn how to deal with those issues in Chapter 9, which focuses on Quality of Service technologies.

The term *switching* can also represent the actions carried out by a network switch. In fact a network switch is a device that performs packet forwarding for packet-switched networks. A switch can forward packets from an incoming port to the necessary outgoing port or ports in order to enable the packet to reach its destination. It is inside of these switches—as well as the routers I'll talk about next—that much of the Quality of Service processing is performed. The switch can extract a frame and determine if it has Quality of Service parameters and, if it does, treat it accordingly. You'll learn much more about switches in Chapter 3.

Routing

Routing is the process of moving data packets from one network to another. A data packet that is transmitted from a computing device may be able to move directly to another device on the same network, or it may need to be forwarded to another network by a router. This is the primary job of a router: to connect otherwise disconnected networks.

Here's a good way to remember the difference between a switch and a router: if you connect multiple switches together, you're just creating a bigger physical network segment. The same is not true with routers. In fact, you should really think of routers as being connected together. Instead routers have two or more interfaces. As seen in Figure 1-20, one interface will be connected to one network and the other interface will be connected to another. This demarcation allows the router to be used as a packet routing device when a device in Network A wants to send a packet to a device in Network B. You'll learn more about routers in Chapter 3 as well as switches and other infrastructure devices.

FIGURE 1-20

Routers routing packets

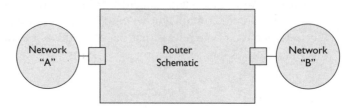

Network Topologies

At this point, you have a foundational understanding of the fact that computers can be connected together to form networks using various connection media and devices such as switches and routers. In addition to this knowledge, it is important that you understand the different types of networks that you can build and the topologies that you can implement within those networks. Of the network types, you will need to be familiar with LANs, MANs, WANs, and GANs for the Convergence+ examination.

LAN

A LAN, or local area network, is usually defined as a group of computing devices connected by a high-speed communications channel that is localized to a campus or single property. A LAN would not be inclusive of the Internet, though it may be connected to the Internet. LANs may be connected together over distances measured in miles or kilometers, but these connected LANs would still be separate LANs, though together they may form a WAN or a MAN.

LANs can be implemented using many different topologies, including bus, star, mesh, and hybrids. The following sections describe each of these.

Bus   The *bus topology*, as depicted in Figure 1-21, requires that all communicating devices share a single bus or communication medium. This bus is usually a coax copper wire that is connected with BNC connectors and BNC Ts and terminators using 50-ohm cables and connectors. The biggest problems with the bus topology are in the maximum device threshold and the single point of failure problem.

FIGURE 1-21

Bus topology

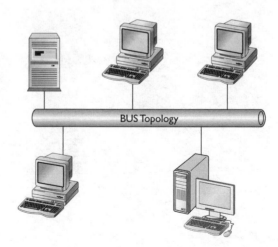

Because all devices share the same bus, you can quickly overwhelm a bus topology. This limitation is because communications occur when a device transmits a signal (frame) on the bus. Only one device can transmit at a time, and this can result in greatly diminished overall bandwidth. For example, there is just 10 Mbps available in common bus topology implementations that use coax cabling. If ten computers were on the bus—even ignoring network management overhead—each computer would only have an average of 1 Mbps available to it when all devices need to communicate. If 100 devices were on the bus, well, you get the picture. The bus becomes saturated very quickly.

Additionally, due to the way the frames are passed up and down the bus and the large number of connectors along the way (each computer introduces a new T-connector), there are many potential points of failure. If one T-connector goes bad, the whole bus shuts down. If one computer is disconnected and the technician fails to couple the bus cable after disconnecting the computer, the circuit is broken and the bus shuts down. As you can see, the bus topology is not the ideal topology for modern networks—at least not by itself.

Star Figure 1-22 shows a star topology. The *star topology* is a hub-and-spoke type of network. All the devices communicate back to the central "hub," and the "hub" passes the information out to the proper "spoke." We rarely use real hubs anymore because they are not as efficient as switches in their utilization of the medium. Hubs receive information from one port and flood it out to all other ports whether they all need it or not. Switches, on the other hand, learn about the devices connected to each port and then only forward information to needed ports as much as possible. Switches or hubs can be used to form a star topology.

FIGURE 1-22

Star topology

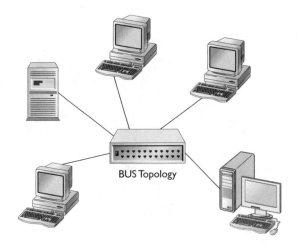

BUS Topology

Of course, you still have a single point of failure. If the switch itself should crash, the star goes down. However, it's much easier to troubleshoot a failed switch that it is to locate the point on the bus where a failure has occurred. For this reason—and the more efficient use of bandwidth—star topologies are much more common in modern LANs.

Mesh   A *mesh topology* is a network structure that includes redundancies for fault tolerance and/or increased bandwidth. Figure 1-23 shows an example of a mesh topology. With this structure, there are multiple routes from and to any endpoint. For example, if router C should fail, the nodes from segment 1 could reach segment 2 through router B. The same is true in reverse: should router B fail the nodes could reach each other through router C.

Hybrids   A *hybrid topology* is any topology that blends bus, star, or mesh. For example, the star-bus (also called bus-star) topology often uses a bus topology to connect the infrastructure and a star topology for connecting nodes to that infrastructure. Figure 1-24 shows an example of a star-bus topology.

FIGURE 1-23

Mesh topology

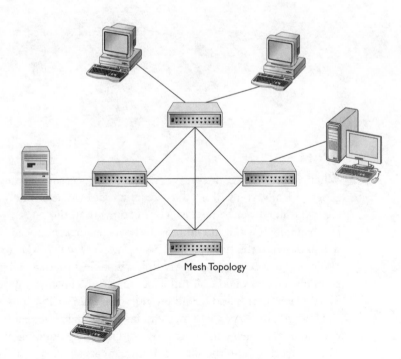

Mesh Topology

Star-bus topology

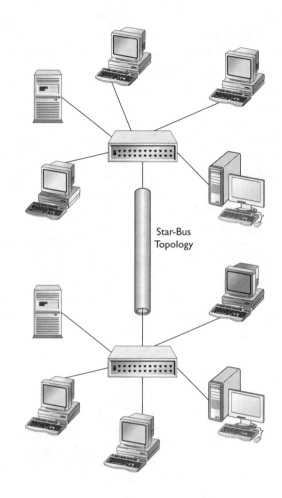

Star-Bus
Topology

WAN

When you need to connect to LANs together that are separated by miles geographically, you will need to create a *wide area network (WAN)*. WANs are created using lower bandwidth connections than those used within the LAN. For example, it is not uncommon to have gigabit core speeds and 100 megabit speeds down to the nodes on LANs today. Very few WANs would support even 100 megabit speeds. Most WAN link will be less than 50 Mbps. This lower speed is simply a factor of cost. A full T3 connection can cost anywhere from $7500 to $12,000 per month and would provide speeds up to 45 Mbps.

However, this speed variance can be overstated. The reality is that most users need to communicate more with users in the same building or location as they are. That's why they're in the same buildings. Years ago, before networks were even considered, the industrial revolution led to large offices for people who managed and administered the workers and products that were being developed in the factories.

The optimization movements of that era led to the collocation of employees who needed to work with each other frequently. This has not changed much today. Even with all of the talk of telecommuting, most people work within a three to five–minute walk of mostly everyone they need to communicate with.

There are, of course, exceptions, like salespeople and others who spend more time communicating with those outside the company, but these are the exceptions and not the rule. This real-world example is why I say that too much emphasis can be placed on "slow" WAN links as opposed to LAN speeds. We don't need as much bandwidth on most WAN links.

With that said, Voice over IP and multimedia over IP can potentially impact this model; however, my experience working in companies with 25,000 people and more tells me that it will still be a percentage factor. WAN links are usually fine as long as they are between 20 percent and 35 percent as fast as the local network.

Many technologies can be used to implement WAN links, including DSL, ISDN, ATM, and Frame Relay. It is beyond the scope of this book to cover these WAN technologies in any greater detail, but you should know that they exist and the general concept of a WAN as opposed to a LAN.

MAN

A *metropolitan area network (MAN)* is a network that is usually established by the local municipality or another service provider. The network will span either an entire city or portions thereof and can act as the carrier for traffic between locations in the city for multiple organizations. Figure 1-25 represents this concept. Notice that the network is used by multiple companies that lease bandwidth or time on the network. This way the companies can simply subscribe to the MAN and do not need to worry about purchasing the components needed to form connections across the city themselves. MANs may be developed using wired or wireless technologies.

FIGURE 1-25

Metropolitan area network

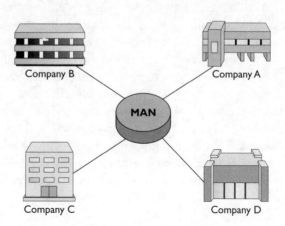

Company B

Company A

MAN

Company C

Company D

If a MAN is implemented by the local telco, it may be wired; however, most MANs that I've worked with have been wireless when implemented by municipalities or private organizations other than the local telco.

GAN

Currently, the Internet is the primary example of a global area network (GAN). However, many refer to the connection between two to LANs as a WAN link or a WAN connection and they refer to the entire corporate network as the GAN regardless of whether it spans the globe or just the eastern United States. Therefore the simplest definition of a GAN is a group of interconnected LANs or WANs that cover an unrestricted geographic area.

CERTIFICATION OBJECTIVE 1.03

Identify the Functions of Hardware Components Used on a Converged Network

The Convergence+ exam expects that you are familiar with various hardware components used in a converged network. These components include servers, in-line power components, wireless devices, and modems. These devices are introduced in this section.

Servers

Something that often shocks many new entrants into the computer networking knowledge domain is the discovery that all network devices are potentially servers and may also be clients. Even a network router is a server. In fact, a network router is nothing more than a computer in a special enclosure without monitor, keyboard, and mouse connections. A router has a processor, memory, and, in some rare cases, a hard drive. In most cases, a router uses non-volatile random access memory (NV-RAM), but the concept is the same: we can store data that will be saved during a power cycle or reboot. The point is that a server is simply a computer that provides services to the network. For this reason, we usually classify a desktop computer as a client instead of a server, but a desktop computer has all the hardware capabilities of being a server. It is simply the intended use of the item that determines if it is a server or a client.

Authentication Servers

In secure networks, devices must be authenticated before they can communicate on the network. This usually requires an authentication server. The authentication server may be a service running on a server that also offers additional services, or it may be a dedicated server or network appliance. An example solution that requires an authentication server is IEEE 802.1X port-based authentication. This authentication method is used mostly in WLANs, but it can also be used and in fact was designed for wired LANs.

Authentication servers can be as simple as Linux computers running a service that uses a flat text file of usernames and passwords to something as complex as IBM Tivoli Identity Manager. The former provides basic authentication and usually only allows access to local resources on the Linux computer or network of Linux computers. The latter can integrate with other systems, such as Windows Active Directory, and provide centralized, policy-based management of user identities. Regardless of the strength and features, authentication servers are an important part of modern networks.

File Servers

Certainly one of the earliest server types seen on computer networks was the file server. In fact, one of the most commonly uttered phrases in the early 90s was "file and print," and it referenced the most common service provided on a network. Today, of course, e-mail and database access far outpaces file and print services in many companies. However, file servers (as well as print servers) are still a staple in most networks.

At stage one, file servers imply provided a place to store files so that they could be accessed by multiple users. It quickly became obvious that security problems existed in this model, and methods were implemented to allow only authorized users into the files and folders (mostly called directories or listings back then). Another capability that was introduced was versioning in order to allow users to recover previous versions of files without needing assistance from administrative staff. Finally, file servers today have the ability to cache and synchronize files with individual client computers and the ability to encrypt data that is stored on the server.

Media Servers

The services which have traditionally been performed by a private branch exchange (PBX) is now frequently performed by *media servers*. Media servers are also called communication servers, call servers, call manager servers, and media gateway controllers. Media servers will be discussed in more detail in Chapters 6 and 7.

Communication and Collaboration Servers

While e-mail servers fall into the category of a communication server, the concept of networked collaboration is more recent. Early solutions included Lotus Notes and Novell Groupwise, and newer solutions include Lotus Notes (still around after all these years), Microsoft Exchange, and Microsoft Sharepoint services. The communication and collaboration servers allow users to communicate with each other via e-mail, chat, and possibly Voice over IP; however, they also allow users to share files, collaborate on the creation of content, and organize projects and priorities. This server category that seems to have gotten much of the attention in the past three to five years.

Database Servers

The final server type that I will cover here is the database server. Oracle, SQL Server, MySQL, and other database server technologies are used in organizations all over the world. The database server acts as a repository for data and provides access to that data. Additionally, modern database servers allow for business logic to be enforced within the database instead of relying on the applications to enforce this logic. Database servers are accessed by clients directly and are heavily used as the back end for Internet web sites.

Inline Power Components

Power over Ethernet (PoE) has proliferated in the WLAN market and is used by many wireless APs. It is also supported by some switches, bridges, and other devices. PoE is a method used to deliver a DC voltage to a device over CAT5 or CAT6 cable. This DC voltage is used to power the device instead of a standard AC power outlet. (Most devices come with transformers that convert AC power to DC power. PoE sends the power directly as DC power.)

CAT5 cables have four pairs of wires in them. Two pairs are used to carry 10BASE-T and 100BASE-TX data. This leaves the other two pairs for other purposes. In the case of PoE, the purpose is to carry power to the device being powered. Some implementations use the same pairs of wire that carry the data to carry the DC voltage, and some implementations use the extra pairs in the CAT5 cabling to carry the DC voltage separate from the data.

One of the most common reasons for using PoE is to power a device where no AC power outlets are available. The other benefit of implementing PoE is the ability to cycle the device being powered from remote. This latter feature is usually only available when the PoE is being provided by a managed switch. The management interface of the switch will allow you to turn off the power on a given PoE-enabled port and then turn it back on. Power cycling is not supported by all PoE-enabled switches.

FIGURE 1-26

PoE with an inline
injector

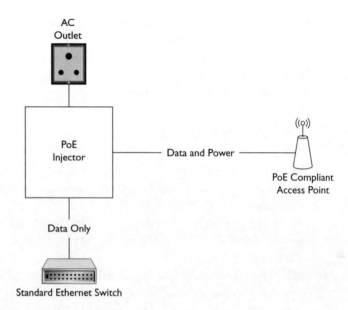

Yet another advantage of using PoE is that a licensed electrician is not usually
required to install it. This is because the voltage is so low that is running across the
CAT5 cabling. Most technicians can run the cables and use PoE. There will likely
be no building codes that will dictate specific guidelines for running the cabling and
powering the end location. Figure 1-26 shows the way PoE would be utilized with an
inline PoE power injector, and Figure 1-27 shows the way PoE would be utilized with
a PoE-enabled switch.

FIGURE 1-27

PoE with a
PoE-enabled
switch

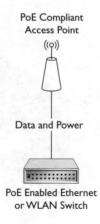

It is important that you know that PoE-enabled switches do not always provide power through all ports. Sometimes half of the available Ethernet ports are PoE-enabled, and sometimes fewer than half can provide DC power to devices. Be sure to check the vendor's documentation to verify the number of PoE ports being provided by the switch you are implementing.

Common Features

There are different types of devices that can provide voltage through CAT5 and CAT6 cables, which power PoE-enabled devices. These types include single-port DC voltage injectors, multiport DC voltage injectors, and PoE-enabled switches.

The single-port PoE injectors will have a single input port and a single output port. The input port is where you connect the Ethernet cable that connects to the network's switch or hub, and the output port is where you connect the Ethernet cable that connects to the device to be powered. Figure 1-28 shows an example of a single-port PoE injector from ZyXEL Technologies (ZyXEL PoE-12). When using a single-port PoE power injector like this one, the power injector itself must be plugged into a standard power outlet. This power requirement means you will likely place the power injector in the closet (or location) with the switch or hub and not closer to the device being powered. Due to the number of power outlets required, single-port power injectors are only recommended when one or two devices need to be powered.

A multi-port PoE power injector is really just a group of Ethernet input ports that pass through a power injection module and then pass on to a matching group of Ethernet output ports. These devices are usually also installed closer to the switch or hub and farther from the powered device. This placement is due to the likelihood of having a power outlet where the switch or hub is located, since the switch or

FIGURE I-28

Single-port
PoE injector

hub will need power as well. Due to the fact that multiport power injectors can power multiple devices while only require one power outlet connection, they are recommended in medium to large installations that require from three to twenty (though this guideline is not an absolute cutoff point) powered devices. When more devices require power, you will likely opt for a PoE-enabled switch.

Large enterprises and networks with more than twenty PoE-powered devices will likely choose to move up to PoE-enabled switches. These switches include power sourcing in the same unit that is the Ethernet switch. It means powering fewer devices through standard power outlets and reducing the number of components that can fail at any given moment. When you use a single-port power injector with a wireless access point, for example, you introduce multiple points of failure. Imagine there are twenty APs that you need to power in this way. You would need 40 CAT5 cables (20 from the switch to the power injectors and 20 from the power injectors to the APs), 20 power injectors, 20 power cords, 20 APs, and at least one switch. These numbers mean a total of 101 individual components could fail and statistically increases the likelihood that you will have a failure at any given time. If you use a switch like the Cisco 3750-E switch that can provide PoE power on up to 48 ports, you reduce the components involved to only 41 components. You've eradicated the need for 20 CAT5 cables, 20 power cords, and 20 power injectors. The likelihood of a failure at any given moment has now been greatly reduced.

In addition to the failure probability reduction, you are gaining the benefit I discussed previously of being able to power-cycle the APs from a central location. With the single-port power injectors, you would still have to go to the physical location where the power injector is located and unplug it and then plug it back in (or flip an on/off switch if it is available). Figure 1-29 shows the Cisco 3750 line of switches with PoE capabilities. The 3750-E model is the one that provides power sourcing as a possibility for all 48 ports; however, it cannot provide the full 15.4 watts of power to all ports concurrently as required by 802.3af.

FIGURE 1-29

Cisco 3750
switch with
PoE capabilities

An additional benefit of PoE-enabled switches is that you do not usually have to enable PoE on all ports. For example, you can use some of the ports for wired devices or non-PoE APs and bridges while you use the other ports for PoE-enabled devices. This provides you with flexibility and is a valid argument for purchasing a switch that supports PoE from the factory or at least purchasing one that can have PoE support added at a later time.

Power over Ethernet (PoE) (IEEE 802.3-2005, Clause 33)

IEEE 802.3-2005 merged the older IEEE 802.3af PoE amendment into the core standard document. The old amendment is now known as Clause 33 in the IEEE 802.3-2005 document. Many, even most, vendors—at this time—are still referencing the standard as IEEE 802.3af, but you should know that it has been rolled into the primary standard now. If you download or access the IEEE 802.3-2005 standard in sections, Clause 33 is in the section two PDF file.

The standard defines a Powered Device (PD) and Power Sourcing Equipment (PSE). The APs we've discussed that support PoE would be examples of PDs. The power injectors and PoE-enabled switches would be examples of PSEs. The clause specifies five elements:

- A power source that adds power to the cabling system
- The characteristics of a powered device's load on the power source and cabling
- A protocol allowing the detection of a device that requires power
- An optional classification method for devices depending on power level requirements
- A method for scaling supplied power back to the detect level when power is no longer requested or needed

The standard then spends the next 57 pages providing the details of this system. You will not be required to understand the in-depth details of PoE for the Convergence+ exam, but this IEEE document can act as your source for more information; however, you should be familiar with the following two terms: midspan and endpoint power injectors.

The standard specifies that a PSE-located coincident (in it) with the switch (technically, the data terminal equipment or DTE in the standards) should be called an *endpoint PSE*. It also specifies that a PSE located between the switch and the powered device should be called a *midspan PSE*. WLAN switches and LAN switches with integrated PoE support would qualify as endpoint PSEs, assuming they are IEEE 802.3-2005–compliant. Multiport and single-port injectors would qualify as midspan PSEs, assuming they are IEEE 802.3-2005–compliant.

Fault Protection

One final note about PoE: fault protection is very important. Fault protection does the work of protecting the devices that are being powered by power injection or that are providing the power injection. A fault occurs when a short-circuit or some other surge in power occurs in the PoE chain. Faults can occur for the following reasons:

- A device does not support PoE and uses the extra two pins used by PoE or for some reason short-circuits the pins.
- An engineer connects an incorrectly wired CAT5 cable.

Due to the nature of things, the last cause seems to be the most common cause. I know I have inadvertently "miswired" a twisted pair cable or two in my time. It's fairly easy to do, since you're dealing with small wires using big fingers and crimpers that haven't been upgraded or improved for a few decades. When a fault occurs, the power source should shut off DC power onto the cable in the path of the fault. Depending on the power injection device, you may need to manually reset the power injector or it may monitor the line and automatically reset when the fault is cleared.

Wireless Access Points

Wireless access points (APs) come in many different configurations, but there are two primary classifications: thick and thin APs. A *thick AP* is an AP that contains all the logic necessary to authenticate and manage wireless client associations. In modern wireless local area network (WLAN) terminology, a thick AP is implemented in a *single MAC model*. This model simply means that all of the MAC layer (remember, the Data Link layer is divided into the LLC and MAC sublayers) functionality is in the AP. A *thin AP* is an AP that depends on a WLAN controller or WLAN switch to provide much of the MAC layer functionality. Thin APs are used when implementing a *split MAC model*. This model means that the MAC is split between the thin APs and the WLAN controllers.

APs basically create virtual switch ports that WLAN clients use to connect to the wired LAN. In most business settings, APs are used in a distribution role allowing wireless clients to gain access to the wired network; however, APs can be implemented in a core role where they function at the core of the network. The latter implementation method, however, is very rare. You'll learn much more about APs in the third chapter.

Modems

A *modem* is a specialized device that converts digital data into analog signals for transmission on regular telephone lines. It also converts analog signals received from the telephone line into digital data for the local device. The word modem is a combination of the words modulator and demodulator. Modulation is the act of encoding data on a carrier, and demodulation is the act of decoding data from a carrier. In the case of a modem, the carrier is the analog signal (sound waves) that is propagated into the phone line as varying electrical currents.

While modems were used mostly by terminals connecting to mainframes in the 1970s and 1980s and then by users connecting to bulletin board systems and the Internet in the 1990s, they are also used to create serial WAN connections over great distances. Of course, modems are still used today in rural areas and even by some on metropolitan areas for Internet connectivity. Since modems can play an important role in a network infrastructure, they will be discussed in greater detail in Chapter 3.

CERTIFICATION SUMMARY

Understanding the OSI model is an important part of preparing for the Convergence+ exam, and it's also an important part of preparing for a career as a network engineer. The reality is that it is a rare piece of vendor literature that does not reference the OSI model in some direct or indirect way. Normally, this documentation will simply state that a certain service is provided at Layer 2 or Layer 3 and will not even mention that it is referencing the OSI model. That's how ubiquitous the model is in network administration and engineering. This chapter introduced you to the OSI model, and you may want to read the actual standard—though you will not need to know any more about the OSI model for the Convergence+ exam—in order to familiarize yourself with it even more.

You also learned the basics of the common cabling types that are used in wired networks today. Coax may be used in rare situations, but UTP is the most commonly used medium. Of course, fiber has become very popular for the network core and even campus area network building-to-building links.

You learned the basic differences between LANs, WANs, MANs, and GANs. The LAN is localized to a property or small area, and the others cover greater areas from metropolitan cities to unlimited global coverage.

Finally, you had a brief introduction to some of the hardware used in networks, including switches, routers, access points, servers, and network appliances. You'll learn more about all of these devices in Chapters 3 and 4. In the next chapter, you'll take the information presented here about the OSI model and apply it specifically to various network protocols that actually make communications happen on your network.

✓ TWO-MINUTE DRILL

Identify the Layers of the OSI Model and Know Its Relevance to Converged Networks

❑ The OSI model is a reference used to design functional networks and is not an actually implemented network technology.

❑ There are seven layers in the OSI model; they are Application, Presentation, Session, Transport, Network, Data Link and Physical from top to bottom. They are also known as Layers 7 through 1 respectively.

Describe Networking Technologies Used in a Converged Network

❑ Unshielded twisted pair cabling has a maximum distance of approximately 100 meters.

❑ Fiber cabling has the longest allowed distances of all cabling types and also provides for the greatest potential bandwidth.

❑ The bus topology is outdated as a stand-alone network solution, though it is beneficial in conjunction with star or mesh topologies.

❑ CAT5 cabling supports data rates of 100 Mbps and sometimes 1000 Mbps. CAT3 cabling only supports 10 Mbps, and CAT4 cabling supports 16 Mbps. For this reason, most implementations have migrated from CAT3 to CAT5, since CAT4 only supports a maximum data rate of 16 Mbps.

❑ The star topology uses a switch or hub as the "center" of the network. All devices send the data to the switch or hub, and the central device forwards it to the proper node or nodes.

❑ Routing is accomplished by using a device called a router or an application that can act as a router in conjunction with compatible routing protocols such as TCP/IP.

❑ There are two primary types of signaling: digital and analog.

Identify the Functions of Hardware Components

❑ Servers provide services to the network such as file storage, communications and collaboration, database access, and call management in Voice over IP implementations.

❑ In-line power equipment uses the Power over Ethernet (PoE) standards to provide power to devices over standard CAT5 and CAT6 cabling.

❑ Switches forward data only to the intended target device as much as possible, and hubs forward data to all ports in the hub.

❑ Modems are used to send digital data across analog phone lines and other analog systems such as wireless media.

SELF TEST

The following questions will help you measure your understanding of the material presented in this chapter. Read all the choices carefully because there might be more than one correct answer. Choose all correct answers for each question.

Identify the Layers of the OSI Model and Know Its Relevance to Converged Networks

1. Which layer of the OSI model is responsible for actually transmitting bits onto the communication medium?

   A. Physical layer

   B. Layer 3

   C. Transport layer

   D. Layer 5

2. At which layer of the OSI model does IP address management occur?

   A. Layer 3

   B. Application layer

   C. Layer 1

   D. Layer 5

Describe Networking Technologies Used in a Converged Network

3. You are implementing a converged network. You need to run cable to a computer that is located 250 meters from the current network. You will need to install either switches or repeaters to implement the connection using standard CAT5 UTP cable. How many switches or repeaters would you need?

   A. 1

   B. 2

   C. 3

   D. 4

4. Which one of the following cabling options provides both the longest possible distances and the highest bandwidth?

    A. UTP

    B. STP

    C. COAX

    D. Fiber optic

5. Which signaling type is identified by sharp contrasts between states rather than gradual variations?

    A. Analog

    B. Digital

Identify the Functions of Hardware Components

6. Which type of server is used to store word processing documents most commonly?

    A. Database server

    B. Media server

    C. File server

    D. Print server

7. You want data packets that are transmitted from client computers on your network to be sent to computers they are intended for but not to other computers. Which combination of topology and device should you implement from those listed?

    A. Bus topology with a switch

    B. Bus topology with a hub

    C. Star topology with a switch

    D. Star topology with a hub

LAB QUESTION

Mark is implementing a network solution that provides Voice over IP. He is running a Voice over IP software package that works on Windows XP and Windows Vista as well as Linux clients. The application specifications state that the computer should be connected to the network with a 100 Mbps connection if any other network applications such as e-mail are to be used during phone conversations.

Conceptually speaking, with what layer of the OSI model will this application speak?

Assuming you're using unshielded twisted pair cabling, what category of cabling must be used at a minimum?

SELF TEST ANSWERS

Identify the Layers of the OSI Model and Know Its Relevance to Converged Networks

1. ☑ **A** is correct. The Physical layer, also known as Layer 1, is responsible for transmitting bits onto the medium.
   ☒ **B, C,** and **D** are incorrect.

2. ☑ **A** is correct. The Network layer, or Layer 3, is responsible for IP address management.
   ☒ **B, C,** and **D** are incorrect.

Describe Networking Technologies Used in a Converged Network

3. ☑ **B** is correct. You will need two switches or routers because the UTP cable can only run about 100 meters. You could run the cable 100 meters to the first switch and then another 100 meters to the second. Finally, you could run a 50-meter patch cable to the computer.
   ☒ **A, C,** and **D** are incorrect. The answer to this question lies in knowing the maximum allowable length of a UTP cable.

4. ☑ **D** is correct. Fiber optic cabling can run for miles or kilometers and can provide many gigabits of throughput.
   ☒ **A, B,** and **C** are incorrect. These are not the cables that provide the longest runs or the greatest bandwidth.

5. ☑ **B** is correct. Digital signals vary sharply and purely. For example, one frequency may represent a 0 and another represents a 1 and these are the only two frequencies used.
   ☒ **A** is incorrect. Analog signals vary gradually and provide great variation in the signal much like the human voice.

Identify the Functions of Hardware Components

6. ☑ **C** is correct. Word documents are files and would therefore, most commonly be stored on a file server.
   ☒ **A, B,** and **D** are incorrect.

7. ☑ **C** is correct. The star topology indicates that data is to be sent to the central device (the switch) and then the central device forwards the data to the intended machine only if that central device is a switch.
   ☒ **A, B,** and **D** are incorrect. The star topology with a hub would not meet our demands, as the data would be sent to all nodes on the star. The bus topology simply will not work, since it does not traditionally use a switch.

LAB ANSWER

The application will communicate with the Application layer, or Layer 7. Remember, the voice application is not the Application layer, but it will use protocols that are at the Application layer such as the Session Initiation Protocol (SIP) or Real Time Streaming Protocol (RTSP). SIP and RTSP could be placed anywhere between Layer 5 and Layer 7 of the OSI model, as they actually operate in the TCP/IP suite at the Application layer of the TCP/IP or DOD model. The Application layer of the TCP/IP model maps closely to the OSI model's Layers 5 through 7.

You will need to use a minimum of CAT5 cabling. This is the first category in the hierarchy that provides 100 Mbps data rates. Certain implementations of CAT5 may also provide data rates up to 1 Gbps; however, you can also implement 1 Gbps over CAT6 cabling.

2

Networking Protocols

> *Do not fold, tear or destroy.*
>
> —*IBM Computer Punch Card Message*

Over the years, industry standards organizations and private companies have developed standard methods of communication that allow computing devices to cooperate and communicate on a network. These standard methods of communication are known as *network protocols*. Understanding these protocols is a very important part of becoming a network engineer or administrator. This chapter will provide you with a detailed introduction to network protocols and information about the functionality of the major protocols used in modern networks. These major protocols include Ethernet, TCP/IP, and various WAN protocols. I'll also review the protocols used to implement Voice over IP solutions, though they will be covered in greater detail in Chapters 6–9.

CERTIFICATION OBJECTIVE 2.01

Describe Protocols

A *network protocol* is a collection of rules, recommendations, and options used to facilitate communications between computing devices. These devices may be directly connected, such as in a bus topology or in the use of a USB device, or they may be connected using internetworking technologies, such as routers, switches, and gateways. It is most important, at this point, that you understand three terms in our definition: *rules, recommendations,* and *options*. Network protocols contain rules to which communicating devices must adhere. They also contain recommendations that may or may not be implemented with careful thought, and they contain optional components that are truly necessary only if those individual services or features are desired.

An example of rules versus recommendations or optional components can be seen in the IEEE 802.11 standards for wireless networking. The IEEE 802.11g amendment to the standard requires the implementation of ERP-OFDM and specifies that the implementation of ERP-PBCC is optional. ERP-OFDM and ERP-PBCC are two different modulation techniques (methods used to transmit digital data on electromagnetic waves) that the amendment specifies. You could say that ERP-OFDM is a rule and ERP-PBCC is a recommendation. The former is required and the latter is optional.

In fact, there is an RFC (Request for Comments) document that specifically outlines the language to be used in RFCs in order to indicate which parts of the protocol or standard are required and which parts are optional. This RFC is number 2119, and you can find it easily by searching the Internet with your favorite search engine for the phrase *rfc 2119*. In this document it indicates that the words "must," "must not," "required," "not required," "shall," and "shall not" all mean that the definition is an absolute requirement or prohibition within that specification. The words "should," "should not," "recommend," and "not recommended" mean that the definition should be omitted from an implementation only after very careful consideration of the impact it will have on the resulting solution. Finally, the words "may" and "optional" indicate that a definition is truly optional and the implementing party (usually a hardware or software vendor) may decide if the component of the standard is useful for their purposes.

Ultimately, RFC 2119 is itself a protocol. It is a protocol that defines how protocol definitions should be written according to RFC standards. By using the language identified in RFC 2119 and indicating that you are using that language, you ensure that the reader will understand the difference in meaning between words like "should" and "may." In common usage, these two words mean the same thing; however, in RFCs they have very different connotations. "Should" is not to be taken as lightly as "may."

Many Internet protocols have been developed through the process of RFCs. RFCs are created by individuals and organizations and may or may not become standards. For that matter, some RFCs are nothing more than jokes—RFC 748 is an excellent example of this. It is an RFC that defines why systems randomly crash and how they must seek user permission to do so. Of course, it is not a real RFC, but it shows how much more open the process of RFC development is than, say, that for IEEE standards and the like. This humor should not be taken as a detractor from the quality of the RFCs that are serious, as they are analyzed by peers and only become standards if they provide value.

Although the first RFC was written in 1969 by Steve Crocker, it is interesting to note that the second RFC actually begins to document how RFCs should be developed. It is also interesting to note the very informal tone of the earliest RFCs such as RFC 6, which is simply titled, "Conversation with Bob Kahn." Of course, these early RFCs were written during the primordial evolutionary phase of the Internet back when it was known as the ARPA Network (according to RFC 1).

e x a m

While you will not be tested on your direct knowledge of the RFCs on the Convergence+ exam, it is these RFCs that define the technical details of many Internet technologies. I would encourage you to read or at least skim the important ones that are mentioned in this chapter and later chapters.

The Need for Protocols

If we did not have standardized protocols, every vendor would implement technology so that their hardware only works with their hardware or their software only works with their software. In fact, this scenario happened for the first decade or so of computing. Have you noticed that you can buy a computer from any computer manufacturer, whether it's an IBM PC compatible or an Apple computer, and they can both talk to the same Internet? This similarity is because they share one or more common protocols. In the case of the Internet, these shared protocols are TCP/IP. In order for two devices to communicate with each other, they must be using a shared protocol or they must communicate through a gateway that performs a protocol conversion for them. It would be much more costly in CPU time to implement the latter, so the preferred solution is to install and use shared protocols on each communicating device.

Today an Apple computer can talk to a Linux machine that talks to a Unix machine that talks to a PC running Windows Vista. Any of these computers can communicate with any other of these computers because of standardized protocols. If you've been working with computers as long as I have and enjoying their great benefits along the way, you will remember the late 1970s and early 1980. We had Apple computers, Commodore computers (the superior devices, of course), Tandy computers, Atari computers, and eventually IBM PCs. These computers each used their own formats for data storage on disks or tapes, and they had no direct way to communicate with each other, since they did not internally support shared protocols. You can be very grateful for the development of standards over the past thirty years that have changed all this.

In RFC 31, apparently penned in February 1968, the authors state:

Network communication between computers is becoming increasingly important. However, the variety of installations working in the area probably precludes standardization of the content and form of intercomputer messages. There is some hope, however, that a standard way of defining and describing message forms can be developed and used to facilitate communication between computers.

From this early communication, it is clear that the computing community understood the need for standardized communication protocols. From the early days of the ARPA Network to the modern Internet, protocol standardization has been important; however, it is even more important today, since millions of nodes communicate with each other on a regular basis.

Network Communication Protocols and the OSI Model

The OSI model, as presented in the last chapter, is a logical model of communications for networked devices and software. The OSI model does not specify the protocols that should be used at each of the seven layers, but it specifies the functionality that should be provided by those protocols. For this reason, you will read and hear statements like "FTP is a Layer 7 protocol" and "TCP is a Transport layer protocol." These phrases are used to indicate the layers within the OSI model where the protocols operate. The reality, however, is that many protocols and network communications in general do not occur in line with the OSI model. TCP/IP is a perfect example. While we can relate the TCP/IP communications model to the OSI model, it cannot be said to meet the OSI specifications or operate in the way the OSI model indicates to exactness.

This concept is important to keep in mind. It is also why some documents will indicate that a certain protocol operates as Layer 6 and another document may indicate that the exact same protocol operates at Layer 7. Since TCP/IP implements only four layers, the Application layer of the TCP/IP model may be said to encompass Layers 5 through 7 of the OSI model. Even though the TCP/IP model has a layer called the Application layer, this layer is not equivalent to the OSI model's Layer 7, which is also known as the Application layer. Instead, you could say that the OSI model Layer 7 functionality is included in the TCP/IP Application layer. In the end, protocols may be said to function according to the OSI model only theoretically, and they may indeed have their own actual communication model that is very different or somewhat different from the OSI model of communications.

In the rest of this chapter, I will look at protocols starting from the bottom of the OSI model—Layer 1—and then work my way up through to Layers 3 through 7.

We'll start by looking at LAN protocols that operate at Layers 1 and 2 and then move on to look at protocols that operate across these LAN protocols such as IP at Layer 3 and TCP or UDP at Layer 4. Finally, we'll consider tunneling protocols that are used to create connections to services or networks by passing data across incompatible or insecure networks using encapsulation techniques.

LAN Protocols

LAN protocols are used to establish and manage communications within localized networks. They facilitate the physical transmission of data from device to device and provide error reporting, control, and correction. The most common LAN protocol is Ethernet, and we'll look at it first. The next LAN protocol, and one that has grown in popularity by leaps and bounds in the past decade, is Wi-Fi. In addition to these two most popular protocols, we'll quickly investigate Token Ring and FDDI, though we will not dive as deeply into their functionality.

Ethernet

The first Ethernet network definition was created in the 1970s and the IEEE 802.3 standard defines and standardizes the Ethernet protocol. The IEEE 802.3 standard was first released in 1985 and supported a 10 Mbps data rate. Since then, amendments have been released that have increased the data rates to 100 Mbps (Fast Ethernet) and 1000 Mbps (Gigabit Ethernet). The IEEE 802.3-2005 rollup document covers the amendments up to that time and now includes the 10, 100, and 1000 Mbps PHY (Physical layer) specification within the single collection of documents.

According to the IEEE 802.3 standard, it defines itself as "a comprehensive international standard for Local and Metropolitan Area Networks (LANs and MANs), employing CSMA/CD as the shared media access method and the IEEE 802.3 (Ethernet) protocol and frame format for data communication." Ultimately this definition means that devices compliant with the IEEE 802.3 standard will use the CSMA/CD MAC (Medium Access Control) sublayer within the Data Link layer of the OSI model and Ethernet frame types at the Physical layer.

exam
watch

The use of CSMA/CD versus CSMA/CA in Ethernet and Wi-Fi networks respectively is a very important distinction. Wired Ethernet networks detect collisions *because they are connected to the wire. Wireless networks avoid collisions as much as possible because they cannot actually detect them occurring.*

Figure 2-1 shows where Ethernet fits into the OSI model discussed in the preceding chapter. You'll notice that the Data Link layer is divided into two sublayers: the Logical Link Control (LLC) and Medium Access Control (MAC) sublayers. Because the MAC and Physical layer (PHY) are the two areas that are most heavily impacted by the IEEE 802 series of standards (including the popular 802.3 Ethernet and 802.11 Wi-Fi standards), you will often see references to the Ethernet MAC and PHY or the 802.11 MAC and PHY. These statements are used as a simple way to reference the MAC specifications and the PHY specifications of the indicated standards.

In order to assist you in implementing telephony services over Ethernet, we'll look at three important details about Ethernet networks:

■ CSMA/CD
■ Supported speeds
■ Frame formats

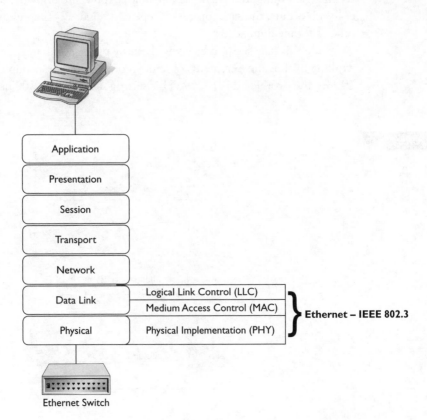

FIGURE 2-1

Ethernet and the OSI model

CSMA/CD   First of all, I want to be very clear. CSMA/CD is not used in many, if not most, Ethernet connections today. This is because most Ethernet connections are between an endpoint (a computer for example) and a switch. The connection between the endpoint and the switch is almost always full duplex, and when it is full duplex, CSMA/CD is not utilized. Now that I've gotten that rant out of the way, let's continue.

Carrier Sense Multiple Access with Collision Detection (CSMA/CD) is the medium access method used in IEEE 802.3 Ethernet LANs that operate in half-duplex mode. It senses whether the medium (remember, this is the physical cabling) is active and, if it's not, it allows transmission of frames. The medium is considered active when a signal is being transmitted. If a frame is transmitted and a collision is detected, a jam signal is sent on the medium to inform all devices that a collision was detected. At this point, all devices back off for a random amount of time, and when their backoff times expire, they may again communicate. As long as an acceptable number of devices share the medium, concurrent moments of expiration (multiple devices trying to transmit after their backoff times expire) are unlikely; however, if another collision occurs, the same process is repeated. Figure 2-2 represents the CSMA/CD process for medium access.

While this may seem to be a very lengthy process, you must remember that we are working in time measurements of nanoseconds (ns). For example, the bit time on a 10 Mbps network card is 100 ns. This setting means every 100 ns a bit can be pushed

FIGURE 2-2

CSMA/CD
process

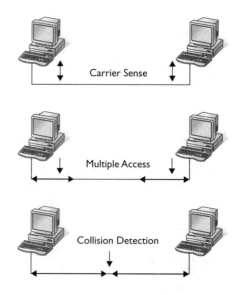

out through the network card. A 100 Mbps device can eject a bit every 10 ns, and a 1000 Mbps network card could eject a bit every 1 ns. To fully understand these time measurements, consider that a microsecond is one millionth of a second, and a nanosecond is 1000 times faster than that. This definition brings us to the topic of supported speeds.

Supported Speeds   The Ethernet standard supports multiple Physical layer speeds. These speeds have different requirements for cabling. For example, you can communicate at 10 Mbps with CAT3 UTP cabling; however, you must use CAT5 cabling to effectively communicate at 100 Mbps using 100BASE-T. Table 2-1 provides a listing of the different IEEE 802.3 Ethernet standards and the cabling and speed factors.

There are a few important things to note about the information in Table 2-1. First, you'll notice that 100BASE-T is a term used to refer to shared MAC extensions that apply to multiple Physical layer entities such as 100BASE-TX for either UTP or STP and 100BASE-FX for fiber. Second, you'll see that I am not referring to amendments when I refer to each of the MAC and PHY layers specified in the IEEE 802.3 standard. This omission is because they have all been rolled up

TABLE 2-1   Ethernet Standards, Cabling, and Speeds

Ethernet Standard	Cabling Requirement	Speed Offered
10BASE-2 (IEEE 802.3-2005, Clause 10)	RG-58 (coax)	10 Mbps
10BASE-5 (IEEE 802.3-2005, Clause 8)	RG-8 (coax)	10 Mbps
10BASE-T (IEEE 802.3-2005, Clause 14)	CAT3 (UTP)	10 Mbps
10BASE-F (IEEE 802.3-2005, Clause 15)	Fiber	10 Mbps
100BASE-T (IEEE 802.3-2005, Clauses 21 through 32)	CAT3 (100BASE-T4), CAT5 (100BASE-TX), Fiber (100BASE-FX)	100 Mbps
Gigabit Ethernet (IEEE 802.3-2005, Clauses 22 through 43)	CAT5 (1000BASE-T), STP (1000BASE-CX up to 25 meters), Fiber (1000BASE-LX and 1000BASE-SX)	1000 Mbps (1 Gbps)
10 Gigabit Ethernet (IEEE 802.3-2005, Clauses 44 through 54)	Fiber (10GBASE-S, 10GBASE-L, and 10GBASE-E), CX4 cabling (10GBASE-CX—this is a copper cabling specially made to allow 10 GB connections of up to 50 feet), CAT6 or CAT6e (10GBASE-T—IEEE 802.3an)	10,000 Mbps (10 Gbps)

into the standard as of the IEEE 802.3-2005 incorporation and they are now properly referenced as clauses within the standard rather than by their temporary working group letter assignment identity. For example, Clause 21 was originally called IEEE 802.3u and was released in 1995. Since it is now simply part of the IEEE 802.3 standard, there is no reason to reference it as IEEE 802.3u any longer. In fact, it can be said that IEEE 802.3u no longer exists and that only Clause 21 remains, since the working group has been disbanded.

Now, you'll notice that CAT6 cabling is required for the 10GBASE-T standard that was introduced in IEEE 802.3an. This amendment was released in 2006, and since it was released after the IEEE-802.3-2005 rollup, it is still referenced here as IEEE 802.3an. It does, however, specify a new clause, which is Clause 55. This amendment also modifies and adds to other existing clauses in the IEEE 802.3-2005 standard such as Clauses 1, 28, 44, and 45. This inclusion is an important thing to understand about the IEEE standards (not just 802.3, by the way). To truly know what the "current" standard "is," you must look at the most recent rollup and then look for any changes made by amendments since that rollup. If an amendment is not yet ratified, it's actually called a draft and not an amendment, and it is not really part of the standard until it is ratified. Currently, the much talked about IEEE 802.11n draft is a perfect example of this. Sometime in late 2008 or 2009 we may see the 802.11n draft become ratified, and then it will be part of the complete IEEE 802.11 standard for wireless local area networking.

One final note about these speeds and standards is that you must remember there are limitations on the length or distance of connections as well. For example, the new 10GBASE-T standard allows connections of only 55 meters for standard CAT6 UTP cabling, but up to 100 meters for enhanced CAT6e UTP cabling. Of course, at this point 10 Gigabit Ethernet is mostly a core or backbone technology and is not being implemented down to the individual network nodes. Most current networks still use either 10 Mbps to the nodes or 100 Mbps to the nodes with either 100 Mbps or 1 Gbps cores. However, 1 Gbps to the nodes and either 1 Gbps or 10 Gbps cores are being implemented. This design would mean that Gigabit Ethernet is used between the switches and the endpoints and 10 Gigabit Ethernet connections are used between the switches and routers on the backbone or core of the network.

e x a m

ⓦatch   *Remember that Ethernet can currently support speeds up to 10 gigabits per second with available technologies, but 100 Gigabit Ethernet is on its way. We'll likely see it sometime between 2010 and 2012.*

Frame Formats   When the phrase "Ethernet frame" is uttered, it can mean
many different things. There are the Ethernet II frame, the IEEE 802.3 frame, the
IEEE 802.3 SNAP frame, and others. The current IEEE 802.3 standard specifies two
frame formats, so I will stick with those here, as they are supported by the majority of
hardware devices that are in use today. These frame formats are:

■   Basic frame
■   Tagged frame

First, a *frame* is nothing more than a collection of well-defined bits that are
transmitted on the physical medium. An Ethernet frame differs from a Wi-Fi frame
in that they are two unique and different networking technologies. However, they
both use frames for the transmission of upper-layer data on the medium.

The Ethernet *basic frame* is supported by all Ethernet devices from 10BASE-T
through to 10GBASE-T. The Ethernet *tagged frame* may or may not be supported by
all devices in a given network. It must be supported by the all devices in the path from
the transmitting node to the receiving node if it is to provide an advantage through the
entire path. In other words, if a network exists similar to that in Figure 2-3, switches A,
B, and C must all support the tagged frames if the benefits they offer (Quality of Service
and VLAN support) are to be realized through the entire path.

Since an Ethernet frame is a series of 1's and 0's (bits), I will represent them in
graphical form from left to right. Think of the left side of Figures 2-4 and 2-5 as
being the first bit transmitted and the right side as being the last bit transmitted.
Figure 2-4 shows a basic frame, and Figure 2-5 shows a tagged frame. Notice that the
difference lies between the source address bits and the MAC client length/type bits.

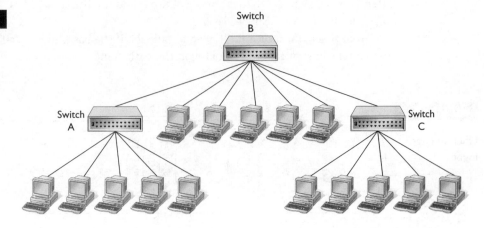

FIGURE 2-3

Typical Ethernet
implementation

FIGURE 2-4

Ethernet basic frame

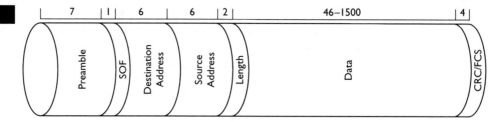

Each frame transmission begins with a preamble and a start-of-frame delimiter (SFD). The preamble is seven bytes (56 bits) long and is used to give the receiver time to "tune into" the signal. Think of this preamble as like saying "hey you" to someone in order to get her attention. You do it to make sure she is listening to you before you begin delivering the important part of your message. Ethernet implements a similar concept by sending 28 1's and 28 0's (in alternating fashion, i.e. 10101010 . . .) in order to allow the receiver to perceive that a frame is coming and prepare to process it. The SFD is a single byte in length and, with Ethernet, is the eight-bit sequence of 10101011. Even if the receiver missed the first 15 or 20 bites, this final eight bit sequence lets it know that the real frame is now beginning. This sequence is true for both the basic frame and the tagged frame.

When the actual frame starts, the first information sent is the destination address, which is followed by the source address. Both addresses are six bytes, or 48 bits, long. These addresses are known as the destination and source MAC (media access control) addresses. They are the addresses, usually burned in, of the network interfaces on the two communicating devices. Keep in mind that, when the true end node is on a remote network, this destination address will be the MAC address of the local router that knows how to get to that remote network and not the actual MAC address of the final destination of the upper-layer data.

From this point forward, I'll need to talk about the basic and tagged frames separately. First, we'll finish looking at the basic frame.

FIGURE 2-5

Ethernet tagged frame

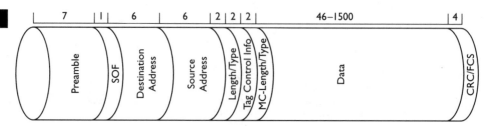

The next field within the basic frame is the Length/Type field. This field can either indicate the length of the MAC service data (the data that has come down from Layers 3 through 7 of the OSI model), or it can indicate the type of protocol the frame represents. If the value in this two-byte (16-bit) field is greater than or equal to 1536, it is specifying the type of client protocol (this size is used by very old Ethernet technologies only), and if it is equal to or less than 1500, it is specifying that actual size of the data payload in the frame. This size is important because the frame's payload will be padded to equal a minimum frame size and this padding must be removed at the receiver in order to properly process the payload within the frame.

Of course the next element of the basic frame is the actual data that comes from Layers 3–7 and is to be transmitted to some final destination. This information will include IP headers, TCP headers, and the actual application data on a TCP/IP network, as you'll learn later in this chapter. For now, just know that it is the data being transmitted to some end location.

The Frame Check Sequence (FCS) is the next field in the Ethernet basic frame. This field is an implementation of a cyclic redundancy check (CRC) algorithm and is used to determine if the received frame has actually been received without error. The sending node generates the four-byte FCS by passing the destination address, source address, length/type, and data fields through a mathematical algorithm that generates a highly unique number. The receiving node processes the same algorithm (which is defined within the IEEE 802.3 standard) and compares the result with the received FCS. If they differ, the frame is corrupted. If they agree, the frame is intact.

Finally, you'll notice that I have a somewhat ghosted field at the end of the frame called the Extension field. Without going into too much detail, the field is only used in Gigabit Ethernet and faster implementations that use half-duplex communications. The reality is that almost no network implements such technology, since nearly all Gigabit Ethernet and higher implementations use full-duplex communications. The field allows for longer cable lengths by increasing the smallest frame sizes through artificial padding. This field is well explained in the book *LAN Technologies Explained* by Philip Miller and Michael Cummins (Digital Press, 2000). I simply could not leave it out of my diagram and explanation of Ethernet frames, since it is part of the actual IEEE 802.3-2005 standard description.

We now bring our focus back to the tagged frame. This frame includes three new fields that are inserted immediately after the source field and immediately before the original Length/Type field, which is now called the MAC Client Length/Type field within the tagged frame. The name change for the basic frame's Length/Type is, of course, because the tagged frame now includes its own Length/Type field, which would be best called an 802.1Q Tag Type field.

The two new fields are Length/Type and Tag Control Information (TCI) in that order from left to right. The new Length/Type field (remember, the 802.1Q Tag Type field) is two bytes long and is used to identify the frame as a tagged frame; it is always equal to the constant 1000000100000000 as identified in the IEEE 802.1Q standard for virtual LAN (VLAN) tagging. The TCI field, which is 16 bits or two bytes long, is broken into three parts: user priority (3 bits—also known as a priority code point [PCP] field), a canonical format indicator (also called CFI and is 1 bit), and a 12-bit VLAN identifier. The 3-bit user priority values are defined in IEEE 802.1p and include eight different values for class of service (CoS) and can be used to provide QoS at the MAC layer or the Data Link layer. The classes are not specified with recommended usage guidelines but can be prioritized according to the needs of the variously implemented systems. The CFI subfield indicates either Ethernet or Token Ring and will be Ethernet in modern implementations. Finally, the VLAN identifier is used to specify the identity of the VLAN to which the frame belongs.

As you can see from these details of the Ethernet frames, Ethernet communications certainly qualify in the definition of a protocol. Ethernet networks use standard frame types for communication of information between communicating nodes.

Wi-Fi

The popular term *Wi-Fi* refers to the technical standards document known as IEEE 802.11 as amended. This standard was first released in 1997 supporting data rates of up to 2 Mbps and has been amended to support 11 Mbps and 54 Mbps data rates. At the time of this writing, the IEEE 802.11n amendment is still in its draft form. When completed, it should offer data rates as high as 600 Mbps.

Like Ethernet, Wi-Fi networks have a standard method used to access the medium (a selected range of RF frequencies). They have speed restrictions, and they have specified frame formats. The following sections will provide an overview of these factors as well as a quick review of the different IEEE 802.11 PHY layers.

CSMA/CA   Ethernet networks (IEEE 802.3) use a form of collision management known as collision detection (CD). Wireless networks use a different form of collision management known as collision avoidance (CA). The full name of the physical media access management used in wireless networks is Carrier Sense Multiple Access with Collision Avoidance, or CSMA/CA. The essence of CSMA/CA is that collisions can happen at many locations in the physical space wherein the RF medium is being utilized at any time during a transmission and may not be detected by the transmitter at its location. Therefore, listening for evidence of a collision, while transmitting is worthless and not a part of the CSMA/CA protocol. The point is that the collision may occur near the remote wireless node, but not be detected at the local wireless node, since the RF signal may not reach it. Since the wireless devices cannot detect collisions with assurance, they must attempt to avoid them.

CSMA/CA is used in wireless networks and was also used in early Apple LocalTalk networks, which were wired networks that were common to Apple devices. Collision avoidance is achieved by signaling to the other devices that one device is about to communicate. This would be like saying, "Listen, for the next few minutes, because I will be talking" in a telephone conversation. You are avoiding the collision by announcing that you are going to be communicating for some time interval. CSMA/CA is not perfect due to hidden node problems, but it provides a more efficient usage of a medium like RF than would CSMA/CD.

Supported Speeds Wi-Fi networks based on IEEE 802.11 support data rates from 1 Mbps through to 54 Mbps at the time of this writing. Once the 802.11n amendment has been finalized, we will have data rates in the hundreds of megabits per second available to use; however, in this text, I will only reference the standards that are currently ratified (actual standards) in any detail.

It is important to note, particularly in Wi-Fi networks, that there is a difference between data rates and throughput. Data rates reference the maximum signaling rates supported by a PHY layer standard implemented in 802.11. Throughput refers to the meaningful data that can be transferred. On a Wi-Fi network with good-quality signals (because there is little interference or other problematic behavior), you will usually achieve between 50 and 60 percent throughput. This means that you really only have about 27–30 Mbps throughput on a 54 Mbps data rate connection.

Of course, the wired side of the wireless connections can also impact actual speeds. If your 802.11g 54 Mbps access point is connected to a 10 Mbps Ethernet switch port, it doesn't really matter that the clients "could" communicate with the access point at the higher data rate (except when they communicate directly with each other). Always make sure your access points are connected to the wired network using 10 Mbps ports at a minimum for 802.11b and 100 Mbps ports at a minimum for 802.11a or 802.11g. Of course, faster ports will be needed when 802.11n is ratified. I am referencing the PHYs by their amendment identifier (e.g., 802.11b or 802.11g) because vendors tend to reference them in this way. Technically, 802.11b is simply the HR/DSSS PHY within 802.11, and 802.11g is the ERP-OFDM PHY within 802.11.

Frame Formats Clause 7 of the IEEE 802.11 standard documents the frame types supported by the IEEE 802.11 MAC. According to this clause, three frame types are supported in IEEE 802.11 networks: management frames, control frames, and data frames. The Type subfield in the Frame Control (FC) field of a general IEEE 802.11 frame may be 00 (management), 01 (control), or 10 (data). The Subtype subfield determines the subtype of frame, within the frame types specified, that is being transmitted. For example, a Type subfield value of 00 with a Subtype value of 0000 is an association request frame; however, a Type value of 10 with a Subtype value of 0000 is a standard data frame. Figure 2-6 shows the relationship of the general IEEE 802.11 frame format to the format of the FC field's subfields. Understanding all of the details about the frame structures and formats is not required of a Convergence engineer; however, it would be of great benefit to you to review Clause 7 of the IEEE 802.11 standard, which defines each frame and each frame field with diagrams similar to the one in Figure 2-6.

While Figure 2-6 represents the general frame format in the IEEE 802.11-1999 standard and even the IEEE 802.11-1999 (R2003) standard, the standard as amended by IEEE 802.11i calls for a change in the FC subfields format. The WEP subfield is now called the *Protected Frame* subfield. There are other changes in the usage of the subfields and their values that have been introduced in both IEEE 802.11i and IEEE 802.11e. Though you will not be tested to this level of depth on the Convergence+ exam, you should be aware of these changes and realize that they constitute IEEE 802.11 as amended. In other words, all ratified amendments have become part of IEEE 802.11 such that when we speak of the IEEE 802.11 standard, we must incorporate these amendments into our conceptualization of the standard. Otherwise, it is important to state that you are referring to IEEE 802.11-1999 and not IEEE 802.11 as amended. Today, if we speak of the IEEE 802.11 standard, this is inclusive of the IEEE 802.11-1999 (R2003) document and all ratified amendments.

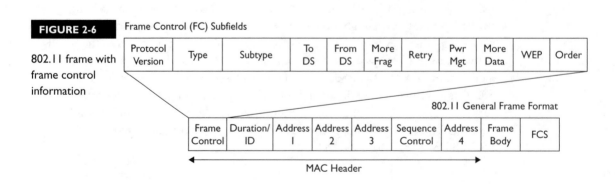

| FIGURE 2-6 | Frame Control (FC) Subfields |

802.11 frame with frame control information

| Protocol Version | Type | Subtype | To DS | From DS | More Frag | Retry | Pwr Mgt | More Data | WEP | Order |

802.11 General Frame Format

| Frame Control | Duration/ ID | Address 1 | Address 2 | Address 3 | Sequence Control | Address 4 | Frame Body | FCS |

MAC Header

IEEE 802.11 PHYs   In today's implementations, there are really only four PHYs in common use: DSSS, HR/DSSS, OFDM, and ERP-OFDM. One of these is also very rare, and that is Direct Sequence Spread Spectrum (DSSS). That leaves three major PHYs (physical layers) in use. In this section, I'll provide you with an overview of these PHYs. In the IEEE 802.11 standard, two additional PHYs are specified: Infrared and Frequency Hopping Spread Spectrum (FHSS). I will not cover these here, as the infrared PHY was never implemented and the FHSS PHY is no longer sold or used (with the exception of very rare existing implementations or hackers). Keep in mind that I am saying FHSS is not used within 802.11 networks anymore. Certainly FHSS is used in Bluetooth devices and other communication systems, but these are not 802.11-compliant.

The HR/DSSS PHY is the High-Rate Direct Sequence Spread Spectrum physical layer specified in IEEE 802.11. This PHY provides data rates up to 11 Mbps, and newer devices that work in the 2.4 GHz spectrum—like HR/DSSS—are backwardly compatible. This means that newer ERP-OFDM devices can interoperate with the older HR/DSSS devices.

The Orthogonal Frequency Division Multiplexing (OFDM) PHY provides data rates up to 54 Mbps, but it operates in the 5 GHz frequency band. Because of this, it is not capable of communicating with HR/DSSS or ERP-OFDM. This PHY is usually called 802.11a though the standards reference it as OFDM.

Finally, the ERP-OFDM or Extended Rate PHY-OFDM physical layer provides data rates of up to 54 Mbps. ERP (the shorthand version of ERP-OFDM) uses the 2.4 GHz frequency band and, therefore, is able to communicate with DSSS and HR/DSSS devices using interoperable communication parameters specified in the standards. It is important to remember that ERP devices support communications with DSSS and HR/DSSS devices, but they do not support communications with OFDM (802.11a) devices. This ERP PHY is usually called 802.11g. Table 2-2 summarizes the different Wi-Fi PHYs and their key features.

In addition to the general fact that ERP-OFDM PHY devices can communicate with HR/DSSS and DSSS PHY devices, you should know that just one HR/DSSS or DSSS device connected in an ERP-OFDM network (a collection of APs and

TABLE 2-2	PHY	Maximum Data Rate	Frequency Band
	HR/DSSS (802.11b)	11 Mbps	2.4 GHz
Wi-Fi Physical Layers and Features	OFDM (802.11a)	54 Mbps	5 GHz
	ERP (802.11g)	54 Mbps	5 GHz

clients that support ERP-OFDM) will slow the other devices down because of the protection mechanism used. Basically, the ERP-OFDM devices all have to send a notification on the wireless medium telling the non-ERP-OFDM devices that they should be silent long enough for the higher-rate transmission to complete. The extra communications slow the entire network down, and this is true any time an HR/DSSS or DSSS device is detected, whether it is part of the network or simply a device near enough to be detected.

Token Ring and FDDI

Token Ring and FDDI are two LAN networking protocols that operate at Layers 1 and 2 of the OSI model, but are not being newly implemented today. However, you may encounter some existing networks that utilize these technologies, so it is only important that you understand very basically what these protocols offer.

Token Ring networks use a token for passing data across a ring topology. A *ring* topology indicates that the network devices are connected together, similar to a bus topology. Instead of terminating the bus at each end, the two ends are connected, forming a ring. A token is passed around the network, and the device that needs to communicate takes the token and passes its data on the token. Each device inspects the token when it arrives at its receiving port to see if the data is intended for it. If it is, it takes the data and either places the token back on the network without the data or adds new data that it wishes to transmit. This description is a simplified overview of how Token Ring networks work. IEEE 802.5 specifies the functionality of Token Ring networks, which support a maximum data rate of 16 Mbps.

The Token Ring "ring" is usually a physical star, but a logical ring. In other words, the cabling usually goes back to a center location, but the token continues to pass around the network in a logical ring manner. A later approved Token Ring standard took the data rate all the way up to 1000 Mbps, but no hardware has every been developed to implement it due to the great popularity of Ethernet.

The Fiber Distributed Data Interface (FDDI) LAN protocol also uses a ring passing mechanism; however, FDDI is fiber based as its name implies. The technology could be implemented on copper, but I know of no actual implementations of CDDI. Much like the Infrared PHY in 802.11, the CDDI implementation was never implemented to my knowledge. Rather than being an IEEE standard, FDDI is an ANSI standard that is derived in part from IEEE 802.4, which is a token bus protocol standard.

In the past, FDDI was used as a backbone technology because it supports up to 100 Mbps data rates; however, Gigabit Ethernet and 10 Gigabit Ethernet are certainly much faster and much less expensive to implement. For this reason, FDDI is—like Token Ring—not being implemented in many new installations today. The lower prices of 100MB Ethernet and the cheaper cost of copper cable has effectively brought an end to FDDI technologies.

WAN Protocols

If you want to connect multiple LANs together, you will need to use a WAN solution. The WAN solution will require a physical medium and the protocols used to communicate across that medium. WAN solutions include HDLC, PPP, Frame Relay, and ATM, among others. In this section, I'll provide a quick high-level overview of these protocols, since they are not covered in depth on the Convergence+ exam.

From the perspective of moving data from LAN to LAN, a WAN may allow various protocols to be passed across it. The good news is that any modern WAN solution will allow the transfer of IP traffic. Certainly without this allowance, the WAN technology would be doomed to failure.

The general process of sending data across a WAN is depicted in Figure 2-7; here are its basic steps:

1. The client computer sends an IP packet to the local router on its subnet.
2. The router forwards the packet to the WAN communications device.
3. The WAN communications device forwards the data across the WAN link/ network.

FIGURE 2-7

Sending data across a WAN

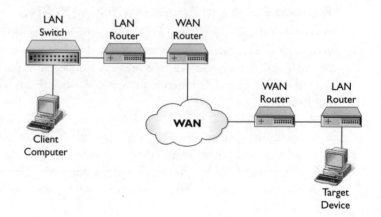

4. The remote WAN communications device receives the data and forwards it to the router on the segment of the target device.

5. The router forwards the data to the target device on the remote WAN.

As you can see from this process, the "inside" of the WAN—often called a cloud—is not really in your control. There are really no exceptions to this concept because WAN links span great distances where you cannot possibly own all points in the link without unbelievable expense. For this reason, we rely on the WAN service provider to grant us the bandwidth and QoS we require. We can only control these parameters—to any great extent—within the LANs on either end of the WAN link. This control becomes very important when implementing Voice over IP solutions that must pass across WAN links.

In order to understand WAN technologies at the level needed to work as a convergence engineer, you'll need to know only the basics of point-to-point WAN technologies and packet-switching WAN technologies. The following sections introduce the two most popular protocols or solutions in each of these two WAN link categories.

Point-to-Point and Packet-Switching WANS

A *point-to-point WAN link* is a WAN connection that is dedicated to just two endpoints. The connection is not shared with other customers of the WAN service provider, and the communications are very simple. Most point-to-point WAN links use either HDLC or PPP for communications, since this is what most WAN routers implement.

A *packet-switched WAN link* is a WAN connection that uses a shared provider network for the connection. The WAN provider's network can be compared to the Internet in that there are many customers connecting to the network; however, it differs in that the WAN provider provides guarantees of bandwidth according to your contract with them. The two most common packet-switched WAN services are Frame Relay and ATM.

HDLC

The *High-Level Data Link Control (HDLC)* protocol is a very simple WAN protocol that implements only a one-byte Flag (this is similar to the SFD in an Ethernet frame in that it indicates an HDLC frame follows), a one-byte Address field (this is not useful in direct links but is a remnant of older implementations), a one-byte Control field, and a four-byte FCS like an Ethernet frame. Between the control field and the FCS field is the actual data being transmitted across the WAN link. Figure 2-8 shows this HDLC frame structure.

It is useful to know that some vendors implement HDLC in a non-standard way. For example, Cisco adds a Type field after the Control field in the HDLC frame. The two-byte Type field is used to identify the LAN protocol being transferred in the HDLC frame. This means that a Cisco router using HDLC can only communicate with another Cisco router using HDLC on the remote end of the WAN link.

PPP

The *Point-to-Point Protocol (PPP)* is a protocol created later than HDLC and one that looks exactly like Cisco's implementation of HDLC from a frame perspective. In other words, the PPP frame looks like the frame in Figure 2-8, except that a two-byte Type field is added between the Control field and the data that is being transmitted. Because of this inclusion, it is one of the most popular protocols used in point-to-point WAN links, since it will work between varied vendors' devices as long as they each support the PPP standard.

PPP is also used in dial-up Internet connections, including ISDN, 56k modems, and other lower modem speeds. In fact, for many years this use has been the most common application of PPP thanks to the many millions of dial-up Internet users around the world.

Frame Relay and ATM

Packet-switched WAN links scale well and can be less expensive than point-to-point WAN links. The two common technologies implemented today are Frame Relay and ATM. While X.25 was popular in the past, I will not address it in this book, since it is a quickly waning technology that has ceased to be useful moving forward.

FIGURE 2-8

HDLC frame format

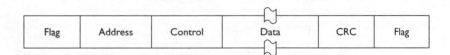

Flag	Address	Control	Data	CRC	Flag

When comparing Frame Relay and ATM, it is first important to know what they share in common and then to know where they differ. Both technologies use packet switching. This term means that they both insert a destination address into a packet and send it through the provider network to that destination address. They both use a concept known as a *virtual circuit* in order to conceptualize the data transfer. Finally, they both transmit data digitally as opposed to being analog like X.25. The dissimilarities are also important. Frame Relay supports service rates of up to 1.544 Mbps, whereas ATM starts at 1.544 Mbps and goes up from there to 622 Mbps or even 2.4 Gbps.

In general, packet-switched WAN connections are used more when many LANs need to be interconnected using a WAN technology. Point-to-point WAN connections are more likely to be used when there are three to four or fewer total LANs needing connectivity. These realities are due more often than not to the cost of equipment and dedicated lines in larger implementations.

TCP/IP

Once you have the physical network connections in place using Ethernet or Wi-Fi or some other protocol, you will need to implement a protocol suite that can provide the functionality for Layers 3 through 7 of the theoretical OSI model. There is no question about the most popular protocol suite in use today, and it is the only suite that I will cover in this chapter. That suite is the TCP/IP protocol suite.

It's important to remember that there is really no such thing as the TCP/IP protocol. TCP is a protocol, and IP is a protocol (though they were the same Transmission Control Protocol from 1974 to 1978, when the single TCP solution was split into TCP for host-to-host communications and IP for network routing). The name TCP/IP indicates that the TCP segments travel as IP packets over the network. In other words, TCP travels over IP just as FTP travels over TCP and SNMP travels over UDP. We're back to the layering concepts again. For this reason, it's important to understand the TCP/IP model as opposed to the OSI model.

TCP/IP Model

Unlike the OSI model's seven layers, the TCP/IP model contains only four layers. The four layers of the TCP/IP model are

- Application layer
- Transport layer

- Internet layer
- Link layer

The *TCP/IP Application layer* can be said to encompass the OSI model's Application and Presentation layers. In other words, Layer 4 of the TCP/IP model performs the services required of Layer 6 and Layer 7 of the OSI model. RFC 1122 specifies two categories of Application layer protocols: user protocols and support protocols. User protocols include common protocols like FTP, SMTP, and HTTP. These protocols are all used to provide a direct service to the user. FTP allows the user to transfer a file between two machines. SMTP allows the user to send an e-mail message. HTTP allows the user to view web pages from a web site. Support protocols include common protocols like DNS, DHCP, and BOOTP. DNS provides name resolution for user requests. For example, if a user requests to browse the home page at www.SysEdCo.com, the DNS support service resolves the web site's domain name to an IP address for actual communications. Because this protocol is an indirect service for the user, it is considered a support service as opposed to a user service. DHCP and BOOTP are both used to configure the IP protocol—and possible other TCP/IP parameters—without user intervention.

The *TCP/IP Transport layer* provides host-to-host or end-to-end communications. Transport layer protocols may provide reliable or non-guaranteed data delivery. TCP is an example of a Transport layer protocol that provides reliable delivery of data (usually called segments), and UDP is an example of a protocol that provides non-guaranteed delivery of data (usually called datagrams). Notice that I did not say that TCP provides guaranteed delivery of data. This is because no network protocol can provide guaranteed delivery of data; however, reliable protocols do provide notice of undelivered data, and UDP does not provide such a notice. Later in this chapter I'll explain why a non-guaranteed data delivery model is very useful and how it plays an important role in Voice over IP networks.

The next layer is the *TCP/IP Internet layer*. The Internet layer is where host identification is utilized to route TCP segments and UDP datagrams to the appropriate end device. The protocol that all Application and Transport layer protocols use within the TCP/IP model is the Internet Protocol (IP). The IP provides the routing functionality that enables the implementation of very large LANs and communications across the Internet. You'll learn much more about this protocol in the section "IP," coming up next. In addition to IP, the Internet Control Message Protocol (ICMP) is considered an integral part of IP, even though it actually uses IP for communications just like TCP and UDP. So ICMP is considered an Internet layer protocol because of its integral use in IP-based communications.

The final or bottom layer of the TCP/IP model is the *Link layer*. This layer is where the upper-layer TCP/IP suite interfaces with the lower-layer physical transmission medium. Some have said that the Link layer is equivalent to the Data Link layer of the OSI model, and of the TCP/IP layers, this description is probably the most accurate linkage of all. In fact, when TCP/IP runs over Ethernet, there is no real Link layer provided by TCP/IP; instead, the MAC and PHY of the IEEE 802.3 protocol provide the functionality of the TCP/IP Link layer to the TCP/IP suite. It is also interesting to note that protocols such as ARP and RARP that actually service the IP protocol are not actually Link layer protocols themselves. Instead, they seem to exist in some nefarious land between the Link layer and the Internet layer. I would suggest that they are simply part of the Internet layer and that you could represent the Internet layer as having an upper management layer (ICMP, IGMP, etc.), a routing layer (IP), and a routing service layer (ARP and RARP); however, this idea is only my thinking and not really part of the standard TCP/IP model.

It has been much debated over the years whether ARP exists at Layer 2 or Layer 3 of the OSI model or at the Link layer or Internet layer of the TCP/IP. I suggest that this debate exists because ARP really works between these layers. In fact, ARP is used to resolve the MAC address (a Layer 2 address) when the IP address (a Layer 3 address) is known. It could be said that ARP provides a service *between* these two layers, and this placement may be the point of argumentation and debate. Wherever you decide to place the protocol, know that you will likely meet opposition to your view.

Figure 2-9 shows a common mapping of the TCP/IP model to the OSI model. Again, keep in mind that this diagram is a mapping for understanding purposes only and that the TCP/IP suite of protocols makes no attempt or claim to mapping in this way. It is simply a helpful way of thinking about the functionality of the suite.

In order to fully understand the working of TCP/IP, it is essential to investigate the core protocols. In the following sections, I'll explain the basic functionality of IP, TCP, and other important TCP/IP protocols for our purposes.

IP

In the earlier section of this chapter that explained Ethernet, you discovered that an address is used to identify each node on the Ethernet. This address is called a MAC address. Now, I'm going to explain another address that each node possesses on a modern network. This is a Layer 3 address known as an IP address. MAC addresses are Layer 2 addresses and IP addresses are Layer 3 addresses. This differentiation will become very important in a moment.

The Internet Protocol (IP) is the Network layer (OSI) or Internet layer (TCP/IP) solution to node identification. This protocol is responsible for addressing, data routing,

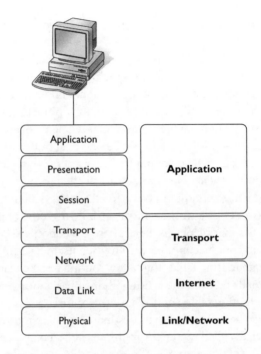

FIGURE 2-9

TCP/IP model
mapped to the
OSI model

and servicing for the upper layers in the TCP/IP suite. The question is simple: why do I need a Layer 3 address in addition to a Layer 2 address?

To understand the answer to this question, you must understand the concepts of network congestion, network segmentation, and network routing. *Network congestion* occurs when the bandwidth on the network is insufficient to meet the demands placed on the network. In other words, you can have too many nodes on a single network. Consider a network with only three computers connected to a 100 Mbps switch. These computers will be able to communicate with each other very rapidly. Now, imagine that you add three more computers to double the size of the network. Each computer will now have an average of half the throughput available to it as was available when there were only three computers on the network. The guideline is that for every doubling of communicating nodes on the network segment (I'll define this in a moment) you halve the average throughput for each node. If we doubled our network four more times, we would have 96 nodes on the network. Each node would have an average of about 1 Mbps of throughput available. This throughput may or may not be enough, and it could indeed lead to network congestion problems, since it is now more likely that a number of nodes will use more than their "fair share" of network bandwidth.

So what can you do when you have too many nodes on a network segment? Create more segments. *Network segmentation* is the act of separating the network into reasonably sized broadcast domains. A *broadcast domain* is a shared medium where all devices can communicate with each other without the need for a routing or bridging device. The most commonly implemented network segmentation protocol is the IP protocol. It can be used, in conjunction with routers, to split any network into infinitely smaller and smaller segments even down to a single node on each segment—which would not be practical or beneficial.

Of course, if you're implementing network segmentation, you must implement *network routing*. Network routing is the process of moving data from one network segment to another in order to allow that data to reach an intended node on a network or network segment separate from the originating node. IP is responsible for the routing of this data. There is a concept known as a routing protocol, but it should not be confused with routing itself. Routing protocols are used to dynamically create routing tables that allow routing to take place. IP uses these routing tables to actually perform the routing. Remember: routing protocols like RIP and OSPF create the routing tables, and IP uses the routing tables to determine the best route available to get a data packet to its destination. These routing tables are stored in memory on devices called routers. A router may be a computer with multiple network interface cards (NICs) acting as a router, or it may be a dedicated routing device.

e x a m
ⓦatch

You should know the basics of IP well. If you do not feel this section has given you enough depth of understanding, you might consider reading a Network+ certification study guide as well. It is *not a prerequisite, but the knowledge a Network+ certified individual would have will be very helpful in passing the Convergence+ exam.*

IP Addresses The IP address itself is a 32-bit address divided into four octets, or four groups of eight bits. For example, the following bits represent a valid IP address:

00001010.00001010.00001010.00000001

This format is sometimes called binary notation, but it is the actual form that an IP address takes. To understand this notation, you'll need to understand how to convert binary bits to decimal, as I'm sure you're used to seeing the previous

IP address shown as 10.10.10.1. These IP addresses are composed of four octets and can contain decimal values from 0 to 255. Let's look at how you would convert an eight-bit binary number into the decimal version that we're used to.

In the earlier section "Ethernet," I pointed out the frame structure and the bits used in each frame. The reality is that any data can be represented by binary bits. Someone simply has to decide how many bits there should be and what those bits should represent.

Ultimately, the smallest element that can be transmitted on any network is a *bit*. A bit is a single value equal to 1 or 0. When you group these bits together, they form bytes. An eight-bit byte is the most commonly referenced byte and is the base of most networking measurements; it is specifically called an *octet* in most standards even though the vendors and networking professionals have leaned more toward the term byte. For example, one kilobyte is 1024 bytes, and one megabyte is 1,048,576 bytes. You will often see these numbers rounded to say that 1000 bytes is a kilobyte or 1,000,000 bytes is a megabyte. The term octet could also be used in these statements; for example, one kilobyte is 1024 octets.

You might be wondering how a simple bit, or even a byte, can represent anything. This is an important concept to understand; otherwise, you may have difficulty truly understanding how a network works. Let's consider just an eight-bit byte. If you have one bit, it can represent any two pieces of information. The 1 can represent one piece of information, and the 0 can represent another. When you have two bits, you can represent four pieces of information. You have the values 00, 01, 10, and 11 available to use as representative elements. When you have three bits, you can represent eight pieces of information, and for every bit you add, you effectively double the amount of information that can be represented. This means that an eight-bit byte can represent 2^8 pieces of information or 256 elements. You have now received a hint about why the numbers 0 through 255 are all that can be used in an IP address octet. (Remember, IP addresses are four octets or four eight-bit bytes grouped together.)

There are standard mapping systems that map a numeric value to a piece of information. For example, the ASCII system maps numbers to characters. Since I can represent up to 256 elements with an eight-bit byte, I can represent 256 ASCII codes as well. A quick Internet search for ASCII codes will reveal a number of sites that provide tables of ASCII codes. For example, the ASCII codes for 802.11 (one of my absolute favorite IEEE standards) are 56, 48, 50, 46, 49, and 49 in decimal form. Since I can represent any number from 0 to 255 with an eight-bit byte, I can represent these numbers as well. Table 2-3 shows a mapping of characters to ASCII decimal codes to eight-bit bytes.

Character	ASCII Decimal Code	Eight-Bit Byte
8	56	00111000
0	48	00110000
2	50	00110010
.	46	00101110
1	49	00110001
1	49	00110001

In order for all this communication to work, both the sender and the receiver of the bytes must agree on how the bytes will be translated. In other words, for information to be meaningful, both parties must agree to the meaning. This concept is the same in human languages. If I speak a language that has meaning to me, but you do not understand that language, it is meaningless to you and communication has not occurred. When a computer receives information that it cannot interpret to be anything meaningful, it sees it as either noise or corrupted data.

To understand how the binary bits in an octet translated to the ASCII decimal codes, consider Table 2-4. Here you can see that the first bit (the right-most bit) represents the number 1, the second bit represents the number 2, and the third bit represents the number 4, and so on. The example, in the table, is 00110001. Where there is a 0, the bit is off. Where there is a 1, the bit is on. We add up the total values in the translated row, based on the represented number for each bit, and find the result of 49 because we only count the values where the bit is equal to 1. This correlation is how the binary octet of 00110001 represents the ASCII decimal code of 49, which represents the number 1 in the ASCII tables.

So how does all this translation apply to IP addressing? IP version 4 addresses are composed of four eight-bit bytes, or four octets. Now, you could memorize and work with IP addresses like *00001010.00001010.00001010.00000001*, or you could work

TABLE 2-4   Converting Bytes to Decimal Values

Bit position	8	7	6	5	4	3	2	I
Represented decimal number	128	64	32	16	8	4	2	1
Example binary number	0	0	1	1	0	0	0	1
Translated	0	0	32	16	0	0	0	1

with addresses like *10.10.10.1*. I don't know about you, but the latter is certainly easier for me; however, the former is not only easier for your computer to work with, it is the only way your computer thinks. Therefore, IP addresses can be represented in *dotted decimal* notation to make it easier for humans to work with. The dotted decimal notation looks like this: *10.10.10.1*.

IP version 4 (which is the current widely used implementation) addresses cannot just have any number in each octet. Remember that there are only eight bits available, so the number in each octet must be an eight-bit number. This means it will be a decimal value from 0 to 255, for a total of 256 valid numbers. There is no IP address that starts with 0. In other words, you'll never see an address like 0.0.0.1 assigned to a device. In fact, the address 0.0.0.0 is reserved to indicate the *default network* and/or the *default device*. In this context, the term default should be understood according to context. For example, the IP address of 0.0.0.23 would refer to host identification 23 in the current (default) network, and flipping the numbers, the IP address of 192.168.12.0 refers to the entire network as a collective. Similarly, 255.255.255.255 is reserved to indicate all nodes or hosts. Of course, on the Internet, 255.255.255.255 would actually refer to every one of the millions of connected devices, and it is not used for any practical purpose.

Another special address that is important for you to know about is the *loopback* address. This is IP address 127.0.0.1. For example, if you use the PING command to communicate with 127.0.0.1, you are actually communicating with your own TCP/IP network stack or protocol implementation. This practice is sometimes called pinging yourself and is used to troubleshoot the local TCP/IP implementation and ensure that it is working properly. Additionally, the Automatic Private IP Addressing (APIPA) process specifies the use of addresses in the 168.*x.x.x* range of addresses. This is now used in Windows systems when a DHCP server cannot be contacted.

An organization known as the Internet Assigned Numbers Authority (IANA) manages the public IP address space. RFC 1166 briefly describes how the IP address space should be partitioned for distribution. It specifies three classes of addresses that are important to our discussion: class A, class B, and class C. Class D addresses are used for multicast implementations, and class E addresses are reserved for testing purposes. Table 2-5 shows the breakdown of the three primary classes of addresses.

TABLE 2-5	IP Address Class	Decimal Representation	Number of Networks
IP Address Classes	A	1.0.0.0–126.255.255.255	126
	B	128.0.0.0–191.255.255.255	16,384
	C	192.0.0.0–223.255.255.255	2,097,152

The IP addressing, however, has been moving from classfull addressing to classless addressing over the past decade. In fact, the IANA even references the classes as "former class B" and "former class C" in many of their web pages, indicating that the age of classfull IP addressing is quickly passing. By doing away with the classes, the IP address space is more usefully managed and distributed. In addition, many companies now use private IP addresses internally and have only a few real Internet addresses for external communication. This is accomplished through something called Network Address Translation, or NAT.

These private IP addresses fall into three different ranges that were originally set aside according to the A, B, and C classes. These address ranges are:

- 10.*x.x.x*
- 172.16.*x.x*–172.31.*x.x*
- 192.168.*x.x*

The class A private address range from 10.0.0.1 to 10.255.255.254 provides about 16.7 million IP addresses and can be divided into thousands of networks using classless subnetting. The class B private address range from 172.16.0.0 to 172.31.255.254 provides about one million IP addresses and can be divided into hundreds of networks. Finally, the class C private address range from 192.168.0.1 to 192.168.0.254 provides 65,556 addresses and can be divided into hundreds of networks as well.

You'll notice that the class B private address range is the only one that provides more than one starting set of decimal values. For example, all of the class A private addresses start with 10, and all of the class C private addresses start with 192.168. The class B private address range can start with any numbers from 172.16 to 172.31.

Ultimately, there is a big difference between a globally assigned IP address and a private IP address. Globally assigned addresses are assigned by the IANA or one of the agencies serving the IANA. Private addresses can be assigned by members of an organization in any way they desire as long as they have implemented a network infrastructure that can support them. The benefit of private addresses is that they are set aside by the IANA and are guaranteed to never be the destination of an actual "on-the-Internet" host. This configuration means that you'll be able to use the exact same IP addresses on your network as those being used on mine and we'll still be able to communicate with each other across the Internet as long as we both implement a NAT solution and possibly port forwarding, depending on the scenario.

Subnet Masks   As you can imagine, a private network that uses the "ten space" (a phrase for referencing the private IP addresses that are in the 10.*x.x.x* range) can be rather large. In order to reduce network traffic on single segments, we can subnet our network and increase performance. To do this task, you will need to implement the appropriate subnetting scheme with subnet masks.

A subnet mask is a binary-level concept that is used to divide the IP address into a network identifier (ID) and a host ID. The network ID identifies the network on which the host resides, and the host ID identifies the unique device within that network. There are two basic kinds of subnetting: classfull and classless.

Classfull (also written *classful*) subnetting simply acknowledges the class of the IP address and uses a subnet mask that matches that class. For example, a class A IP address would use the first eight bits for the network ID, and therefore the subnet mask would be:

11111111.00000000.00000000.00000000

Notice that the portion of the IP address that is the network ID is all 1's, and the portion that is the host ID is all 0's. For example, if the IP address were 10.12.89.75 and we were using classfull subnetting, the subnet mask would be 11111111.00000000 .00000000.00000000, which is represented as 255.0.0.0 in dotted decimal notation. As one more example, consider a class C IP address of 192.168.14.57. What would the classfull subnet mask be? Correct. It would be 11111111.11111111.11111111.00 000000. This difference is because a class C IP address uses the first 24 bits to define the network ID and the last 8 bits to define the host ID. This network ID would be represented as 255.255.255.0 in dotted decimal notation.

Most configuration interfaces allow you to enter the IP address and also the subnet mask in dotted decimal notation. This notation makes configuration much easier; however, if you want to perform classless subnetting, you will need to understand the binary level where we now reside in our discussion.

Classless Inter-Domain Routing (CIDR) is the standard replacement for classfull addressing and subnetting. Classless subnetting allows you to split the network ID and the host ID at the binary level and, therefore, right in the middle of an octet. For example, you can say that 10.10.10.1 is on one network and 10.10.10.201 is on another. How would you do this separation? Let's look at the binary level. Here are the two IP addresses in binary:

00001010.00001010.00001010.00000001 (10.10.10.1)
00001010.00001010.00001010.11001001 (10.10.10.201)

In order to use CIDR and indicate that the final octet should be split in two so that everything from 1 to 127 is in one network and everything from 128 to 254 is in another, we would use the following subnet mask:

11111111.11111111.11111111.10000000

You might be wondering how this subnet mask works. In order to understand it, consider it in Table 2-6. The first row (other than the bit position identifier row) of the table is the IP address of 10.10.10.1, and the third row is the IP address of 10.10.10.201. The second row is the CIDR subnet mask. If you count the columns carefully, you'll see that the first 25 positions have a 1 and the last 7 positions have a 0. Where there is a 1, the IP addresses must match or they are on different network IDs. As you read across the rows and compare the first row with the third row, you'll notice that they are identical until you read the 25th bit position. In the 25th bit position, the first row has a 0 and the third row has a 1. They are different and therefore are on different network IDs. This is CIDR in action.

Because of the fact that CIDR subnetting allows subnet masks that mask part of an octet, you will see subnet masks like 255.255.255.128 (which is the decimal equivalent of the subnet mask row in Table 2-6). Instead of representing the subnet mask in decimal notation, it is often simply appended to the end of the IP address. For example, the IP address and subnet mask combination in Table 2-6 could be represented as 10.10.10.1/25. This representation, which is sometimes called Variable Length Subnet Mask (VLSM) representation or CIDR representation, is becoming more and more common. It indicates that the IP address is 10.10.10.1 and the network ID (sometimes called the subnet or subnetwork) is the first 25 bits of the IP address.

IP Routing   At this point you're probably beginning to wonder why all this subnetting really matters. Other than the fact that it can be used to reduce the size of a network segment, which may or may not be a benefit, depending on the infrastructure type you've implemented, it allows IP routing to function. To simplify the process down to the level that you really need to know in order to work with converged networks, the local TCP/IP implementation on a device needs a method to determine if it can send the data directly to the end IP address or if it needs to send it through a router.

TABLE 2-6   Subnetting with CIDR

B 1	B 2	B 3	B 4	B 5	B 6	B 7	B 8	B 9	B 10	B 11	B 12	B 13	B 14	B 15	B 16	B 17	B 18	B 19	B 20	B 21	B 22	B 23	B 24	B 25	B 26	B 27	B 28	B 29	B 30	B 31	B 32
0	0	0	0	1	0	1	0	0	0	0	0	1	0	1	0	0	0	0	0	1	0	1	0	0	0	0	0	0	0	0	1
1	1	1	1	1	1	1	1	1	1	1	1	1	1	1	1	1	1	1	1	1	1	1	1	1	0	0	0	0	0	0	0
0	0	0	0	1	0	1	0	0	0	0	0	1	0	1	0	0	0	0	0	1	0	1	0	1	1	0	0	1	0	0	1

Going back to the example in the section "Subnet Masks," imagine that IP address 10.10.10.1 is attempting to send a packet to IP address 10.10.10.201. How does the machine at 10.10.10.1 know if it needs the router (also called the default gateway) or not? The answer is that it determines the network ID of its own address and looks at the destination address to see if it has the same network ID. If the network IDs match, the Address Resolution Protocol (ARP) can be used to discover the MAC address of the destination IP address because they are on the same network. If the network IDs do not match, ARP is used to discover the MAC address of the router and the IP packet is sent to the router. The local device assumes that the router knows how to get to any IP address in the world.

Inside the router, it must find the best path to the destination IP. Once this path is determined, the router discovers the MAC address of the nearest router in that path and forwards on the IP packet. This process continues until the target end node is reached. However, it all started when that first device evaluated the IP address against the subnet mask and determined that it needed the help of the local router or default gateway. Figure 2-10 shows this process as described.

TCP

While IP is used to move data around on internetworks until that data reaches its intended target, the Transmission Control Protocol (TCP) is used to both provide reliability in those deliveries and determine the application that should process the data on the receiving device. The reliability is provided by segmenting, transmitting, retransmitting, and aggregating upper-layer data. The application determination is accomplished through the use of TCP ports.

TCP Segments The TCP protocol takes data from the upper layers and segments that data into smaller units that can be transferred and managed. Because IP sometimes drops IP packets due to congestion and because packets can travel different routes to the destination, some TCP segments (which are sent across IP) may arrive at the destination out of order or not arrive at all. For this reason, TCP provides resequencing when data arrives out of order and resends data that doesn't

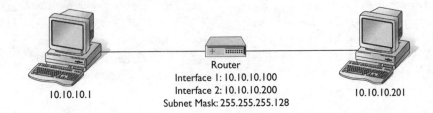

FIGURE 2-10

IP routing from a local machine to a remote node

10.10.10.1

Router
Interface 1: 10.10.10.100
Interface 2: 10.10.10.200
Subnet Mask: 255.255.255.128

10.10.10.201

make it to the destination. This service provides reliable delivery of data and makes TCP very useful for such solutions as file transfers, e-mail, and NNTP (the Network News Transfer Protocol). Each of these applications need reliable delivery, and this ability means TCP is a prime candidate.

TCP Ports   In addition to reliable delivery, TCP uses ports to determine the upper-level applications that should receive the arriving data. Some port numbers are well-known, others are registered, and yet others are unassigned or private. Common ports include 21 for FTP, 80 for HTTP, and 25 for SMTP. Knowing which port a service uses has become very important in modern networks due to the heavy implementation of firewalls. Firewalls often block all internetwork traffic except certain ports. If you don't know the port number the service is attempting to utilize, you won't know what exception to create in your firewall.

INSIDE THE EXAM

Why Is TCP/IP Important?

TCP/IP is the foundational protocol of the Internet. Modern private networks primarily use this protocol as well. If you do not know the basics of IP, TCP, and other protocols in this suite, you will not be able to administer a modern network.

The success of TCP/IP has been largely due to its utilization on the Internet. If you wanted to browse web pages, download from FTP sites, or read text at Gopher sites, you had to be running the TCP/IP protocol. Many early networks ran TCP/IP as well as some other protocol like IPX/SPX or Banyan Vines. First, network vendors began by supporting TCP/IP alongside their proprietary protocols, and eventually they moved their systems not only to support the TCP/IP suite, but to rely upon it. Today, Novell, Microsoft, Unix,

Linux, and Apple computers all use TCP/IP as the primary communication protocol.

At this moment, a very gradual transition is happening in relation to the IP protocol. IPv6 has been available for a number of years, and operating systems have slowly incorporated it into their available protocols. Windows Vista has very integral support of IPv6, as does Windows Server 2008. Unix and Linux machines have supported it for some time, and Apple's Mac OS X also supports it. Once IPv6 support is available on the vast majority of computers, we'll likely see it used more and more on our networks; however, the Convergence+ exam will not test your knowledge of this protocol according to the stated objectives as of early 2008.

RTP

The *Real-Time Transport Protocol (RTP)* is a protocol designed for moving audio and video data on internetworks running IP. Unlike TCP, RTP is not focused on reliability but rather on rapid transfer. It is used when the data being transferred must arrive in a timely fashion or it provides little value. For example, an e-mail message can take two or three minutes to arrive, and once it has arrived, you can read it just fine; however, a five-second statement made into a Voice over IP phone cannot take two or three minutes to arrive. That five-second statement would have been broken into many packets and they must arrive very quickly. If there are large gaps in their arrival, the call will either be lost or the listener will not be able to intelligently hear the message. RTP runs over UDP, which is discussed in a later section.

RTCP

The *Real-Time Transport Control Protocol (RTCP)* is used with RTP. It is used to gather and report on performance information about the network communications. Administrative applications have been developed that allow support professionals to monitor media communications issues like latency, jitter, and packet loss. These issues will all be addressed in later chapters.

UDP

Since RTP and RTCP use UDP, it would be a good idea to know what this protocol does. UDP is responsible for connectionless communications. In other words, it just drops the data on the wire and says, "I hope you get there." At first glance, you might think there would be no use for such a protocol; however, this protocol is very important to our converged networks, as it allows for timely delivery of information and ignores reliability completely. UDP doesn't care if the data gets there and in fact doesn't even expect an acknowledgment from the receiver.

Here's why this is a good thing. Since UDP does not have to process an acknowledgment, it can just keep the stream of data flowing. The receiving end doesn't have to pause to acknowledge receipt either. The end result is that there is less processing at both the sender and the receiver. Another benefit is that network congestion is reduced, since fewer packets are being transmitted. Finally, the packets that are transmitted are smaller, since the UDP header is smaller than the TCP header. The UDP header is 8 bytes and the TCP header is 24 bytes. This larger size means that 16 bytes of extra information has to be transferred with TCP, and this difference would be detrimental to voice communications where latency and packet loss must be kept to a minimum.

Other TCP/IP Protocols

There are actually dozens of common protocols used in TCP/IP networks. I'll cover just a few of them here, as they can have an impact on converged networks.

ICMP   The *Internet Control Message Protocol (ICMP)* is used for administration and management purposes within the TCP/IP suite. For example, when you PING another computer on the network, you are actual performing an ICMP echo request and the remote host is sending back an ICMP reply. Many modern firewalls have prebuilt exceptions for ICMP. If you enable these exceptions, you are allowing PINGs to come through the firewall.

FTP and HTTP   The *File Transfer Protocol (FTP)* and the *Hypertext Transfer Protocol (HTTP)* are probably still the two most popular application protocols in the world. Every time you use a web browser to search the Internet, read news, or access an update from a vendor's web site, you are using HTTP. Much of the time, when you download files from the Internet, you are using FTP. These protocols are extremely useful and still have a long life ahead of them.

DNS, DHCP, and BOOTP   The *Domain Name System (DNS)* or *Domain Name Service* is the protocol that really made the Internet take off. Do you think the Internet would have been as successful as it has been if people had to visit 32.15.17.89 for news and 165.145.209.217 for weather? I don't think so. However, the speed with which someone can type in www.SysEdCo.com and quickly arrive at my company's web site has made the Internet easier to use. DNS provides domain name to IP address resolution. This resolution is very important because all Internet communications use IP addresses, but we humans like to have something simpler to remember.

An additional benefit of DNS is that it allows for less intrusive IP address changes. For example, if the IP address of a server is 10.10.10.18 and the host name is server1.company.local, you can change the IP address to 192.168.12.17 and—as long as the host name stays the same—the users will never need to know the difference. They can keep right on connecting to server1.company.local and not worry about the underlying IP address.

The *Dynamic Host Configuration Protocol (DHCP)* and *Boot Time Protocol (BOOTP)* are used to dynamically configure the IP protocol on client machines. These protocols help prevent the long lists of IP addresses that were so popular on early networks in the 1990s. I remember having a text file with more than 700 IP addresses in it. Any time an address changed or a machine was added, I had to update that file. In modern networks with DHCP, this problem is no longer the case. The client device receives an IP address and keeps it for some period of time. Once that period of time expires, the client can attempt to renew the address or may simply acquire a new one.

In addition to dynamic IP configuration, modern networks support dynamic DNS (DDNS). DDNS allows hosts to have a consistent host name with changing IP addresses. Despite the changes to the IP address we can still reach them easily since the host name does not change and the host name–to–address mappings are maintained dynamically. The client computer registers its host name with the DNS service either each time the computer is started or each time the IP address changes.

CERTIFICATION SUMMARY

This chapter introduced you to the concept of networking protocols. It is very important that you understand the basics of the protocols that are used in a converged network so that you can both implement them and troubleshoot them when you are not experiencing the results you desire. We investigated Layer 1 and Layer 2 protocols first. These protocols include Ethernet, Wi-Fi, and various WAN technologies. Ethernet is the most widely used wired LAN protocol, and IEEE 802.11 is the most widely used wireless LAN protocol. It is essential that you understand the basics of these lower-layer protocols in order to ensure you've implementing the right technologies for your converged network.

Next, you learned about TCP/IP in general and the model that is implemented in TCP/IP. This model was compared and contrasted with the OSI model. You evaluated the Internet Protocol (IP) in some depth, as it is the foundation of the TCP/IP suite. You also learned about the TCP, RTP, and UDP protocols at an overview level. Finally, you learned about the additional protocols in the TCP/IP suite such as ICMP, FTP, HTTP, DHCP, and DNS.

TWO-MINUTE DRILL

Protocols

❑ Ethernet is a Layer 1 and Layer 2 protocol that is used on wired networks.

❑ Wi-Fi, or IEEE 802.11, is a Layer 1 and Layer 2 protocol that is used on wireless networks.

❑ Common Ethernet implementations provide data rates from 10 Mbps to 10 Gbps.

❑ Common Wi-Fi implementations provide data rates from 1 Mbps to 54 Mbps.

❑ IEEE 802.11n has not been ratified at this time, but will provide much higher data rates of up to 600 Mbps.

❑ Ethernet uses CSMA/CD (collision detection), and Wi-Fi uses CSMA/CA (collision avoidance).

❑ TCP/IP uses a different model than the OSI and divides communications into four layers: Application layer, Transport layer, Internet layer, Link layer.

❑ The Internet Protocol (IP) provides addressing and routing for the TCP/IP suite.

❑ The TCP protocol provides reliable transfer, and the UDP protocol provides timely transfer.

❑ Subnet masks are used to define the network ID and the host ID in an IP address.

❑ IP uses the subnet mask to determine when the router is needed to reach the destination address.

❑ The following IP address ranges are reserved for private use and are not used on the global Internet: 10.$x.x.x$, 172.16.$x.x$ through 172.31.$x.x$, 192.168.$x.x$.

❑ DHCP and BOOTP are used to dynamically assign IP addresses to client devices on a network.

❑ DNS is used to resolve domain names to IP addresses. This protocol allows for the use of host names for communications instead of IP addresses.

SELF TEST

The following questions will help you measure your understanding of the material presented in this chapter. Read all the choices carefully because there might be more than one correct answer. Choose all correct answers for each question.

Protocols

1. You are implementing a new Voice over IP solution. The client has asked you to provide both wireless and wired access to the voice system. Which of the following are Layer 1 and Layer 2 protocols that can be used to implement this solution?
   A. Ethernet
   B. Wi-Fi
   C. TCP
   D. UDP

2. Why has Ethernet largely replaced FDDI?
   A. Ethernet is faster than FDDI.
   B. Ethernet is newer than FDDI.
   C. FDDI is no longer available.
   D. FDDI must run on fiber.

3. What is the difference between a frame and a packet?
   A. The term frame usually refers to the data at Layer 2 that is ready to be transmitted.
   B. The term packet usually refers to the data at Layer 2 that is ready to be transmitted.
   C. Frames are encrypted, and packets are not.
   D. Packets are encrypted, and frames are not.

4. Which of the following technologies are used in Ethernet networks?
   A. CSMA/CA
   B. Frames
   C. CRC
   D. OFDM

5. Which of the following technologies or concepts are used in Wi-Fi networks?
   A. CSMA/CD
   B. Frames

    C. Radio frequency

    D. OFDM

6. Which of the following are protocols used to dynamically configure the IP addresses on IP devices?

    A. DHCP

    B. DNS

    C. FTP

    D. BOOTP

7. Which of the following is a default subnet mask when using classfull IP subnetting?

    A. 255.255.255.128

    B. 255.255.128.0

    C. 255.0.0.0

    D. 255.255.255.0

8. You are implementing a Voice over IP solution. You want to make sure that your data is delivered in a timely fashion to the receiving device. Which of the following two protocols is the right choice?

    A. TCP

    B. UDP

LAB QUESTION

The XYZ corporation is implementing a converged network. As the network engineer, it is up to you to select the core protocols that will run on the network. You must implement both wired and wireless solutions. There will be about 300 devices connecting to the wired network and 30 devices connecting to the wireless network. The wireless devices need at least 4 Mbps of bandwidth each. Which of the following protocols will you be implementing: how and why?

- Ethernet
- Wi-Fi
- TCP/IP

SELF TEST ANSWERS

Protocols

1. ☑ **A** and **B** are correct. Both Ethernet and Wi-Fi are Layer 1 and Layer 2 protocol solutions. They define a Layer 2 solution that is usually called a MAC layer and a Layer 1 solution that is usually called a PHY layer.

    ☒ **C** and **D** are incorrect. Both TCP and UDP are Layer 4 protocols in the OSI model, and they are Transport layer protocols in the TCP/IP model.

2. ☑ **A** is correct. When FDDI was first released, it was a very fast solution for the backbone of modern networks. Today, Ethernet is just as fast or faster thanks to Gigabit Ethernet and 10 Gigabit Ethernet.

    ☒ **B, C** and **D** are incorrect. Ethernet has actually been around longer than FDDI. FDDI is still available, and you can run a version of FDDI on copper, though it is usually called CDDI.

3. ☑ **A** is correct. The term *frame* is usually used to reference the data that has been serviced by the Data Link layer and is ready to be transmitted on the wire. The term *packet* refers to the Network layer data (the TCP/IP model called this the Internet layer) that is managed by the IP protocol in most networks today.

    ☒ **B, C** and **D** are incorrect.

4. ☑ **B** and **C** are correct. Both frames and CRC (cyclic redundancy check) are used in Ethernet implementations.

    ☒ **A** and **D** are incorrect. CSMA/CA is used in Wi-Fi networks, and so is OFDM.

5. ☑ **B, C** and **D** are correct. Frames are used in Wi-Fi networks, although they have a different format than Ethernet frames. Radio frequency is also used, as it is the medium on which the Wi-Fi signals travel. OFDM is the modulation scheme used in IEEE 802.11a for 54 Mbps data rates.

    ☒ **A** is incorrect. CSMA/CD is used in Ethernet networks.

6. ☑ **A** and **D** are correct. The Dynamic Host Configuration Protocol (DHCP) is used to dynamically configure the IP protocol on devices, and the BOOTP protocol does the same.

    ☒ **B** and **C** are incorrect. DNS is used to resolve domain names to IP addresses, and FTP is used for file transfer.

7. ☑ **C** and **D** are correct. 255.0.0.0 is the default subnet mask for a class A address, and 255.255.255.0 is the default subnet mask for a class C address. Remember that classfull addressing is the old way and that it has been replaced by Classless Inter-Domain Routing (CIDR).

    ☒ **A** and **B** are incorrect.

8. ☑ **B** is correct. The User Datagram Protocol (UDP) provides connectionless communications and does not offer reliability; however, it is very timely, as it sends data faster and has less overhead than does TCP.

☒ **A** is incorrect. The Transmission Control Protocol (TCP) provides reliable delivery, but it does not provide timely delivery. Therefore, it should not be the foundation of Voice over IP implementations.

LAB ANSWER

Answer may vary, but here is a possible answer:

You will use Ethernet for the wired network. 100BASE-T should be sufficient between the end devices and the switches, and Gigabit Ethernet will be used between the switches and to the Internet-connected router. Additionally, you may choose to install one or two internal routers to segment the network. This design will require you to implement the appropriate IP subnet mask configuration.

You will need to implement Wi-Fi for the 30 wireless clients. Since they each need 4 Mbps of bandwidth, you'll need to provide more than 120 Mbps of total bandwidth. You'll probably want to implement a minimum of six IEEE 802.11g access points. This design gives you an aggregate data rate of 324 Mbps, but only about half of this will be actual data throughput for the users. The resulting throughput of approximately 170 Mbps exceeds the demands of the specification.

Of course, you will implement TCP/IP across both the Ethernet and Wi-Fi layers, since Voice over IP is required. You may need to consider roaming issues for wireless Voice over IP client devices as you'll learn in later chapters.

3

Infrastructure Hardware

> *We must look for some method which allows us to use our most sophisticated equipment as much as possible as if we were connected directly to the remote computer.*
>
> —Steve Crocker, *Request for Comments (RFC) 1, 7 April 1969*

From the early days of computer networking the goal has been the same: to allow users access to networked resources in a timely fashion and to provide the best experience possible in balance with available funds. This concept is illustrated well by the Steve Crocker quote that begins this chapter. He was dealing with issues like latency and response times in the network when he wrote RFC 1, and every evolutionary move in the computer networking arena has continued along this path. We are continually increasing the data rates, reducing latency and response times, and improving availability. This task is accomplished through the utilization of many different network infrastructure components, and that is the topic of this chapter. You will learn about the major components used to build a network in order to serve the end nodes as well as possible. In the next chapter, you will learn about those end nodes and how they use the infrastructure that we discuss in this one.

You will begin by learning about the various network design models that are available. This discussion will include centralized versus decentralized computing and edge versus core technologies. You will then learn about the specific devices used to form the network. These devices include switches, routers, firewalls, network appliances, wireless devices, and more.

CERTIFICATION OBJECTIVE 3.01

Recognize Network Models

Networks can be physically and logically designed in many different ways. In Chapter 1, you learned about the bus and star topologies and how these topologies can be combined to form many different physical implementations. As an example, Figure 3-1 shows a diagram of a network that effectively uses a bus topology at the core and then implements a number of interconnected star topologies for end-node connectivity. The point is that you can design your physical network in many different ways, and it is up to you to implement the structure that best serves your organization's needs.

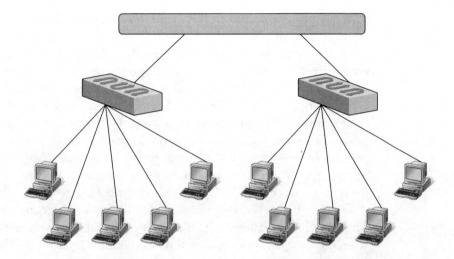

FIGURE 3-1

A combination topology

In addition to these physical implementation models, we have many different logical ways to think about the network. When considering logical network design, a layered approach is usually used. A single device can operate at one layer or in multiple layers at the same time. It is only for the sake of thought processes and administrative boundaries that we conceptually place the logical model over the physical model. In this section, I'll review both physical design considerations and logical design philosophies that have become very important in converged networks.

Network Design Models

When it comes to the actual physical implementation of your network, you have two primary options. The centralized network has been used for more than thirty years, and the decentralized network has evolved over the past twenty to become the most common network implementation type.

Centralized Versus Decentralized

The common example of a centralized network model is the traditional mainframe implementation. In this model, all the network resources and processing power are in a centrally located, powerful computer. The access nodes were usually dumb terminals, and they were given this name because they had no local intelligence for any processing other than the display of the information sent back from the mainframe or IBM AS/400. Eventually, some companies began installing desktop PCs with terminal emulation software installed that allowed the users to run powerful programs on the local PC while also accessing the centralized information on the mainframe.

Modern centralized networks often employ multiple servers at the center of the network, but all communications come into the center from a surrounding "ring" of networked devices. This design is also sometimes called centralized computing today. Often the ring is only a virtual ring, but all of the network resources sit behind one entry point, usually a router. An example of this is depicted in Figure 3-2. This model may be better called centralized resource networking, as all the resources are on a single segment, but it is often referenced as centralized networking. In the strictest sense, centralized networking refers to the data and processing existing at the center and access provided through a distributed model.

Decentralized networking indicates that the network resources are close to the point of need. For example, the file server used by the accounting department may be on the same network segment as the accounting department. The database server used by the engineering department may be on the same segment as the engineering department. This concept is depicted in Figure 3-3. This design is probably the more common model being implemented today. No unnecessary data must traverse across the entire network. Users in one segment can certainly access resources in another segment, but the resources they need the most are closest to them—on either their segment or a close neighbor.

FIGURE 3-2

Centralized resource networking

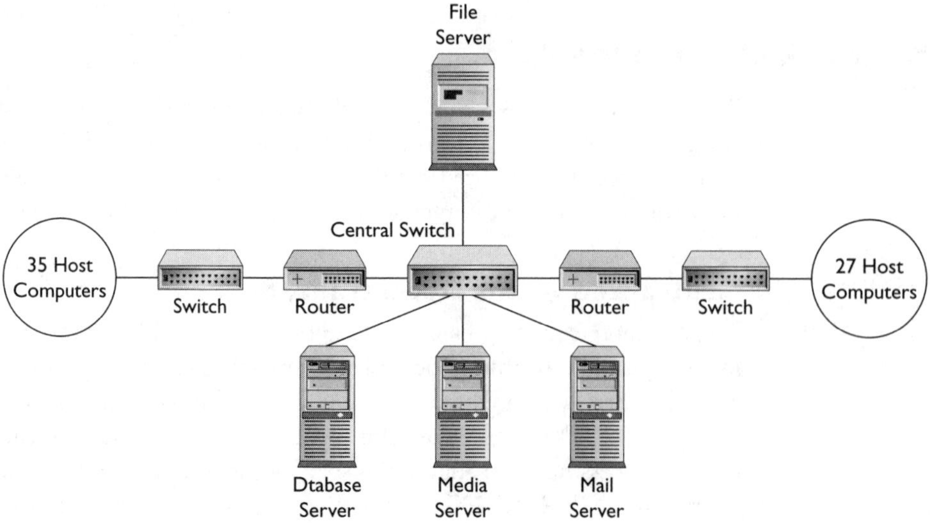

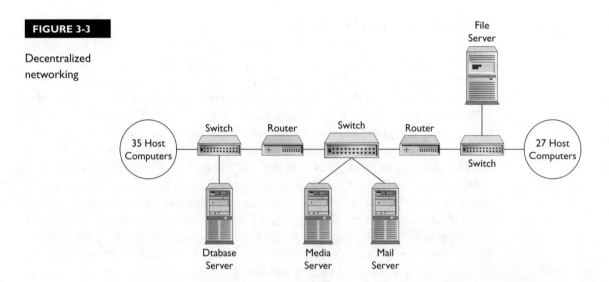

FIGURE 3-3

Decentralized
networking

Configuring Converged Resources   The concept of centralized versus
decentralized networking becomes very important in a converged network.
Voice data must travel as rapidly as possible from the sender to the receiver.
Delays in data transfer can cause problems with voice communication, such as
poor quality or dropped calls. Multimedia communications, such as streaming
video, can also be impacted by your decisions here. The best practice is to keep
communications as close to the network segment where they originate as possible
when it comes to your non-converged or traditional data (e-mail, file transfer,
printing, etc.). This design keeps the backbone or Distribution layer "pipelines"
open so that the voice or video data can move across it as quickly as possible. The
reality is that voice data packets are very small, so they will move across the local
segment very quickly once they reach it, even if they are in contention with other
data packets. This concept is particularly true if Quality of Service mechanisms are
implemented as discussed in Chapter 9.

However, dozens or even hundreds of calls may need to be routed across the
backbone or Distribution layer of your network as quickly as possible. Keeping
unneeded traditional data off those layers can be very helpful. This segregation will
be most easily accomplished if you implement a decentralized model where the most
needed resources are on the same segment or at least within the same distribution
group or workgroup.

Flat Versus Tiered Networking

When implementing a decentralized network, you can choose between a flat or a tiered networking model. In some ways, these two models are logical in nature in that a device in a tiered or layered model can exist in two tiers at the same time. This method results in the tier being more conceptual or functional than physical. The same physical device may perform functions in two tiers. Therefore, the actual physical implementation may not mirror the logical behavior of the network.

The flat network model is represented in Figure 3-4. Here we see that all of the switches and routers function in a similar way. They all implement access control lists and policies and most of the switches have both routers and end nodes and other switches connected to them. In this model, there is no "fast" backbone because all of the devices perform security verifications and similar functions.

Now notice the difference in the model in Figure 3-5. Certain functions only take place within certain tiers or layers. In fact, the most common model has three layers: the Core layer, the Distribution layer, and the Access layer. Because each layer performs different functions and the Core layer does little more than move data, the overall performance of the network is improved. Let's look at each of these layers individually.

Core Layer   As shown in Figure 3-5, the Core layer (sometimes called the network backbone) is responsible for moving a lot of data as fast as possible. This name is used because all global data access moves across this layer. In other words, when users in one office want to access resources in another office, that data will most likely pass across the Core layer. The Core layer may or may not include WAN links, and a network core can exist even in a single building implementation. For example, you may choose to implement a series of Gigabit Ethernet routers that route between three or four major sections of your network at the Core layer. Figure 3-6 illustrates a configuration like this one. Notice that there is a Gigabit Ethernet connection from the Distribution layer routers to the Core layer routers and there is a Gigabit Ethernet connection between each of the three Core layer routers.

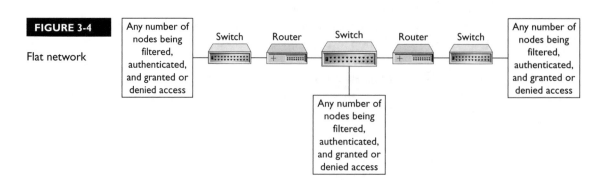

FIGURE 3-4

Flat network

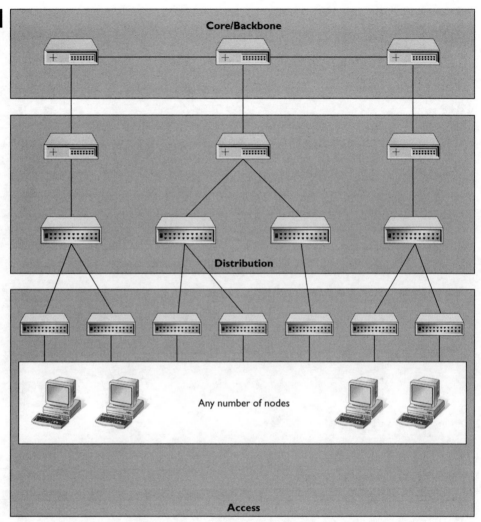

FIGURE 3-5

Tiered or layered network

Here at the Core layer, there is no intensive packet analysis for security purposes. The data is simply moved along as quickly as possible. Notice in Figure 3-6 how each major section of the network has its own printers and file storage, as well as database servers. This placement keeps traditional data off the Core layer as much as possible and allows converged data to move across this backbone very quickly. You can never prevent traditional data from moving across the Core layer completely, but this design will greatly improve your overall converged data performance.

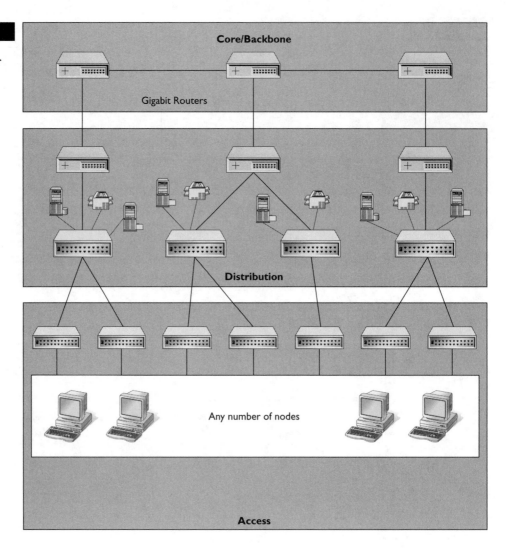

FIGURE 3-6

Fast Core layer
implemented

Distribution Layer The Distribution layer is where access control lists (role-
based access control, network-based access control, etc.) would be implemented as
well as other policies. This layer acts as the intermediary between the Access and
Core layers. It is at this layer that the decision is made as to whether data should

pass across the Core layer or not, and it is this layer that directs data from the Core layer to the appropriate end devices. You will usually find firewalls, packet filters, queuing devices, multiple routers, and switches at this layer. You may also choose to implement WAN links at this layer, though they may also be implemented at the Core layer. The Distribution layer is where the majority of network activity takes place, but it takes place in multiple separated networks all residing at the Distribution layer and interconnected by the Core layer.

Access Layer   The Access layer will contain direct access devices such as hubs, switches, and wireless access points. This layer is where your desktop computers are connected to the network and your laptop computers as well. This layer is where your Voice over IP phones are connected and associated with the network at large. You may also further segment your network at this point. For example, you may have multiple segments within the Distribution layer as indicated in Figure 3-6, but you may create additional segments within each Distribution layer segment at the Access layer.

You may also have additional network access policies managed at the Access layer or in a shared management model with the Distribution layer. For example, you may have wireless access points at the Access layer configured to authenticate wireless clients using an authentication server that resides in the Distribution layer. With an implementation like this one, authentication is shared between the Access and Distribution layers.

Branch or Edge Network Solutions

Some network solutions are referenced as branch or edge devices in vendor literature and common vernacular among network engineers. This terminology arises from thinking about network solutions in terms of a tree analogy. The core of the network is often considered the *trunk*, while the distribution devices are considered the *branches* and the end nodes are considered the *edge* of the network.

Some devices are considered edge devices by default. For example, a firewall is usually referred to as an edge or perimeter device. This name is used because it is installed at the edge of a network and allows controlled communications with the outside world or the remote networks. Other devices like spam filters, e-mail gateways, and certain Voice over IP (VoIP) gateways may also be considered edge devices.

CERTIFICATION OBJECTIVE 3.02

Identify the Functions of Hardware Components

Now that you know the basics of network engineering or design, you need to understand the various hardware components that make up the network. Remember that we are building a foundation that will help us implement, manage, and troubleshoot a converged network. Before we can understand how a network functions when data and voice become converged, we must understand how a data network functions alone. These first four chapters are providing you with that foundation, and then Chapters 5 and 6 will give you a foundational understanding of voice or telephone networks. The convergence of the two will begin in Chapter 7. If you are a Network+ certified individual or a CCNA, you will most likely already know the information covered in the rest of this chapter. However, you may choose to review this information again in order to have it fresh in your mind as you prepare for the Convergence+ exam.

Routers

The first device that I will cover is the router. A *router* is a network device that is capable of moving data from one network to another using different algorithms, depending on the network protocols implemented. The most common type of router used within LANs is an IP router. Additionally, routers are used as interfaces to WAN service providers so that they have an interface operational on the LAN and another interface operational on the WAN. When data needs to traverse to the remote network, it will pass in through the LAN interface and out through the WAN interface. On the remote network, the data will pass in through the WAN interface and out through the LAN interface. Of course, when the communications are reversed, so is the interface utilization.

In order to help you better understand routers, I will document them from four perspectives:

- Functionality
- Common features
- Physical installation options
- Configuration process

In addition to these perspectives, I'll provide information about routing protocols, since they are so important to the functionality of routers. At the same time, the basics of IP routing will be presented.

Functionality

To help you understand what a router really is and does, consider that a router is nothing more than a computer. If you were to install two NICs in a single computer and then connect one NIC to one network and the other NIC to another network, your computer could be configured to route between the two networks. The Windows operating system has had routing capabilities in it since the early Windows NT days, and Linux systems have this capability as well. In fact, there are a few routers on the market that actually run a scaled-down version of the Linux operating system.

While most computers have hard drives, memory chips (RAM), and a processor, most routers have non-volatile random access memory (NVRAM), memory chips (RAM), and a processor or set of special processors. Computers use the hard drive to store permanent information that needs to be retained between boots, and routers use the NVRAM for this purpose. This difference allows the routers to boot quickly and, probably more important, reboot quickly. It also reduces moving parts and therefore common points of failure. In comparison to computers, network routers very rarely fail. Even a consumer-grade router, such as one from Linksys, will usually work for well over ten years; however, most computers do well if they make it four or five years without needing a hard drive replaced at a minimum. Notice what it is that is most likely to fail: the hard drive. This problem is why the NVRAM is so beneficial.

on the
job

The reality is that things are changing somewhat. Hard drives seem to be outlasting processors and memory these days. I'm not sure if it's just an illusion of my experience or if it's true, but it seems that the great heat in the average computer case is causing failures in areas less frequently seen in the past.

You can also use a computer to perform the functions of a router. The computer may be running Windows 2000 Server and routing can be enabled across two networks by using two network cards. When the computer with the IP address of say 10.10.1.5 needs to communicate with the computer at say 10.10.2.7, it must communicate through the computer. The Windows 2000 Server receives the communication on the NIC at 10.10.1.1 and sends it out of the NIC at 10.10.2.1 so that it can reach the destination of 10.10.2.7. Since this routing process is taking place on a computer running Windows 2000 Server, the routing may be slower, since the server is not likely to be a dedicated router. This server may be providing DHCP services, DNS services, domain services, or any other service supported by the Windows Server. This additional overhead is why we usually use dedicated devices as routers.

A dedicated device has at least two major benefits. First, the processing will most likely be faster, since it is dedicated to the process of routing. Second, the up time will most likely be greater, since you will have to perform fewer upgrades and you will experience fewer hardware failures (remember, non-moving parts). On the first point, the processing will not only be faster because the entire device is dedicated to routing, but also because the software is optimized for that purpose. With a regular PC running an operating system that supports routing, the operating system is most likely doing many unnecessary things in relation to the intention of routing.

Routers, in most cases, route IP traffic. Where does the IP protocol operate in the OSI mode? At Layer 3 or the Network layer. This tells you that a router is a Layer 3 device. Routers are most commonly used to connect switches, which are Layer 2 devices in most implementations, together to form larger networks than could be otherwise created. It is important to know that some routers can perform switching with added components and some switches can perform routing. However, for our purposes here, we'll treat the two as completely separate devices and ignore the customized modern routers and switches offered by today's vendors.

Common Features

Regardless of the vendor, routers share a common set of features, which include:

- CPU
- Memory
- NVRAM
- ROM or BIOS
- Operating system
- Interfaces
- Management methods

CPU   Processor speeds vary in routers from less than 100 MHz to greater than 1 GHz. Keep in mind that the router is dedicated to routing, so a speed of, say, 266 MHz is not as slow as it sounds by today's standards. However, enterprise-class routers will have both faster processors and more memory than consumer-grade routers in most cases. Additionally, many consumer-grade routers are hard-coded to disallow data from the Internet that is not based on a previous internal request and this feature simply cannot be disabled. This feature is unacceptable in enterprise networks.

Memory   The newest routers support 1 gigabyte or more of RAM for massive processing capabilities. Again, keep in mind that these dedicated devices do not have the 100–500 megabytes being consumed by the operating system as a PC does. Most of this RAM is being utilized for the work of routing. Older and consumer-grade routers may have as little memory as a few megabytes. Those with less than 1 megabyte of memory are of little use today.

NVRAM   The NVRAM in routers and other network devices is usually used to store the configuration settings for the device. In addition to storing the configuration in NVRAM, you can usually upload the configuration to an FTP or TFTP server or you can save the configuration to a local PC when connected via the console port on the router or an HTTP web-based configuration interface.

ROM or BIOS   Just like in a computer, the ROM or BIOS in a router contains the bootstrap program used to get the device up and running. This may include initial system checks known as the power on self test (POST), and it may include features related to customizable components in the router. The ROM or BIOS is often updated using a flash mechanism and downloaded modules from the router vendor's web site.

Operating System   Again, like a PC, a router has an operating system. The famous Cisco Internetwork Operating System (IOS) is used on most all newer Cisco routers (and other Cisco devices for that matter) and is probably the most well-known router OS in the world. However, each vendor typically uses its own proprietary OS, since that gives the vendor a competitive advantage. Even consumer-grade routers have an OS; it's usually just much less powerful than those in the enterprise-class routers. In fact, sometimes the only difference between a consumer device and an enterprise device is the software that it's running. This difference is similar to the way that home PCs ran Windows 98 in the late 90s while many enterprise computers ran Windows NT Workstation. The consumer PCs could have well run the enterprise operating system, but it simply offered more complexity than the average home user desired.

Interfaces   Routers typically come with one or more built-in interfaces and the ability to add more interfaces through add-on modules. Each vendor refers to these add-on modules with differing terminology, but you can think of them like PCI cards for a desktop computer. Just as you can add a PCI wireless network card to a computer and—poof—the computer now has wireless capabilities, you can add a new card to a router and provide additional capabilities.

These add-on interfaces are usually used to route between Ethernet and some other serial technology such as HDLC, PPP, or ISDN. You may, for example, need to route between your local network and the EVDO network provided by a cell provider. You can usually purchase an add-on to a router that will allow it to route internal users onto the Internet across the EVDO provider's network. Some routers come with two Ethernet ports and have no feature available for exchanging or upgrading ports. In these cases, you'll be forced to use the router as is, which means you can only use it to route from one Ethernet subnet to another. In many cases, these types of routers are used to route from your local network to the Internet using business-class DSL or cable Internet service. I use a TP-LINK TL-R480T router in my home office. It is shown in Figure 3-7. With street prices of less than $150, this device is a very powerful router for SOHO (small-office/home-office) installations with anywhere from 5 to 50 nodes. You'll notice in the picture that this is one of those devices that is both a router and a switch. It has a four-port switch for a local LAN and a single Ethernet port for the WAN.

Management Methods   When it comes to managing routers, the options are nearly endless. You can manage most enterprise routers in any of the following ways:

- **Console**   This connection uses a serial interface and a terminal emulation program allowing for command-line management of the router.
- **Telnet**   This application gives you the same options as the console (as long as Telnet is enabled), only you manage your router across the network.
- **Web-based**   Using a web browser, you can connect to the router and configure it using a graphical management interface.

FIGURE 3-7

TP-LINK
TL-R480T

- **Custom Applications**   Some vendors provide custom applications that run on Windows, Linux, or the Mac OS and can be used to configure the router. Third-party companies also sometimes provide such applications.
- **SNMP**   The Simple Network Management Protocol may be able to be used with some routers to configure them on a large scale.

The reality is that these are just the most common configuration and management options. You may also be able to use SSH, SFTP, and other methods with the routers that you implement. The key is to know the most secure and efficient methods. If you implement an insecure management solution, such as Telnet across an unencrypted channel, the administrative account will be exposed and hackers can find their way easily into your network. You'll learn more about this in the last two chapters of this book when you study security.

You should consider downloading the manuals for two or three routers from two or three vendors to expose yourself to the various configuration options available. Usually, just reading the sections on initial installation and configuration are enough to expose you to the basics of how you would interact with that device. When selecting your reading materials, make sure you get the documents from enterprise hardware vendors, as their configuration options are very different from those for consumer-grade devices. Most consumer-grade routers are really broadband Internet routers, and they are very limited in their configurability and in their management interfaces. In most cases, if you want to experience a command-line interface (CLI) to the router, you'll need to use enterprise-class routers.

Physical Installation Options

Most enterprise-class routers are designed to fit into rack mounts, since rack mounts are commonly used in data centers and wiring closets. SOHO-class routers may simply be shelf devices, meaning that they rest on a shelf, but even these devices usually come with mounting hardware (or it can be purchased separately) to mount them to the wall or even in a rack. The big items to consider when installing them are:

- The distance from the router to the switches or Internet connections or other routers. You must ensure that you can run a cable without incurring signal loss because of increased length.
- The power source. There must be a source of power where the router is installed. Some devices may accept PoE, but most routers will not, since they are core infrastructure devices.
- Ease of access. If you need to test a port, change a cable, or replace an interface, you will want to be able to access the router easily.

FIGURE 3-8

Cisco 3825 series
router

Figure 3-8 shows a typical router. This is a Cisco 3825 series router, which has been a very popular device for many years. It is not powerful enough for the most intensive modern operations, such as voice QoS and Gigabit Ethernet data rates, but it is still one of the most commonly used examples of a router.

Configuration Process

Since every vendor's routers will be configured differently, it is not feasible to provide detailed step-by-step instructions here. However, there is a basic process that should be followed when installing and configuring any infrastructure device, including a router. That is to configure it offline and then connect it to the network. Here's the basic flow:

1. Unpack the router and place it on a stable surface for initial configuration.
2. Connect the router to a power source.
3. Connect to the router using the appropriate mechanism (console, Ethernet, etc.).
4. Power on the router.
5. Update the router's software if necessary.
6. Perform the basic configuration of the interfaces so that they will function appropriately on your network.
7. Perform any security configuration steps required.
8. Save the configuration.
9. Power off the router.
10. Install the router in the production location and power it on.

At this point, you have configured and installed the router, and it should be performing as configured for your network. You'll want to test the network and ensure that this is correct. Can you reach the network on the other side of the router from each side? Can only the nodes that should be able to pass through the router indeed do so? These and other factors should be verified.

IP Routing and Routing Protocols

As I stated previously, routers perform their most important tasks at Layer 3. This layer is where the IP protocol operates and in today's networks IP routing is the primary function of a Layer 3 router. It is very useful for you to understand how a router works its magic. It all begins at Layer 1 and it ends at Layer 1 as well. To understand this concept, consider Figure 3-9.

The work of a router can be summarized as follows:

1. Receive incoming frames on each interface.

2. Extract the IP packet from the incoming frame.

3. Evaluate the IP header in order to determine the destination of the packet.

4. Look in the routing table to determine the best route to the destination.

5. Encapsulate the IP packet inside a new frame and transmit it on the interface that connects to the next step in the route.

6. Process the next received frame.

FIGURE 3-9

The router at work

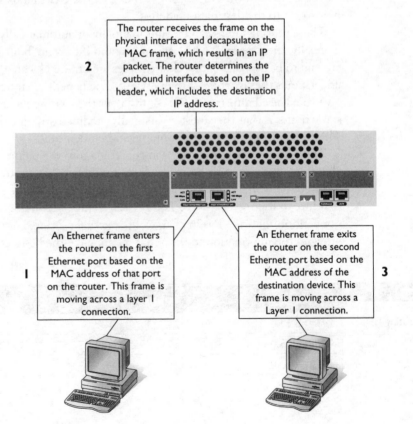

2 The router receives the frame on the physical interface and decapsulates the MAC frame, which results in an IP packet. The router determines the outbound interface based on the IP header, which includes the destination IP address.

1 An Ethernet frame enters the router on the first Ethernet port based on the MAC address of that port on the router. This frame is moving across a layer 1 connection.

3 An Ethernet frame exits the router on the second Ethernet port based on the MAC address of the destination device. This frame is moving across a Layer 1 connection.

As you can see, the process is really quite simple. The router must remove the preamble and MAC frame header and the FCS from the Ethernet frame, which results in the original IP packet. This original IP packet will remain the same as it moves from source to destination as long as no dynamic tagging is used. The header of the IP packet contains the destination address as well as the source address. The router can use the destination address to determine the best way to reach that network on which that destination address exists. To do this task, it will use its routing table.

The router's *routing table* is a listing of known networks and the routes to those networks. The simplest routing table may look something like Table 3-1. Each entry will contain an IP address and a subnet mask. These two values are used to determine a destination network. The same IP address can be listed multiple times with different subnet masks and would result in different networks based on the subnet masks. The Via column in the sample table represents the "way to the destination" network or host. For example, based on this routing table, if the router received an IP packet destined for 192.168.15.73, it would forward that packet on to 192.168.5.2. Now considering the subnet mask, we know that 192.168.15.73 is not on the same network as 192.168.5.2, but that node (which is another router) knows how to get to the destination address.

These routing tables can be built manually or automatically. If they are built manually, they are said to be *static routes*, and if they are built automatically, they are said to be *dynamic routes*. Static routes are entered by an administrator who understands the structure of the network. The benefit of static routes is that they give you, the administrator, full control over the routing process. The problem with static routes is that they must be manually modified anytime the network changes. This task can become time consuming and burdensome.

This point is where routing protocols come into the picture. Don't get confused about the phrase routing protocol. A *routing protocol* is a protocol that discovers the neighbor networks around a router and dynamically builds the routing table for IP to utilize in routing decisions. The key is to remember that a routing protocol does not perform routing. IP is in charge of the actual routing, but the routing protocol provides the information to IP so that it can make the best decision.

TABLE 3-1	IP Address	Subnet Mask	Via
IP Routing Table Sample	192.168.13.0	255.255.255.0	192.168.5.1
	192.168.15.0	255.255.255.0	192.168.5.2
	192.168.20.0	255.255.255.0	192.168.5.2

There are many routing protocols, but the most popular are:

- BGP
- IS-IS
- OSPF
- IGRP
- EIGRP
- RIP

Routing protocols are often categorized as either interior or exterior. Of those listed, only the Border Gateway Protocol (BGP) is considered an exterior routing protocol. BGP is used for routing on the Internet and is a distance-vector routing protocol. Distance-vector protocols choose the best route based on how many hops or routers the packet will have to pass through in order to reach the destination.

IS-IS (Intermediate System to Intermediate System) is an interior routing protocol (interior routing protocols are used within local networks) and is a link-state protocol as opposed to a distance-vector protocol. Link-state protocols actually look at the state of a connection. For example, is the link up or down? Additionally, link-state protocols can usually measure the quality of the link and the speed of the link to truly find the best route. For this reason, in enterprise networks, link-state protocols are often preferred over distance-vector protocols.

Consider the following scenario. Imagine that three routers are interconnected such that router A is connected to both routers B and C and routers B and C are both connected to each other. This design forms a logical triangle. Now, further assume that all three routers are connected to local subnets on another interface. This configuration is represented in Figure 3-10. Notice that the link between routers A and B is a 128 Kbps link. Now notice that the links between A and C and C and B are actually 1.5 Mbps links. When a user on subnet A wants to communicate with a user on subnet B, what is the fastest route? Well, distance-vector would say to use the link from router A straight to router B because the hop count is the lowest; however, the reality is that the fastest route is to add hops on much faster connections going through router C to get to router B. A link-state routing protocol may catch this and give appropriate preference to that route entry in the routing tables.

OSPF (Open Shortest Path First) is another link-state interior routing protocol. It borrows some of its features from IS-IS and is probably the most popular link-state protocol in use on modern networks.

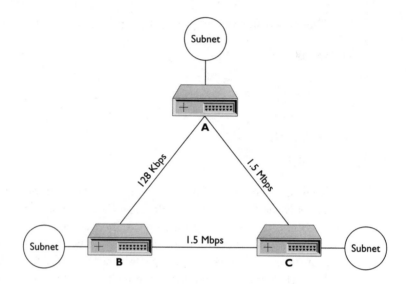

FIGURE 3-10

Distance-vector
versus link-state

Both IGRP (Interior Gateway Routing Protocol) and EIGRP (Enhanced IGRP) are distance-vector routing protocols that were developed by Cisco. IGRP was created in the 1980s by Cisco to overcome some of the limitations of the RIP protocol, which was and is limited to having 16 hops in a route. This limitation affected the overall size of the network. Additionally, RIP supported only a single metric: hop count. IGRP added new metrics such as internetwork delay and load. This addition makes the route calculation similar to a link-state protocol. EIGRP is simply an enhanced version of IGRP that was created in the 1990s to improve efficiency. The biggest change is in the fact that EIGRP does not send out a periodic update to all neighboring routers of its routing table. It instead discovers neighbors and communicates with them directly, greatly improving network efficiency.

RIP, the Routing Information Protocol, is one of the oldest distance-vector routing protocols still in use today. RIP and RIPv2 are excellent solutions for small networks with two or three routers. The big problem with using them in larger networks is that they do send broadcasts to all neighboring routers, whether anything has changed in the routing tables or not. This design is not very efficient. Also both versions are limited to 16 hops in a route. This number limits the size of the network to medium-sized organizations anyway. Those medium-sized organizations would be much better served by OSPF or EIGRP and should avoid RIP.

Switches

In order to help you understand the benefit an Ethernet switch brings to your network, let's review the method used to access the medium in Ethernet networks. Remember that Ethernet uses CSMA/CD, or Carrier Sense Multiple Access with Collision Detection. Just as we have rules of etiquette for human communications (though they are sometimes assumed and not really taught), Ethernet networks have rules for communicating.

Every Ethernet device complies with the rules of CSMA/CD. These devices need to be able to detect activity on the medium before they attempt to use it for their own communication. This method is like being in a meeting and using your ears to listen for other conversations before you begin speaking yourself. In addition, the Ethernet device needs to have a method for detecting a collision even if it begins communicating. In other words, it is possible for two Ethernet devices connected to the same medium to begin communicating at exactly or almost exactly the same moment. This situation will result in a collision. When this collision happens, a jam signal is sent on the medium letting all the devices know that a collision has occurred and that they should all begin the backoff operation before attempting to communicate again. Relating again to our meeting of humans, this event is like your beginning to speak at the same moment as one of the other attendees. You will both sense this "collision," and according to many possible parameters, one of you will back off and let the other speak.

Here's the question: If there are 200 people in a room, is it more likely that two people will begin talking at the same moment, or is it less likely? The answer is clearly that it is more likely. This scenario is also true on your Ethernet network. When you have more nodes connected to a shared medium, you are more likely to see collisions on that medium. The goal is to reduce the number of nodes on the medium. This task can be done by using routers to implement smaller collision domains, but there is also another way.

What if you could implement a network where there were never any collisions? You can, and that network is a network that uses switches. This desire is why hubs have all but been removed from enterprise-class networks and switches have been implemented in their place. A *switch* is defined as a network device that filters, forwards, or floods Ethernet frames based on the destination MAC address of each frame. To best understand switches, you should understand the differences between unicast, multicast, and broadcast traffic.

■ *Unicast* traffic is traffic that moves from one point (the source) to another point (the destination). The traffic or frame is intended for a single endpoint.

- *Multicast* traffic is traffic that moves from one point to multiple specified points. The traffic or frame is intended for multiple endpoints that are defined or listed.

- *Broadcast* traffic is traffic that moves from one point to all other points in a broadcast domain. The traffic or frame is intended for all endpoints rather than a list of endpoints or a single defined endpoint.

It is essential that you understand these three types of traffic, as they are all processed by switches. Switches can handle broadcast and multicast traffic, but their great power is in how they handle unicast traffic. A hub, now an outdated device, receives frames on each port and floods those frames out all other ports. Some devices were created known as switching hubs, but they were really simple switches for which—for some reason—people felt a great need to keep the term hub in the name. A switch receives frames on each port and then analyzes the frame to see if it is a unicast, multicast, or broadcast frame.

I'll explain what happens next in the following section, "Functionality;" however, it is important that you understand this one guideline: a switch implements a number of segments (at the Data Link layer) equal to the number of ports it provides, and these segments experience no collisions. Of course this guideline assumes that you are using full-duplex communications. This assumption is because full-duplex communications use one pair of wires to send data to the switch and another pair of wires to receive data from the switch. Since this configuration is a pair of one-way streets, there will be no collisions and CSMA/CD is not used. This design greatly improves actual data throughput as opposed to management overhead. It also allows you to grow network segments (though they are logical, since each full-duplex connection is like a segment in itself as far as unicast data is concerned) much larger. Many enterprise networks have segments as large as 500 nodes on the segment. I certainly wouldn't recommend segment sizes any larger than that.

Figure 3-11 shows an example of a network switch. This model is a 3Com switch; other popular switch vendors include Cisco, HP, Nortel, and Foundry Networks.

FIGURE 3-11

3Com switch

Functionality

So how does the switch work its magic? The first thing that you need to know is that a switch is a learning device. As data comes in and out of the switch, it notices the MAC address of the sending device as it transmits data through a particular port. Since the device sent data to the switch through that port, the switch knows that it can reach the device (or its MAC address) through that same port. This learning process is repeated again and again, and it forms a database in memory that tracks the various MAC addresses and the ports through which they can be reached.

Now when a frame comes into the switch destined for a known MAC address, the switch forwards that frame to the appropriate port. When a frame comes into the switch destined for an unknown MAC address, the switch floods the frame to all ports. In the end, a switch is effectively a multiport bridge. The traditional and now obsolete bridge had two ports in most implementations. One port existed on one network, and the other port existed on another. Each port learned the MAC addresses on that side of the bridge, and the bridge only forwarded frames from one side to the other that were actually destined for a device on the other side. Switches implement the same basic functionality, only there are multiple "virtual bridges" within the switch. In fact, most switches state that they support the IEEE 802.1D standard, which is not a switching standard, but is rather a bridging standard. Cisco indicates that their switches implement the IEEE 802.1D Spanning Tree Protocol (STP).

Just like routers, and all other computing devices, a switch is a computer.

Common Features

Network switches support many common features, of which the most common are listed here:

- Autosensing
- Autonegotiating
- ASICs processors
- LEDs
- Managed/unmanaged
- VLANs

An *autosensing* switch is a switch that automatically determines the speed for each port. Some switches are 10/100, and others are 10/100/1000, while still others are only 100/1000. The point is that any port can accept any of the valid speeds.

Most switches do not fall back to the slowest connected device, but some do and this is an important consideration.

When a switch is *autonegotiating*, it means that the switch can determine if communications are to be half-duplex or full-duplex. Half-duplex communications use the same pair of wires to both send and receive, and the result is a reduction in data throughput. Full-duplex communications use two wires to send and two wires to receive, resulting in potential full use of the available throughput. In a full-duplex installation, CSMA/CD is not used and management overhead is therefore reduced. The switch basically creates a single LAN between each endpoint and the switch port.

Application-Specific Integrated Circuits (ASICs) are special processors designed for specific purposes. The routers tend to use general-purpose processors just like a desktop PC. Switches may use general-purpose processors, but most of them use ASICs processors. Since the processor is actually created for the purpose of switching, it is much faster at what it does. In fact, many switches are able to offer line speed switching. What this ability means is that—assuming there are not other operations being performed in the switch that would delay processing—the frame can come in the port, be processed, and be sent out the proper port at the same speed as it would have traveled in a direct crossover cable from node to node. Of course, line speed switching is a concept and rarely a reality, since multiple frames may be coming into the switch from multiple ports concurrently.

Most switches also have *light emitting diodes (LEDs)* that indicate the operation of the switch. Each port may have an LED that indicates communications, speed, duplex, and other parameters. Additionally, there is usually a collection of LEDs that can be used to determine many factors about the health and utilization of the switch. For example, you can usually see how much of the switch's capacity is actually being utilized by simply looking at the LEDs.

However, the greatest management and administrative power comes through the switch management interface. Some switches are *managed,* and others are *unmanaged.* Consumer-grade switches that you would buy at the local computer store are traditionally unmanaged switches. This difference simply means that there are no configuration options for the switch other than a possible uplink or duplex toggle button on the physical unit itself. Indeed there is rarely a software-based configuration interface. There are exceptions to this rule, but as a general guideline, the less expensive switches do not have management features.

Enterprise-class switches are usually managed switches. This function means that they have software-based management interfaces. This interface may be a web-based interface that is accessed through a standard web browser, or it may be a custom management application. Some switches also support centralized management through a custom application. These switches can usually be managed en masse and can sometimes even be grouped together to form one extremely large (hundreds or thousands of ports) virtual switch for management purposes.

Finally, *virtual LANs*, or *VLANs*, may be supported. A VLAN is a logical LAN that exists only in the memory of the switches and/or routers. They can exist independent of the physical LAN implementation. For example, a node that physically exists in one building and is three routers away from a node that physically exists in another building can be on the same VLAN as that other node. They will have the same IP network ID and have different host IDs. VLANs are used for management purposes. You can apply policies, restrictions, and more to VLANs. For example, you can say that a user that exists on a particular VLAN can only access the Internet but cannot access any internal servers. This ability to contain traffic is very useful for guests on your network.

Physical Installation Options

Switches are installed in much the same way as routers. They usually exist in the Distribution and Access layers of a tiered network model, though they may exist at the Core layer, since the distinction between switches and routers is being blurred by modern Layer 3 and Layer 4 (or higher) switches.

Configuration Process

Configuring a switch can be very simple or equally complex. In a simple installation, you just plug the cables into the right ports and power on the switch. You're done. In a complex installation, you'll need to configure access control lists, policies, VLANs, Quality of Service, and other parameters. You'll learn about these different parameters in more detail in later chapters.

Firewalls

A *firewall* is a device that controls or filters traffic between networks. It may be located between sections of your internal network, or it may be located between your network and an external network such as the Internet. The firewall is most frequently deployed between a private network and a public network. The most important thing to remember in a converged network is that most existing firewalls will not be configured to allow incoming Voice over IP traffic when you first start to merge the voice and data networks. You'll need to open the appropriate TCP and

UDP ports in the firewall to allow this traffic through. You'll learn much more about firewalls in Chapter 13 when you learn how to secure your converged network.

WAN Devices

When creating the modern converged network, you'll need to set up WAN links as well as local network links. To do these tasks, you'll need to understand a few terms related to these devices. However, keep the fact in mind that, unless you work as a WAN engineer for a service provider, the service provider will assist you with the initial connection of your network to the WAN. For this reason, you will not need to be an expert in relation to what happens inside the mystical service provider's network, but you will need to understand the basics of the hardware that exists at your site. This hardware is usually called customer premises equipment (CPE).

The basic hardware used at your site will include one or more of the following items:

- Modems
- CSU/DSU
- NT1

Modems

You learned the basics of how a modem operates in Chapter 1. You also learned that they can be used with ISDN and DSL lines, though these are usually called digital modems. You can run the PPP protocol over a modem connection and then route IP traffic across that connection. This connection is one way to form a WAN link.

CSU/DSU

When using T1 or E1 (or higher) lines, you'll need a channel service unit/data service unit (CSU/DSU). You could say that a CSU/DSU is like a modem for T and E lines. The CSU and DSU are supplied in one piece of networking equipment and they are therefore usually called a CSU/DSU device. The CSU/DSU sits between your network router and the T or E line. The DSU is responsible for converting the T or E line frames into frames that the local network can process and local frames into frames for the T or E line. The CSU is the actual entry point to the WAN local loop or service provider. It also provides testing services for the provider so that they can determine if a problem exists within the customer premises or between the CSI and the central office (CO).

NT1

A *Network Termination 1 (NT1)* is used with ISDN connections. The NT1 converts the two wires coming in from the service provider to the four lines needed by the ISDN equipment; it also is used for line health monitoring and maintenance operations. In many cases, the NT1 is build right into the ISDN terminal device, removing the need for a separate component. In the United States the NT1 is to be provided by and located on the customer premises, but in Europe it is provided by and located on the provider's network.

Wireless Networking

In recent years, wireless networking has become very important to converged networks. More companies are seeking to deploy Voice over WLAN than ever before, and you must understand the fundamental operations of a WLAN in order to effectively implement a converged network that includes wireless devices. For this reason, I will cover WLAN technologies in some detail in this section. While you may not be heavily tested on your knowledge of WLANs (the objectives mention only that you need to understand access points), the reality is that your real-world activities will almost certainly require you to understand this information.

e x a m

ⓦatch   ***Make sure you understand
the portion of this section that covers
access points well. You are not likely to be
tested on the other portions in great depth,***   ***but you will need this knowledge as you
begin working with converged technologies
in wireless networks.***

Wireless Design Models

WLAN design models will have a tremendous impact on the performance of your converged network. Some models provide very fast communications, and others focus on centralized management and control, while still others blend the best of both worlds. Many WLAN design models exist, and the one you choose to implement will depend on the needs discovered during a site survey and customer interviews. This section provides an overview of these WLAN design models.

Site-to-Site Connections When using WLAN technology to form site-to-site links, you will create either point-to-point (PtP) or point-to-multipoint (PtMP) links. This section describes both.

A PtP WLAN connection is a dedicated connection between two wireless devices. These two devices are usually bridges that allow for the bridging of two otherwise disconnected LANs. These wireless connections allow for the creation of large-scale campus networks and may even be used to create metropolitan networks that span cities. They provide the benefit of connecting disconnected LANs over some distance without the need for leased lines or running cable when the connection is created within a large campus or otherwise owned area. Figure 3-12 shows a PtP connection.

These PtP connections will use semidirectional or highly directional antennas to form the connection. These antennas, unlike the more common indoor omnidirectional antennas that are seldom aimed at anything but rely on reflections to get the job done, do focus the signal mostly in a desired direction so that more amplitude is available in that direction.

FIGURE 3-12

PtP and PtMP
wireless
connections

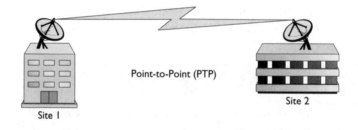

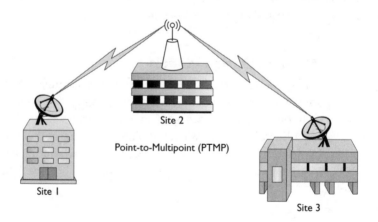

A PtMP wireless link is created when more than one link is made into a central link location like that represented in Figure 3-12. An omni- or semidirectional antenna is usually used at the central location, and semi- or highly directional antennas are used at the other locations. This design is a kind of hub-and-spoke configuration. It is similar to a star topology in a wired network.

When creating outdoor or indoor bridge links, you will have to decide between these two topologies. When there is only one connection needed, you will usually choose the PtP model and when there is a need for multiple locations to link back to a central location, you will usually choose the PtMP model. However, there are times when multiple PtP links may be justified instead of using the PtMP model. Specifically, this design may be needed when you cannot accept the throughput constraints imposed by having a single antenna positioned centrally that is accessed by all remote locations.

While PtP and PtMP links are mostly considered when creating bridge links, the truth is that an access point (AP) is a bridging device accessed by laptops and desktops in a typical WLAN installation. The common WLAN implementation of multiple stations accessing a single AP is a PtMP model. If each station had its own AP, this design would be a PtP model. I've seen this PtP implementation scheme used in situations where shared access to an AP would not provide the needed throughput, though this configuration is both costly and extremely rare.

It is important to note, however, that "real" IEEE 802.11 bridge devices implement a full or nearly full IEEE 802.1D feature set between 802.11 and 802.3. Doing the full IEEE 802.1D processes does require significantly more RAM and CPU power, and hence the greater price of dedicated WLAN bridges.

WLAN Models

In the common WLAN PtMP model, there are two primary implementation methodologies: the single MAC model and the split MAC model. The single MAC model is also known as an *edge* or *intelligent edge* model, and the split MAC model is also known as a *centralized* model.

When a single MAC model is used, it means that the APs contain all of the logic within them to perform MAC layer operations (remember, the MAC layer is a sublayer within the Data Link layer of the OSI model). In other words, all IEEE 802.11 services reside within the AP with the possible exception of security services when IEEE 802.11i is implemented. The single MAC model is the oldest and is still very popular in small and medium-sized WLANs. There are both costs and benefits of the single MAC model.

Single MAC model costs:

- Decentralized administration may require more ongoing support effort.
- APs may be more expensive, since they have more powerful hardware.
- Each AP may be able to handle fewer client stations.

Single MAC model benefits:

- There is no single point of failure. If one AP goes down, the others continue to function.
- Less wired network traffic is required to manage the wireless stations.
- More features are offered within the APs themselves.

The split MAC model is called such because portions of the MAC layer operations are offset to centralized controllers and other portions remain in the AP. These types of APs are often called thin APs, since they do not perform as many functions as the traditional APs (fat APs). The split MAC model is very popular in large networks today and is becoming more popular in smaller networks as well. There are costs and benefits associated with the split MAC model too.

Split MAC model costs:

- There is a possible single point of failure at the WLAN controller.
- Increased wired network traffic is required to manage the wireless stations.
- Fewer features are offered within the APs themselves when using truly thin APs.

Split MAC model benefits:

- Centralized administration may reduce ongoing support efforts.
- APs may or may not be less expensive, since they can have less memory and processing power.
- Each AP may be able to handle more client stations, since the AP doesn't have to handle management processing overhead.

You may have noticed that, in a large way, the benefits of the split MAC model are the costs of the single MAC model and the benefits of the single MAC model are the costs of the split MAC model. While there are certainly more details involved than this choice, it is important to understand that you will be giving up something regardless of the model you choose. The key is to determine what is best for the organizational and technical needs of the organization in which you are implementing the WLAN. You will learn more about this decision in the section "Wireless Site Surveys" later in this chapter when you learn about site surveys and WLAN network planning.

Wireless Mesh Networks   Another wireless networking model is the wireless mesh networking model. Earlier you learned about the PtP and PtMP models. In the database world, you have a one-to-one relationship model, and this design is like the PtP model in WLANs. You also have a one-to-many relationship model, and this design is like the PtMP model in WLANs. However, database theory also presents a many-to-many relationship model, and this design is much like the mesh networking model in WLANs. Therefore, you could say that mesh networking is like a multipoint-to-multipoint (MPtMP) model.

In a mesh network, all APs can connect to all other stations that are turned on and within range of each other. Additionally, data travels through each node so that each node is both a router/repeater and an end node at the same time. The benefits of a mesh networking model include:

- Communications within areas that would normally have many LOS obstructions
- Data routing redundancy

The first benefit is seen because mesh nodes are placed close enough to each other that a path will always be available around obstructions that would normally prevent wireless links. Figure 3-13 illustrates this benefit. Notice that data can travel from node A to node B and then to node C and finally to node D. If this were not a mesh network, there would be no clear path from node A to node D.

The second benefit is also seen in Figure 3-13. If the route mentioned previously (A to B to C to D) was to become unavailable, there is data routing redundancy in that the route from A to F to E to D could be utilized.

The IEEE 802.11s amendment is currently in development and will specify a standard for wireless mesh networking. The normal backbone or distribution system (DS) for a WLAN is an Ethernet LAN. However, the IEEE standard leaves the specification open so that a wireless distribution system (WDS) could also be used. The IEEE 802.11s amendment is aimed at detailing just such a WDS. This change means that our future could see networks that are entirely wireless without a single Ethernet cable (or other wired standard) anywhere. Right now, it seems that the more wireless we implement, the more wires we install; but this problem could

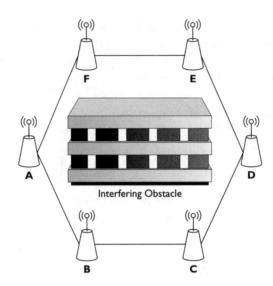

FIGURE 3-13

Solving LOS
issues with mesh
networking

change with evolving modulation schemes, frequency distribution, and powerful processors at lower prices. This migration will be aided by both the IEEE 802.11n amendment for a MIMO PHY and the 802.11s amendment for a mesh-based WDS, but there is still plenty of work to do and plenty of uses for those wires. While we are years and more likely decades from an entirely wireless infrastructure (and some suggest it will never come), the potential is exciting.

Where We've Been and Where We're Going in the Industry

To put the pieces together, this section will present the WLAN models that have evolved over time. I will start with the first model that was implemented using IEEE 802.11 technology and then progress through the evolutionary stages of WLAN design models. While the models did not necessarily evolve in a precisely sequential order as presented here, the adoption of the differing models does seem to have followed a path much like this one.

Intelligent Edge (Distributed)   The first devices to be released to the market were the standard fat APs that are still used heavily today. This AP contains the entire logic system needed to implement, manage, and secure (according to the original IEEE 802.11 specification) a WLAN. The benefit of this type of WLAN is that implementation is very quick when you are implementing only one AP. The drawback to this type of WLAN is that implementation is very slow when you are implementing dozens or hundreds of APs. There are many networks around the world that have more than 1000 APs. You can imagine the time involved if you have to set up each AP individually. At stage one, intelligent edge, this design was your only choice. The APs implemented in this model are also known as autonomous APs.

WLAN Network Management Systems (Centralized Management/ Distributed Processing)   When we arrive at stage two in the evolution of WLAN management, we encounter centralized configuration management with distributed intelligence. The devices and software that provide this functionality are known as a WLAN network management system (WNMS). This stage provided much faster implementations of traditional fat APs and worked using SNMP or other proprietary communication protocols to configure the APs across the network. The WNMSs usually supported the rollout of firmware so that the APs could be updated without having to visit each one individually. This model provided scalability, but did not reduce the cost of the APs and did not offset any processing from the APs so that they could handle more stations at each AP. In this model, autonomous APs are still used.

Centralized WLAN Architecture (Split Mac)   That brings us to stage three: centralized WLAN architecture. This networking model utilizes lightweight or thin APs and depends on a wired network connection to the WLAN switches. The WLAN switch contains all the logic for processing and managing the WLAN. This design allows the APs to handle more client stations and provides for simple implementation. For example, most of these systems allow you to connect the lightweight AP (sometimes called an access port to differentiate them from an access point) to the switch that is connected to the WLAN controller, and the AP and controller will automatically synchronize without any intervention from the engineer. Of course, there is still the requirement of initial setup and configuration of the controller, but moving forward it can be automatic. The things that are automatically configured may include the channel used by the AP, the encryption methods used, the SSID, and more.

Hybrid WLAN Architecture   The hybrid WLAN architecture uses a WLAN controller like the centralized architecture and represents stage four. The difference is that hybrid APs are used instead of lightweight APs. A hybrid AP is an AP that can perform some or all of the functions needed within a BSS and can also allow for some or all of these functions to be managed by the central controller.

Unified WLAN Architecture   The stage is now set for another evolutionary move where the wireless controlling functions are simply integrated into the standard wired switches used within our network cores. This design would mean that the switches that provide wired network functionality to wired clients will also have the capability to serve the needs of wireless APs so that specialty wireless switches/controllers are no longer needed as separate devices. Today's centralized and hybrid solutions usually depend on a connection from the wireless controller to a wired switch that actually has connections to the APs. The future may see more development of multiport switches that have wireless controller functionality built in, reducing the need for an extra wired switch. In fact, many vendors already have such capability available in their newest switches.

Access Points

There are really two main devices that help you install and manage a wireless network: access points and wireless controllers. WLAN controllers are beyond the scope of this book, but they are covered in great detail in one of my other books, CWNA *Certification Official Study Guide* (McGraw-Hill, 2007). Here, we will focus on the access point (AP) in some detail, as understanding it is an objective of the Convergence+ exam.

APs are the most frequently installed infrastructure (non-client) devices. They provide access to the WLAN and may also bridge to a wired LAN. APs provide a point of access to the WLAN and derive their name from this functionality. WLANs are built from basic service sets (BSSs), and each BSS has one and only one AP. A BSS is defined as a group of wireless stations, including the AP, that have synchronized with each other and share an SSID. When multiple APs work together to form a larger network throughout which clients may roam, they form an extended service set (ESS).

In most cases, an AP will provide connectivity to a wired LAN or WAN for wireless client stations (STAs); however, this configuration does not have to be true. APs are often used at construction sites to form controlled and secure networks that are entirely wireless (with the exception of the power cords connected to the APs) as just one example of the use of APs where access to wired networks is not the intent. They may also be used in meeting spaces and conference facilities to set up wireless networks that do not connect to any outside wired networks.

Autonomous access points are APs that contain the software for total management of the WLAN processes within themselves. These were the only kind of APs in early WLANs. *Lightweight access points* are APs that contain limited software and depend on centralized WLAN switches or controllers to provide the remaining functionality. Today, many wired switches either include the functionality to act as a WLAN controller or they support it through add-on modules. There is no standard for implementing lightweight versus autonomous APs and the way in which they are implemented varies from vendor to vendor. Autonomous APs are sometimes called fat or think APs where lightweight APs are also called access ports (as opposed to access points) or thin APs. Figure 3-14 shows a network implementation using autonomous APs and Figure 3-15 shows the use of lightweight APs.

Some APs can act as either an autonomous or lightweight AP, depending on the configuration determined by the WLAN administrator. When used as an autonomous AP, all the AP software features are enabled. When used as a lightweight AP (or access port), many of the AP software features are disabled or they are simply controlled by the centralized WLAN switch or controller.

When lightweight APs are brought online, they are automatically configured by the WLAN controller or switch. This configuration may include the automatic installation or update of firmware (internal software used to run and manage the AP). Many vendors ship their lightweight APs with no firmware loaded; the firmware is installed when it first connects to the WLAN controller. Symbol (now owned by Motorola) does this with their 5100 series WLAN switches and access ports.

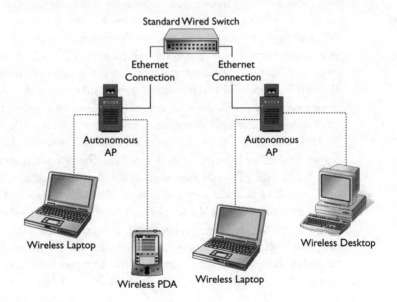

FIGURE 3-14

Autonomous AP
implementation

FIGURE 3-15

Lightweight AP
implementation

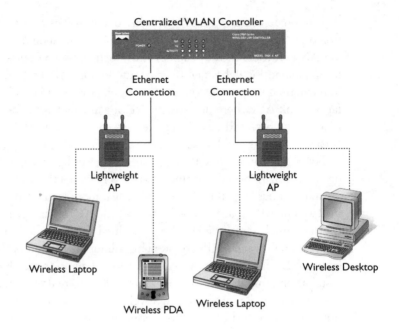

An AP, like the other devices we've evaluated, is basically a small computer
that includes one or more radios and usually one Ethernet port. Inside the AP are
a processor and memory. In fact, one of the big differences between enterprise-class
APs and those designed for SOHO implementations is the processing power and the
amount of memory available in the AP. Many WLAN administrators are surprised
when they first learn that many APs either run a flavor of Linux or can run Linux
through flash updates. It is important to remember that you may lose support from the
device vendor if you flash the device with an operating system that is not supplied by
the vendor. For example, firmware is floating around on the Internet that converts
Linksys WRT54G WLAN routers into more enterprise-like devices with advanced
features usually only provided in WLAN switch/AP combination installs. These
features include VPN endpoint support for client connections, more powerful filtering,
and centralized management and control. Again, if a WLAN administrator chooses to
install such firmware, he/she will likely lose all support from the hardware vendor.

APs, both autonomous and lightweight, come in many shapes and sizes. Some have
antennas built in, and others use external antennas. They come in round, rectangular,
and other shapes. Some are designed for mounting on walls or ceilings, and some are
designed to be placed on desktops or shelves. Figure 3-16 shows multiple APs in the

Symbol product line, and Figure 3-17 shows examples of Cisco APs. Figure 3-18 provides examples of SOHO-class APs from Linksys, and Figure 3-19 shows a sample Netgear AP.

APs come with common features and require various configuration processes. The following sections document each of these important factors. First, the common features will be covered, although it is important to note that, while these features are common, they are not available in all APs. Second, I will walk you through the basic installation and configuration of an AP.

FIGURE 3-18

Linksys APs

Common Features By common features, I mean features that are commonly seen in APs and not necessarily features that are common to all APs. Some APs will have all of the features listed here and more, while others may lack one or more of the listed features. Features that will be covered include:

- Operational modes
- IEEE standards support
- Fixed or detachable antennas
- Filtering

FIGURE 3-19

Netgear APs

- Removable and replaceable radio cards
- Variable output power
- Ethernet and other wired connectivity
- Power over Ethernet support
- Security capabilities
- Management capabilities

The IEEE 802.11 standard defines an AP only as a STA that provides access to the distribution services via the wireless medium for associated STAs. It does not define the three common operational modes that are found in APs. These modes (root, bridge, and repeater) are specific implementations of a WLAN STA for varied purposes, and in some cases, they may be proprietary rather than matching an IEEE standard. For example, in bridge mode, an AP is implementing a network functionality that is not directly stipulated in the IEEE 802.11 standard. Root mode is the closest to the IEEE 802.11 standard, and many APs meet the IEEE 802.11 standard exactly when running in root mode.

The first mode offered by most APs is root mode. An AP operating in root mode provides wireless clients with access to the WLAN and possibly a wired network. Root mode is the default mode of operation for all WLAN devices sold as APs. Some WLAN bridges are really APs that come with the operating mode set to bridge mode, and they are nothing more than a standard AP operating in bridge mode. Full-function WLAN bridges will implement a complete 802.1D bridging feature set. When APs operate in root mode, they may still communicate with each other, but the communications are not related to bridging. In root mode, inter-AP communications are usually related to the coordination of STA roaming. Figure 3-20 shows a typical installation of an AP in root mode.

Bridge mode is used to create a link between two or more access points. When only two APs are used, a point-to-point link is created. When more than two APs are involved, a set of point-to-multipoint links are created. In a bridge mode implementation, the APs involved usually associate only with each other and do not accept client STA associations. Exceptions to this rule exist, but not as the normal implementation, since it would reduce the throughput available for the bridge link connection. Figure 3-21 shows a typical installation of a set of APs in a point-to-point bridge mode implementation.

Figure 3-21 shows an implementation of bridge mode that reveals one possible scenario where it may be beneficial. The AP in the Administration building is associated with the AP in the Research building. The two otherwise disconnected LANs are merged into one via the WLAN bridge link created using bridge mode of the APs.

AP implemented
in root mode

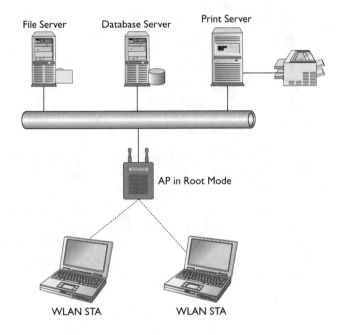

APs implemented
in bridge mode

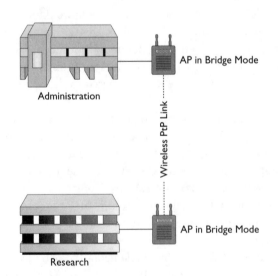

The final mode, repeater mode, is used to extend the range of a WLAN beyond its normal usable boundaries. The repeater AP acts as the AP for clients that would otherwise be out of range of the distant AP operating in root mode. Where a root AP is the connection point for many clients and is a client to no other APs, the AP in repeater mode is a client to the AP in root mode while also accepting connections from client stations itself.

Repeater mode in a WLAN AP should not be confused with the functionality of an Ethernet repeater. Ethernet repeaters regenerate the received signal in order to allow it to travel farther than it would otherwise travel. They do not decapsulate and encapsulate data as a WLAN repeater will. The AP running in repeater mode will decapsulate the data frames received from the clients and encapsulate them for transmission to the root mode AP. In other words, the WLAN AP in repeater mode will receive data from the WLAN clients associated with it and then retransmit that data to the root mode AP with which it is associated. Figure 3-22 shows an AP operating in repeater mode to provide access to remote clients.

Keep in mind that an AP operating in repeater mode must be able to communicate with the clients associated with it, as well as the root mode AP with which it is associated. Because of this requirement, the repeater mode AP will usually have to implement a basic service area (BSA) that overlaps with the BSA of the root mode AP by at least 50 percent. This design reduces the overall coverage area that may be provided if each AP were operating in root mode and forming an ESS; however, Ethernet connectivity is not always available to provide for the preferred implementation, and repeater mode may be used in these scenarios.

APs on the market today support a wide range of IEEE 802.11 amendments, but it is difficult to find hardware that supports some of the older PHYs such as FHSS. Most equipment supports ERP, HR/DSSS, DSSS, or OFDM. The vendors usually report this support as 802.11g, 802.11b, 802.11, or 802.11a respectively. Many devices are said to be 802.11b/g devices. This distinction simply means that the devices implement the ERP PHY, which is capable of communicating with HR/DSSS PHY devices as well.

In addition to the PHYs that are supported, you should consider the standards-based security features that you may require. Some APs support IEEE 802.11i, and some do not. Some still support only WEP encryption, but thankfully these devices are becoming harder to locate. Most modern APs will support both WPA and WPA2 with preshared keys (PSK) at a minimum, and many will support WPA and WPA2 Enterprise, which utilizes a RADIUS authentication server.

Another standards-based feature to consider is Quality of Service (QoS). If you need support for QoS extensions, you should ensure that the AP has support

FIGURE 3-22

AP in repeater
mode

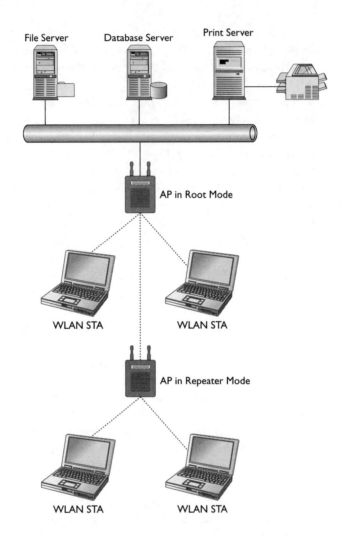

for IEEE 802.11e or the Wireless Multimedia (WMM) certification by the Wi-Fi Alliance. These QoS features will be very important if you intend to support VoWLAN or video conferences over the WLAN.

Newer APs tend to support the newer IEEE standards while also supporting older standards. One of the benefits of a newer ERP PHY–based device is that it can communicate at the 54 Mbps data rate with other ERP PHY devices and it can also communicate at the 11 Mbps data rate with older HR/DSSS PHY devices. Of course, the ERP protection mechanism kicks in whenever an HR/DSSS PHY device is associated with the ERP AP. This design means that the AP will transmit a frame that can be understood by the HR/DSSS machine(s) before transmitting the frame that can only be understood by the ERP machine(s). This first frame is used to cause a backoff timer to kick in on the HR/DSSS machines so that they will not interfere during the ERP frame transmission. This setup reduces overall throughput. The moral of the story is that one HR/DSSS device associated to your ERP AP will cause the entire BSS to slow down to some extent.

In addition to the benefit of backward compatibility with the HR/DSSS PHY, ERP PHY devices are able to support more data rates than HR/DSSS devices, so as the data rate changes, it does not necessarily drop by half at a single step the way an HR/DSSS device does when it goes from 11 Mbps to 5.5 Mbps in one step.

Finally, APs may not support utilization in every regulatory domain. You should be sure to verify that the APs you are purchasing are authorized for use within your regulatory domain. IEEE 802.11h specifies support for European nations, and IEEE 802.11j specifies support for the regulatory domain of Japan. For more specific information regarding your regulatory domain, check with the regulation management organization in your country.

Very few enterprise-class APs do not support detachable antennas. Some SOHO APs may have built-in antennas with no external antenna connectors, but this design is rare and even these less-expensive devices usually support replacement antennas. Detachable antennas are beneficial from at least two perspectives: the physical location of the antenna and the selection of a different antenna type.

The ability to move the physical location of the antenna to a different location than that of the AP is a valuable one. You can use RF cabling to move the antenna to a location that is more practical for the transmission and reception of RF signals and locate the AP itself closer to power outlets. This setup can be advantageous when you do not have power outlets closer to the RF signal transmission and reception locations.

The second benefit is that of replacing the antenna with a different antenna type. You may want to provide coverage down long, narrow corridors (patch or panel antennas), or you may want to provide coverage in an area horizontally with as little RF energy propagating upward and downward as possible (higher-gain omni antennas). Whatever the motivation, a detachable antenna provides you with the capability to better control how the RF energy is radiated from the antenna and therefore how the AP provides coverage in the BSA. Figure 3-23 shows an AP with a detachable antenna.

FIGURE 3-23

D-Link AP with
detachable
antenna

Most APs offer two kinds of filtering at a minimum. The first kind is MAC address filtering, and the second is protocol filtering. Filtering functionality provides the WLAN administrator with the capability to limit which STA frames can pass through the AP according to the hardware configuration of the STA (MAC address) or the protocol being used, such as HTTP.

MAC filtering has often been referenced as a security solution, but it should not be thought of as such. It may be useful from the perspective of making it harder to accidentally associate with the wrong AP, but MAC filtering should not be considered a security solution in WLANs. This is because MAC spoofing (stealing a MAC address from a valid STA) is easy to do and step-by-step instructions are readily available on the Internet. The only common value seen from MAC filtering today is its use in specific association limitation scenarios. For example, a training center near my home office uses laptop computers in the training rooms. They do not want the laptop computers to be moved from room to room, but instead want them to stay in designated rooms. The simple solution was to use MAC filtering in the AP in each room. Each room's AP contains the MAC addresses of the laptops that are supposed to be in that room. The AP's output power is throttled back to reduce the coverage area provided. Now, if someone takes a laptop from the designated room to another room, the laptop will have to associate with an AP with a very weak signal in the remote room. Throughput suffers, and in most cases, the laptops cannot connect in such scenarios because the rooms are far enough apart. Again, if this configuration was being done as a security solution, it would be a very bad idea. Any moderately skilled cracker can spoof a MAC address very quickly. I cannot emphasize enough that MAC filtering should not be considered a security solution.

Protocol filtering can be used to disallow specific protocols or only allow specific protocols. This feature usually allows for filtering of both the frames arriving through the radio and those reaching the Ethernet port. You may also filter only the radio-side (wireless) frames or only the wired frames, depending on the AP and vendor. Some APs can filter out frames based on the actual file extensions the user or machine is trying to access on the Internet. For example, if the user attempts to access a WMV file and the WLAN administrator has chosen not to allow access to such streaming media for performance reasons, the AP can disallow such requests. Most APs can blindly block all HTTP requests or FTP requests and other such Internet protocols as well.

An additional kind of filtering, though less common, is that of wireless STA to wireless STA filtering. Some APs will allow you to create virtual APs (VAPs) within one physical AP. You can then determine if wireless STAs associated with one VAP can communicate with wireless STAs associated with another VAP (inter-VAP filtering). You can also determine if wireless STAs can communicate with other wireless STAs associated with the same AP (intra-VAP filtering). Finally, you can disallow all client-to-client communications and only allow the STAs to use the AP for access to the wired medium. This type of filtering can be useful when you want one physical AP to service public and private clients. The public clients may have limited access to the network and therefore to the private clients. The private clients may have normal access to the network. In this way, one AP effectively provides access to both internal users and public guests.

Some APs are designed to support one PHY only, while others are designed to allow for multiple radios and therefore multiple PHYs. These multiple-radio APs are usually called dual-radio APs because one radio is needed for the OFDM PHY and another is needed for the HR/DSSS or ERP PHY. The latter depends on whether the device is an 802.11b and 802.11a device or an 802.11a and 802.11g device. It is important to remember that all devices claiming to be IEEE 802.11g compatible must also allow associations with IEEE 802.11b devices. This requirement is because the ERP PHY may provide for associations with devices that are using the HR/DSSS PHY. These devices may provide a feature for disabling HR/DSSS PHY associations, and this task is often accomplished by only allowing associations that support data rates of more than 11 Mbps.

Many APs, like the Cisco 1200 series, provide for replaceable radio cards. This ability allows you to upgrade the device for future standards by upgrading the firmware or operating system and the radio cards. Figure 3-24 shows the 1200 series AP. The antennas shown include the OFDM PHY antenna (the square antenna) and the ERP or HR/DSSS antennas (the dipole or rubber ducky antennas). In the

FIGURE 3-24

Cisco 1200 Series
AP with multiple
radios and
antennas

case of the 1200 series AP from Cisco, the 2.4 GHz PHYs (HR/DSSS and ERP) are supported by a built-in radio card and the 5 GHz PHY (OFDM) is supported by an add-on radio.

Many APs support replacement radios through the use of PCMCIA or CardBus WLAN NICs. In these cases, the replacement radio cards usually have to be purchased from the vendor that created the AP. This limitation is due to the limited cards supported by the software running within the AP.

These APs that support replacement radio cards may support two modes of use. The first is to act as a single AP that is reached using multiple PHYs such as OFDM and ERP. The second is to have each radio card configured as if it is a separate AP. In this case, both cards will likely use the same PHYs and will simply operate on different channels. For example, one card may operate on channel 1 and the other card may operate on channel 11. This configuration allows the WLAN administrator to service twice as many clients in the coverage area while still using a single AP.

Variable output power provides the WLAN administrator with the capability of sizing cells more accurately. Remember, this feature should not be considered a security solution by itself because a remote client with a powerful WLAN card and the right antenna can often still pick up the signal of the WLAN and also transmit data to the WLAN. However, as an RF management philosophy, cell sizing makes a lot of sense.

As an example, consider a facility with the need for four different WLANs (for security reasons or otherwise) that must coexist in a fairly small space. Throughput is not a paramount concern, since the users of the WLAN perform minimal data transfers, though these data transfers happen several times per hour. Figure 3-25 shows a simplified floor plan of this facility. In order to implement the four distinct WLAN BSAs (cells), APs can be installed in areas A and D that use antennas that directs most of the RF energy inward. These antennas could be mounted on the walls near areas B and C and facing away from them. In areas B and C, APs could be installed centrally to the areas and use standard omnidirectional antennas. These APs could have their output power settings lowered to ensure that there is minimal overlap into areas that are not intended for coverage by these APs.

Of course, a scenario like this one can be implemented to provide unique configuration parameters for each BSA; however, you must remember that this type of cell size reduction does not of itself equal security, but it would help in RF spectrum management in small areas that need different types of WLAN access such as that depicted here.

Some APs provide variable output power management based on percentages, and others, based on actual output power levels. For example, an AP may allow you to specify that the output power be 25 mW, 50 mW, or 100 mW. Other APs may only allow you to state that the output power should be at 25 percent, 50 percent, or 100 percent. These are just examples, but it is important to know what you're looking for when you enter an AP's configuration interface. Figure 3-26 shows the variable output power management (Transmission Power) interface for a 3Com 8760 AP. You can see that this device provides percentage-based management of the output power.

Unless an AP is providing WLAN services and access to a wireless-only LAN, the AP must have some interface through which it can connect to a wired LAN. In most APs, this connection will be an Ethernet connection. Depending on how old the AP is and the model of the AP, it may support only 10 Mbit Ethernet.

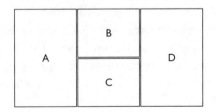

FIGURE 3-25

Simplified floor plan needing four distinct cells

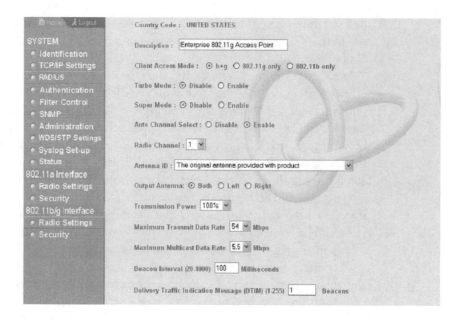

FIGURE 3-26

3Com 8760
Transmission
Power
management

Newer models should support 100 Mbit and even Gigabit Ethernet. With an OFDM or ERP PHY, you should ensure that the AP provides at least a 100 Mbit Ethernet connection. This way, the wired side can keep up with the wireless side. If the device supports a 54 Mbps PHY (which will likely give up to 26 Mbps data throughput) and a 10 Mbit Ethernet connection, the wired side will fail to keep up with the wireless side and it will give the illusion of poor wireless performance. In multi-radio cards with more than two radio cards, you will want to seek out a WLAN AP with a Gigabit Ethernet port. Of course, the switch to which the AP is connected must also support Gigabit rates, and you may have to analyze other links in the chain from the AP to the common service providers users will be accessing. This point is where data flow analysis can benefit you in your planning of the WLAN.

It is also important to remember management overhead that will be incurred on the wired side of the AP. Most centralized management systems, whether in a WLAN switch or controller or in a computer-based application, will perform their management through the Ethernet connection. This design prevents the management activity from interfering with wireless activity; however, it may also utilize measurable portions of the Ethernet connection and may be enough reason to warrant the use of 100 Mbit Ethernet ports as a minimum. I would certainly not buy a brand new AP today that only has a 10 Mbit Ethernet port.

In addition to standard CAT5 or CAT6 cabling, some APs may support 100Base-FC fiber connections. Since fiber is rated for longer cable runs, it may provide a solution to a scenario where the AP needs to be located more than 100 meters (the limit of CAT5) from the switch port. Of course, this design means the switch as well as the AP must support fiber.

It seems more enterprise-class APs support Power over Ethernet (PoE) than not. Support for PoE allows for the installation of APs in areas where no power outlets reside, but where you can run network cables to carry the power. While PoE is very popular for WLAN devices, because it can provide extra features such as power-cycling the device as well as powering the device in the first place, it is sometimes more cost effective to run the power to the area rather than to use PoE. This situation is usually the case when only one location needs the power outlet and the power run would be only a few feet.

Consider the implications of PoE carefully before deciding against it. You often hear that the primary benefit of PoE is the ability to install APs where there is no AC power outlet; however, it is certainly a major benefit to be able to power-cycle (stop and start the device) an AP that is installed in the ceiling and plugged into a power outlet there. Many PoE switches support the stopping and starting of power injection on the PoE ports using the graphical management tools the vendor provides. This inclusion means you can restart an AP from your desk, even if you cannot get into the management interface of the AP and even if the AP has stopped responding to other management interfaces that communicate with the device through the network layers. To me, this ability is an equally valuable benefit to that of being able to place an AP where there is no power outlet.

PoE support is usually not found in SOHO APs like those from Linksys or Netgear. Most enterprise APs do support PoE, but check with your vendor to ensure you purchase a model that supports it if you need it. While more and more enterprise-class APs do support PoE, some still do not.

Modern APs often provide a mesh networking function. The function allows the AP (AP1) to act as a client to multiple other APs (AP2 and AP3 for example) and treat the individual associations with these other APs as ports across which it can bridge traffic for the STAs associated with it (AP1). When a client needs to reach a destination that is reachable through AP2 but that client is associated with AP1, AP1 will bridge the packets across the association with AP2 on behalf of the client.

There is a limit to the number of associations these APs can make. For example, the Symbol AP-5181 AP can create up to three mesh associations with other APs. The Symbol AP-5181 calls these connections client bridges or client bridge mode. At the same time, the device can act as a base bridge and accept incoming client bridge connections from other AP-5181 APs. With these capabilities, a somewhat dynamic mesh network can be built over time across which client traffic may be directed. All of the associations, in the Symbol APs, are based on the SSID (called the ESSID in Symbol's documentation, though this is not IEEE standard terminology). In other words, the mesh network is built dynamically based on the SSID and the other APs in client bridge mode or base bridge mode, or else both are discovered through beacon scanning.

Figure 3-27 shows a network implementation using APs that support a mesh networking mode. In this case, MU1 is associated with AP1 and MU2 is associated with AP2. Since AP1 is a client bridge to AP2 and AP2 is a client bridge to AP3 while being a base bridge to AP1, both MU1 and MU2 can access the files on the file server. This ability is possible even though AP1 may not be connected to an Ethernet port. The association AP1 has with AP2 becomes the port across which it bridges network traffic destined for the file server.

More and more of the newer APs are coming equipped with hotspot support. This support usually includes walled garden support and may also include connectivity to online payment processing services if you are providing a for-pay hotspot. Having this support built in is also useful when you simply want to provide a "guest" network for visitors to your organization's facilities.

APs support a large pool of common security capabilities. These include:

- MAC address filtering (a common item in vendors' lists of security features)
- IEEE 802.1X port-based authentication
- IEEE 802.11i

FIGURE 3-27

Mesh networking mode implemented

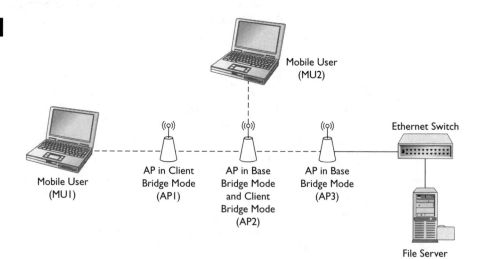

- SSH and SSH2 for management access
- HTTPS access to web-based management
- Legacy WEP (we shouldn't even call this a security capability, but vendors do)
- WPA/WPA2
- SNMP v3 for secure SNMP management
- Various EAP types (some are secure; some are not)
- Built-in firewalls
- Support for VPN tunnel endpoints and pass-through
- Content filtering

Your role as a WLAN administrator or converged network engineer may include the selection of APs that support the security technologies required by your security policies. Today, these policies will likely specify that you cannot implement an AP that uses WEP for data encryption and you must therefore select an AP that supports WPA-PSK at a minimum. More likely, in an enterprise implementation, you will be implementing full IEEE 802.11i support from this point forward—until a newer and better security technology comes along. This last statement is not meant to indicate that IEEE 802.11i is insecure, only that it will be some day. That day may be ten years down the road, but it will come. By that time, the IEEE will likely have developed newer security recommendations and standards, and for that matter, IEEE 802.11 may not even be the common WLAN standard. Things change.

APs will provide different methods for configuration and management of the devices. These methods will vary from vendor to vendor and from model to model within a given vendor's product lines. However, there are common methods utilized. These common methods include:

- Console (serial)
- Telnet
- SSH
- SNMP
- Custom software applications
- Web-based interfaces

Console or serial interfaces are usually only provided on enterprise-class hardware. For example, Cisco, Proxim, Symbol, and 3Com devices are likely to come with console interfaces for configuring them. Linksys, Belkin, D-Link, and Netgear

FIGURE 3-28

Netgear WG302
AP

devices are less likely to come with such an interface. This exclusion should not be taken as a given; for example, the Netgear WG302 AP (see Figure 3-28) supports a console port as well as most of the other common management interfaces mentioned in this section. Many vendors that were once known as only SOHO vendors are beginning to attempt to cross over into the enterprise market.

When using a console interface to configure an AP, you will usually connect a serial cable from your computer to the AP. You may also use a USB-to-serial converter such as the one seen in Figure 3-29. Once connected, you will use a terminal program such as HyperTerminal, in Windows, to connect to the device. Alternatively, you could use the CLI (command-line interface) provided by the vendor. Each vendor's CLI will be somewhat different, and sometimes they will be wildly different. This difference is one of the major arguments for using consistent hardware throughput your organization: you only have to learn one set of CLI commands rather than a varied set. The good news is that the CLI is usually used at initial configuration or for device reload, and the other graphical interfaces are usually used for ongoing maintenance and configuration support.

FIGURE 3-29

USB-to-serial
converter

The Telnet and SSH or SSH2 interfaces will be similar to the console management method in that the CLI will be utilized. The difference is that the CLI is utilized across the network rather than through the console port and a serial cable. When using these management methods across the network, you should be careful to ensure that some form of encryption is in use. Otherwise, with Telnet for example, the commands being transmitted from your machine to the AP are being sent in clear text that is easily readable in any common Ethernet packet analyzer.

SNMP is widely supported among WLAN devices. Due to security vulnerabilities in earlier versions, you should choose only devices that support SNMP v3 and—eventually—higher. SNMP provides for centralized mass configuration management. SNMP is not a proprietary technology, so one centralized application can often manage multiple vendors' APs.

Custom software applications may come with the AP and are usually provided on a CD-ROM when they do. These applications are usually designed to run on Windows clients, since these clients are so popular in enterprises. The applications may provide first-time configuration only, or they may provide for ongoing configuration management. Due to the proprietary nature of these applications, they provide limited value to very large-scale installations.

Finally, web-based configuration interfaces take advantage of built-in web server software in APs to allow for remote configuration through the Ethernet interface. While you may be able to enable web-based management through the WLAN interfaces, I do not recommend it. This setting means that an attacker can try to guess the password and then manage the WLAN device across the WLAN. He or she will not even need to gain access to your physical network. For this reason, if you enable the web-based administration interface at all, it should only be enabled for the Ethernet port. Web-based management interfaces are provided on nearly all APs whether they are built for enterprise or SOHO use. Earlier, Figure 3-26 shows a portion of the web-based interface on a 3Com AP.

In addition to the configuration features mentioned here, most WLAN APs also allow you to save the configuration to a file that can be downloaded from the device to a disk. This feature allows you to quickly and easily reload the configuration at a later point. It also provides for quick changes from one configuration to another. Some APs also provide onboard storage of multiple configurations among which you can switch.

Configuration Process Many new APs will come out of the box with the antennas detached. In this case, you will need to first attach the antennas before the AP will be able to radiate the RF signal. You may wish to wait until after configuration to attach the antennas, but this step is really optional, since you will not be connecting the AP to the LAN until you have configured it properly.

As the last sentence suggested, you should configure the AP before connecting it to the actual wired LAN to which it will provide access. This step helps to remove the potential for wired-side access before the AP is properly configured and reduces the likelihood that you will provide an entry way into your LAN—though only for a short time—during the configuration window. Most APs come from the factory with little or no security set, so they can certainly provide a point of vulnerability by default.

After the AP is properly configured according to your security policies and configuration standards, you will need to connect the AP to the wired LAN via the Ethernet port. You may also need to connect the antennas if you did not connect them before configuration, or if you disconnected them during configuration for security reasons.

Finally, you should test the AP to ensure that you can connect to it with a client configured with the appropriate security and configuration standards that match with the AP. If you are using an AP model for the first time, you may also want to perform some load testing to verify whether the AP works as advertised (in relation to throughput and concurrent connection) or not. You may need to adjust the number of installed APs according to real-world performance with some devices.

AP Summary   This section has been rather long and for good reason. You must understand APs if you are to successfully implement VoIP on WLANs. In the end, APs come in many different shapes and sizes, as seen in Figure 3-30. The devices in Figure 3-30 all come from one vendor, and yet they are very different in form factor and capabilities. APs usually support a common set of IEEE standards, security capabilities, and mounting options. Common management interfaces include console, Telnet, and web-based interfaces among others. Most APs that are used in enterprise installations today support SNMP for centralized management and may support custom software provided by the AP vendor. As a WLAN administrator, it is important that you understand these options and be able to choose among them effectively.

Wireless Site Surveys

In order to determine the right hardware and placement of that hardware for your wireless LAN, you'll need to perform a site survey. I'll walk you through a typical site survey for a small business. In most small businesses, the primary focus is on reducing interference sources because the space being covered is usually relatively small. Site surveys for larger installations are more complex than what is described here.

FIGURE 3-30

Aruba WLAN AP
product family

The basic process of performing the site survey for a small business can be broken into three main steps:

1. Get an overview of the organization.
2. Perform the physical site survey.
3. Document your findings.

Get an Overview of the Organization

An organizational overview consists of understanding where the organization is currently, where its members hope to go, and what their intentions are for the wireless network. You'll interview managers and users and gather the appropriate documentation. I suggest providing a summary of this information to the decision makers before performing the physical site survey. This summary verifies that your understanding is correct and provides an opportunity for the decision makers to change the intended plans.

Interviewing   Understanding the organization's business processes is an essential first step. Interview the management staff for each department impacted by the wireless network to determine their current processes and future needs. You may also want to survey some of the end users of the current network and future wireless network. Determine what applications they run, how they obtain the information they enter into the applications, what their primary responsibilities are, and any other information that helps you better plan the wireless network.

Interviewing methods vary, but here are a few techniques you can consider using:

- Focus groups
- One-on-one meetings
- Surveys
- Observation

While all of these techniques can reveal beneficial information, I've found observation to be effective and often essential. Have you ever had someone ask you how you do a certain portion of your daily work? If you have, you probably found it difficult to recall every step along the way. In my training classes, I often ask an attendee to tell me the step-by-step process he goes through to get to work every morning. Without fail, he forgets that he starts the car before he backs out of the garage or he forgets that he closes the garage door before driving away. The point is that little steps are often forgotten. The users of the future wireless network are going to have the same difficulty. By observing their actions, you ensure that the steps you are documenting are the actual steps taken to perform the work.

Gathering Documentation Next, you want to gather blueprints and floor plans for pre-evaluation and for documentation use later on. In addition, acquire any available network diagrams. These diagrams might show the services running on the network, which helps you determine bandwidth needs for the wireless network. They can also be used to discover available Ethernet ports if they are sufficiently detailed. Figure 3-31 shows a simple network diagram for a small business network.

This diagram does not provide any information on available ports in the switches or routers, which is not uncommon, but at least you know where the devices are located. You can also discover bandwidth needs based on the fact that there is a SQL Server and three file servers. You need to ask the network administrator how heavily the SQL Server is used and how often the file servers are accessed.

Perform the Physical Site Survey

When performing the physical site survey itself, you're likely to encounter other wireless networks in the same area. It's not uncommon to see information similar to that in Figure 3-32 when using NetStumbler to perform an existing wireless networks analysis.

If you're surrounded by organizations using all three channels for 802.11b/g in an area, you might need to plan for an 802.11a implementation. An 802.11a network

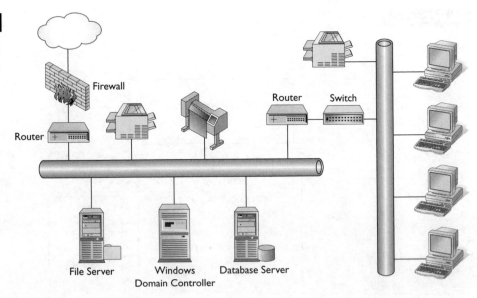

FIGURE 3-31

Network diagram for small business

is more expensive, and for this reason, you might choose to attempt negotiations with the organizations surrounding the intended implementation area. Figure 3-33 illustrates a scenario where you might be able to negotiate a modification of settings in another organization's implementation.

In this figure, the organization north of your facility is using 802.11g devices on channel 6. The organization south of your facility is using 802.11g devices on

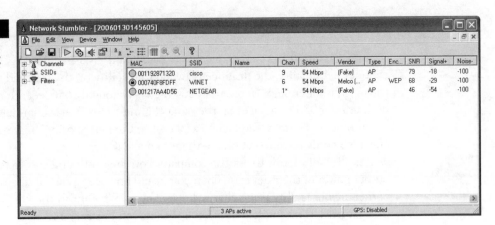

FIGURE 3-32

Multiple existing networks

Negotiation
scenario

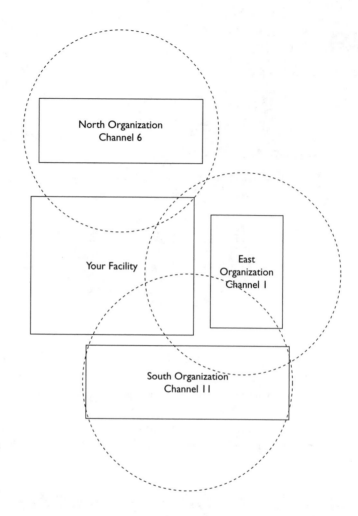

channel 11, and the organization east of your facility is using 802.11g devices on channel 1. In reality, the distance between the facility on the east and the one on the north is far enough (more than 200 meters or roughly 656 feet) that they could both utilize 802.11g devices on the same channel. This frees a channel for your environment. There would be no extra cost to the neighboring organizations, and in effect, you only have to get one of them to make the change.

Another solution is to ask the companies on the north and east to reduce the output power of their devices. Then you install an access point in the southeast corner of your facility using channel 6 and an access point in the northwest corner on channel 1. This implementation is depicted in Figure 3-34.

Implementation after adjusting power levels

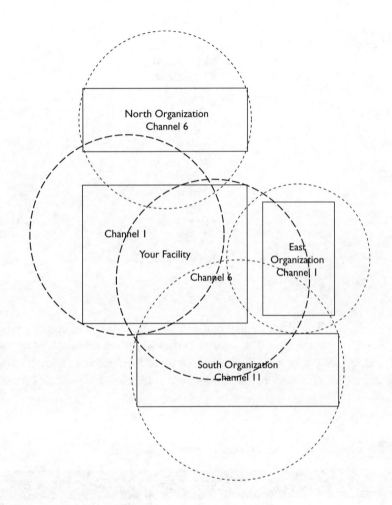

Document Your Findings

The final step of the site survey is to document your findings. This document can be a 3 to 5 page recommendation or several hundred pages bound with a full installation and maintenance plan. In most cases, you can provide rough sketch drawings, similar to the one in Figure 3-35, that will suffice for small businesses. You can take this drawing back to the office and recreate it using Visio or another diagramming tool for the client presentation.

In addition to the drawing, you want to provide a list of devices and their settings required for the installation. Table 3-2 is an example of such a document; however, it's not intended to be a recommended list. Each situation varies and demands individual analysis.

FIGURE 3-35

Site survey
drawing

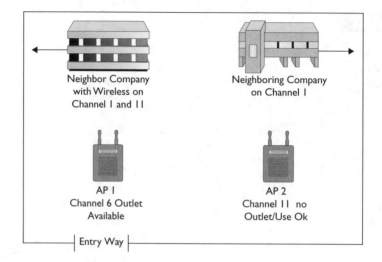

Neighbor Company
with Wireless on
Channel 1 and 11

Neighboring Company
on Channel 1

AP 1
Channel 6 Outlet
Available

AP 2
Channel 11 no
Outlet/Use Ok

Entry Way

The ultimate goal of a wireless site survey is to ensure that you can implement a license-free WLAN solution that will meet the needs of the organization. The result should be a detailed plan that includes a listing of equipment and the installation (both location and configuration) instructions for that equipment.

TABLE 3-2   Site Survey Recommended Devices and Settings

Device	Installed Location	Settings	Requirements
AP (Linksys WAP54GP)	North closet	Channel 6, WPA	WPA, variable output power, PoE
AP (Linksys WAP54G)	South storage room	Channel 11, WPA	WPA, variable output power
AP (Linksys WAP54G)	Front office	Channel 1, WPA	WPA, variable output power
Router (Linksys WRT54G)	IT center	Channel 11, WPA, VPN Pass-through	WPA, variable output power, VPN pass-through
35 PCMCIA Cards (Linksys WPC54G)	Throughout facility	Based on location	WPA, 802.11G

Obsolete Devices

Some devices are just not as common in newer networks. This is usually because they've been replaced by better technology. The two devices that come to mind are hubs and bridges. Hubs are multiport devices that allow multiple computers to communicate with each other, usually with UTP cabling. The problem with hubs is that they flood every incoming frame out on every port. This design effectively creates a large broadcast domain, and this negative has been overcome by modern switches.

In the process of fixing that which was broken in the hub, switches also made bridges all but unnecessary. A switch, as you learned earlier in this chapter, is really a multiport bridging device. Since switches perform the same operations as bridges, they have mostly replaced bridges at this point. However, because these devices just keep on working (remember, no moving parts?), you are still likely to see them hanging around in very old networks. If the network has existed for more than ten years, there's a good chance you might run into a hub or a bridge. In fact, this inclusion may be one of the main things you need to deal with as you prepare the network for the modern age of convergence. Ridding your network of hubs will give you a tremendous performance gain for converged data.

CERTIFICATION SUMMARY

This chapter introduced you to the infrastructure hardware used on both data and converged networks. You learned about routers, which provide network segmentation and are used at the Core and Distribution layers of the network. You also learned about switches, which provide non-CSMA/CD Ethernet for full-duplex connections and greatly increase the performance of your network. You also learned about basic WAN hardware and wireless access points, which are a very important part of modern converged networks, since so many networks will support wireless VoIP client devices.

You started the chapter by gaining an understanding of different network design models, including flat and tiered models. The tiered model includes three layers. The Core, Distribution, and Access layers are part of the tiered, or layered, model.

TWO-MINUTE DRILL

Recognize Network Models

❑ The Core layer provides fast data routing and forwarding.

❑ The Distribution layer provides access control and policy management.

❑ The Access layer provides connectivity to the network infrastructure.

❑ Centralized networking keeps the data and processing at the center of the network, and decentralized networking distributes both closer to the endpoints.

Identify the Functions of Hardware Components

❑ Routers are used to connect networks and route or move data among them.

❑ Switches are used to provide full-duplex non-collision connectivity to nodes on a network.

❑ Switches have mostly replaced bridges and hubs.

❑ Access points are used to provide wireless access to wired or wireless networks for compatible clients.

❑ Access points may be fat or thin, meaning they may have the complete operational logic in them or in an external provider.

❑ Access points should be installed carefully so that they provide the needed access in the right locations and do not interfere with each other.

SELF TEST

The following questions will help you measure your understanding of the material presented in this chapter. Read all the choices carefully because there might be more than one correct answer. Choose all correct answers for each question.

Recognize Network Models

1. You have implemented a network solution that includes a single server that is accessed by forty client devices. The client devices simply display the information sent back from the server, and all processing happens at the server. What kind of networking model have you implemented?

   A. Tiered

   B. Flat

   C. Decentralized

   D. Centralized

2. Which of the following are layers in a common tiered network model?

   A. Core

   B. Access

   C. Presentation

   D. Distribution

3. You are implementing a firewall. On what portion of the network model will it most likely be implemented?

   A. Core

   B. Edge

   C. Data Link

   D. Transport

Identify the Functions of Hardware Components

4. You need a device that will allow you to move data throughout your network based on IP address information. Which of the following devices are you most likely to implement?

   A. Switch

   B. Router

   C. Access point

   D. CSU/DSU

5. You are implementing a WAN connection. Which of the following devices is unique to WAN links and may be needed during your implementation?

    A. Switch

    B. Router

    C. Access point

    D. CSU/DSU

6. An NT1 device is needed with which of the following types of WAN technologies?

    A. T1

    B. E1

    C. ISDN

    D. DSL

7. When you implement an access point in a wireless network, you are creating what logical structure?

    A. Basic service set

    B. WAN

    C. Branch

    D. CPE

LAB QUESTION

You are implementing the network represented in Figure 3-36. This network has been designed by an individual who is not aware of a tiered network model. How could you redesign this network—while still providing the services needed by the end nodes—so that a tiered model is in place?

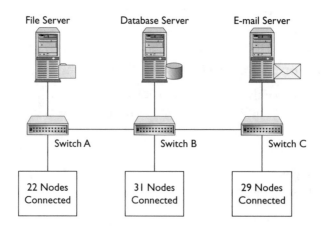

FIGURE 3-36

Flat network design needing to be rebuilt

SELF TEST ANSWERS

Recognize Network Models

1. ☑ **D** is correct. This scenario describes a centralized networking model.
☒ **A, B,** and **C** are incorrect. A tiered or flat network model has to do with how the data gets from node to node, and either could technically support either centralized or decentralized models.

2. ☑ **A, B,** and **D** are correct. The most common layered network design model is the three-layer model consisting of Core, Distribution, and Access layers. Do not confuse this with the OSI model, which is a network communications model. The OSI model of communications operates over a flat or tiered network design model.
☒ **C** is incorrect. The Presentation layer is a layer of the OSI model and not a layer in the network design model.

3. ☑ **B** is correct. Firewalls are most frequently implemented at the edge.
☒ **A, C,** and **D** are incorrect. Firewalls are not usually implemented at the core because the core needs to move data as fast as possible and this would not allow filtering. Data Link and Transport are layers of the OSI model, and while a firewall may operate at these layers, they are not related to where the firewall would be placed on the network.

Identify the Functions of Hardware Components

4. ☑ **B** is correct. A router is a Layer 3 device and will move data based on IP addresses.
☒ **A, C,** and **D** are incorrect. Switches work at Layer 2 based on MAC addresses and access points also work at Layer 2. The CSU/DSU is a WAN solution.

5. ☑ **D** is correct. A CSU/DSU is unique to WAN links.
☒ **A, B,** and **C** are incorrect. These devices are all used within local networks.

6. ☑ **C** is correct. ISDN links require the use of an NT1.
☒ **A, B,** and **D** are incorrect. These are all WAN technologies, but they do not require the use of an NT1.

7. ☑ **A** is correct. A BSS is created when you install an AP.
☒ **B, C,** and **D** are incorrect. WANs are created using special hardware and a service provider's network. A branch is similar to the Distribution layer in the three-tier model. CPE stands for customer premises equipment and may include things like NT1s and CSU/DSUs.

LAB ANSWER

Figure 3-37 shows a possible solution. Notice how the network now uses core routers at the Core layer and distribution switches that are connected to Access layer switches for end node access. Notice that the servers are in the Distribution layer and are positioned closest to the users who need access to them.

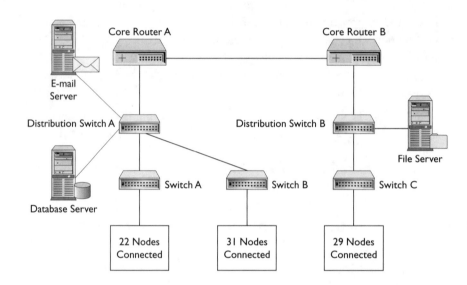

FIGURE 3-37

Tiered model
after redesign

4

Client Devices

> *Things should be made as simple as possible, but not any simpler.*
>
> —*Albert Einstein*

The contents of this chapter form a necessary foundation for understanding the endpoints used on a converged network and the networking technologies used on the same. Without a fundamental understanding of these concepts, the detailed specifications of soft phones, PC phones, and Wi-Fi phones will not make much sense. Since these topics fit well in the category of our traditional data networks, I'll cover them here. They will be extended in later chapters when specific voice and media devices are explained.

There are two major topics covered in this chapter: client device connectivity and TCP/IP implementations. In the first topic, you will learn how a client device becomes part of the network. Wired devices will be covered first and then wireless. The second topic will look at specific implementations of TCP/IP and their possible features and limitations on various operating systems and devices. This knowledge will prove very useful as you begin to explore the Quality of Service features in later chapters. You will be able to relate that knowledge to this foundational information. I'll keep it as simple as possible, but not any simpler. Thanks, Albert.

CERTIFICATION OBJECTIVE 4.01

Identify the Various Endpoints Used in a Converged Environment

In Chapter 7, you will learn about IP phones, soft phones, and other converged endpoint devices. This section covers the basic network functionality of a computing device and allows you to better understand the chapters ahead. If you've read the previous three chapters or have the knowledge they contain, you will be well prepared to understand the topics covered in this chapter. Remember that our focus is on the networking foundations needed to understand convergence technologies. The network must be there in order for voice traffic to traverse it. This discussion of endpoint or client device connectivity is very important because it explains how a computing device joins, participates in, and disconnects from the converged network.

Client Device Connectivity

There are very literally two ways you can connect to a network because there are only two kinds of media: wired and wireless. Some would prefer the terms cabled and uncabled, since fiber cabling does not necessarily consist of wires. Whichever you prefer, you have these two connection types. In the previous three chapters, you learned about the different wired and wireless connection types. In this chapter, you will learn about the way these devices actually become part of the network. We want to answer these three fundamental questions:

- How does a device join the network?
- How does a device participate in the network?
- How does a device remove itself from the network?

Understanding the answers to these three questions is crucial to your understanding of how client devices function on a converged network. For example, you will need to know how a wireless client roams from one access point to another, and you will need to know how wired and wireless clients authenticate to the network. These questions will be answered in this chapter, and the answers will help you understand the specific devices that you learn about later in the book.

Wired

When it comes to wired networking, from the perspective of the installer, joining the network seems to be more about the physical hardware being in place than anything; however, it is important that you differentiate between the physical medium across which network traffic passes and the network itself. You can actually run multiple networks across a single network medium. This concept is an important distinction. In order to participate as a wired device in a wired network, a device must be connected to the physical network medium and it must be configured so that it may communicate as part of a logical network that operates on that medium. It is for this reason that both wired and wireless devices can participate in a shared logical network. They may have different physical network connections, but they can communicate at Layers 3 through 7 in much the same way.

For this reason, the major focus of this section and the wireless section to follow will be on how the physical devices access and communicate on the medium. However, since it makes logical sense to talk about it here, the process of joining the logical network will be addressed in sequence. In this section, you will see the step-by-step process from power on to full communications ability that a wired client utilizes to join and participate in the network. The next section will go through the same process for wireless devices. Finally, you'll see how a network device removes itself from the network, though a device will sometimes simply disappear without announcement (the user powers it off).

The first question that must be answered is this one: How does a device join the network? The answer to this question begins by analyzing a piece of hardware known as a network interface card (NIC). In some devices there may be a network interface chipset as opposed to a NIC. This design is common on modern desktops, since many of these desktop computers have built-in network adapters that are on the motherboard rather than an add-on card. Other devices, such as protocol analyzers and line testers, which do not actually join the network, usually have dedicated hardware for performing the work for which they are designed.

Since network connectivity begins with the NIC, you should understand the different types of NICs that are used in modern networks. These network cards include the following:

- ISA, PCI
- PC-Card
- Mini-PCI
- SD and compact flash
- USB
- Onboard

The oldest type of NICs still in use today are the ISA (Industry Standard Architecture) adapters. ISA NICs came in either 8-bit or 16-bit versions and may still be lingering in some older computers on networks. I'm continually amazed at the lifetime of some devices. In the last quarter of 2007, I was working in an organization on their new WLAN when I noticed a cutting table that was used in their manufacturing process. To my surprise, an old 386 computer was being used to manage the cutting table and this computer was connected to the network. The good news is that you are not likely to see any ISA cards used for converged technologies. Most video conferencing done from computers will be on computers that were manufactured in the last few years. This younger age means they will more than likely use built-in modern NICs, or PCI/PCI-X NICs.

The Peripheral Component Interconnect (PCI) standard was first available for implementation at version 2.0 in 1993 and has evolved since that time. Most NICs use PCI, which is a 32-bit bus; however, some newer gigabit NICs are implemented as 64-bit devices and support the newer PCI-X standard. The Mini-PCI standard was

FIGURE 4-1

PCI network
interface card

developed for use in laptop and notebook computers. Mini-PCI is basically PCI 2.2
implemented with a different physical form (usually called the form factor). Figure 4-1
shows a typical PCI NIC.

Many hand-held devices such as Windows Mobile and Palm devices support
compact flash (CF) and secure digital (SD) interfaces. These interfaces can be used
to add network connectivity. Multiple manufacturers have released CF-wired NICs
for portable devices, but most SD components are used for wireless connectivity
instead of wired.

The Universal Serial Bus (USB) standard has become very popular in the past
decade and is widely used for network connectivity. Of course, it is important to
keep the limited speed of USB 2.0 (the currently implemented standard) in mind.
Since USB 2.0 can only support maximum data rates of 480 Mbps, it certainly
cannot keep up with the data rate of a gigabit Ethernet adapter. However, the reality
is that 480 Mbps is faster than 100 Mbps, so a USB 2.0 Gigabit Ethernet adapter
will be faster than a 100 megabit Ethernet adapter. Figure 4-2 shows an example of a
USB Ethernet adapter.

FIGURE 4-2

USB Ethernet
adapter

Finally, and true for most modern desktops, onboard network adapters are very popular. They come in 100 Mbps and gigabit speeds and can be disabled in order to install an add-on NIC. If your system has only a 10 Mbps built-in adapter, you should probably consider upgrading it to an add-on 100 or 1000 Mbps adapter in order to keep up with converged demands. Granted, the voice and video traffic alone will not consume anywhere close to the 10 Mbps available, but that also doesn't leave as much room for the integrated data communications.

It is important that you understand the impact of these different types of connectors or form factors. It's more than just choosing the right NIC for the available slot in your computing device. Sometimes it's about choosing the right computing device that provides the needed slot. For example, you cannot connect an ISA card to Gigabit Ethernet and achieve gigabit speeds. This fact is because the bus is simply not capable of this serial speed. However, a PCI or PCIe card (or a PCI-X card) could keep up with Gigabit Ethernet, and the newer PCIe multilane cards can even keep up with 10 Gigabit Ethernet with the right configuration.

Now that you understand the basic NIC types that are available for portable and fixed computing devices, you can begin to investigate how these NICs interact with the network and the local machine. Remember that data travels on the computer bus in parallel. In other words, a 32-bit bus can be said to have 32 parallel lanes of communication. A NIC, however, is serial in nature. Data travels in series on a single lane. This is why full-duplex NICs are so beneficial. They provide one pair of wires for outgoing communications and another pair for incoming. This design provides bidirectional traffic, though it is still serial from point A to point B.

In order to perform this serialization of internal data, the NIC will usually buffer the data and then transmit it as quickly as the medium and interface type will allow. The data is transmitted on the medium, and the NIC relinquishes control of it at that point. Of course, the NIC cannot travel with the data. This is why protocols such as TCP have built-in error checking. The receiving machine will transmit acknowledgments back to the sending machine in order to provide proof of receipt. If the sending machine does not receive an acknowledgment within a given amount of time, it will resend the data or time out and cancel the transmission.

When you power on a device that communicates on the network using a wired interface, the NIC comes alive almost immediately and begins sensing the medium in order to determine if connectivity is available. Every NIC has an identifier that is, or should be, unique to that NIC. This identifier is known as a MAC address and also as a:

- Layer 2 address or L2 address
- Hardware address or hardware ID

- Physical address
- Burned-in address (BIA)

The Medium Access Control (MAC) address is often called a Layer 2 or L2 address because it is how the MAC layer (remember, this layer is a sublayer of the Data Link layer) entity is identified as unique on the network. Since the address is burned into the memory of the NIC in most cases, it is also known as a burned-in address (BIA), a hardware address or ID, or a physical address. These latter names are all based on the fact that a BIA will always go with the NIC regardless of the machine in which it is placed or the location in which it operates. That address belongs to that card.

When a switch receives an Ethernet frame from a NIC in a computing device, it documents the MAC address in its internal memory with the port through which the communication arrived. This mapping allows the switch to direct traffic to that MAC address through that port. You could say, then, that a computing device joins the network or makes the network aware of its existence the first time it transmits a frame on the network. It is at this point that the switch is aware of this MAC address as existing on the network.

Most NICs provide external LEDs for analysis of the state and operation of the card. The following are common LED indicators used by NICs:

- Green connected (link or connection LED)
- Amber collisions (Activity LED)

So how do you read the green and amber LEDs? In most cases a solid green LED is a good connection. This solid green LED will usually show within 15–30 seconds or less of initial power on. If the green LED is blinking, it usually indicates intermittent connectivity. The NIC could be failing or the cable, switch port, or patch panel may be bad. When the green LED is simply not on, it's usually a very bad indication. The physical wire is not being detected by the card. It may not be seated completely, or it or another part of the connection chain may have completely failed.

The amber LED is usually used to indicate collisions. If this light is blinking, it's completely normal; however, if this light is solid, it may indicate in inordinate number of collisions and is cause for investigation.

So far you've learned that a NIC translates the parallel computing within a computer to the serialized data streams sent on the network. You've also learned that, at power on, the NIC detects the medium and reveals its discoveries through LEDs that are usually visible on the outside of the computing device or NIC itself. At this point, the operating system that is used by the device will begin to load.

At some point, this operating system will need to communicate with the NIC, and this point is where drivers come into play.

NICs require drivers in order to operate. If you've ever had a computer with a NIC installed that would not communicate on an attached network even though the LEDs showed connectivity, you've either seen a situation where a driver has not been installed or where the higher-layer network configuration was improperly managed. Just as in any other hardware device, the operating system must know how to talk to the NIC and the NIC must be able to pass data to the operating system.

The interaction between hardware and the operating system is handled differently by each operating system, but the concept is the same. A device driver is a piece of software that knows how to receive requests from the operating system and submit those requests to the controlled hardware. This software also knows how to receive attention signals (interrupts) from the hardware and pass incoming information to the operating system. In the end, device drivers are nothing more than small software programs that perform these operations for you.

Since device drivers are really just software applications, it's important for you to keep the following software truism in mind:

Applications can perform poorly even though they meet functional requirements.

What is meant by this statement? Functional requirements state *what* a system must do, but they do not state *how* the system must do it. For example, the NIC might have a functional requirement of being able to communicate on 100 Mbps Ethernet networks. That's the what. The reality is that a NIC could communicate on a 100 Mbps Ethernet network, but only provide data rates of between 30 and 35 Mbps, as an example. This NIC would meet functional requirements in that it operates on 100 Mbps Ethernet networks, but it performs poorly. Sometimes this poor performance is actually a problem with the NIC itself, but I've found that it is just as likely to be a problem with the device drivers. This possibility is why it is so important to test the NIC's performance as well as functionality and then be sure to check for updated drivers if the performance is subpar.

Once the operating system loads the device driver, the computing device is officially "on" the network from a communications perspective at OSI Layer 1 and Layer 2; however, it is not yet on the internetwork. In order for this step to happen in modern networks, the device must have a proper IP configuration and identity on the network. The identity on the internetwork will be the IP address, and the IP configuration will include subnet mask, default gateway, and possibly other parameters. On most modern networks, this identity means acquiring an IP configuration set from a Dynamic Host Configuration Protocol (DHCP) server. Figure 4-3 shows a diagram of the process used to acquire an IP configuration set

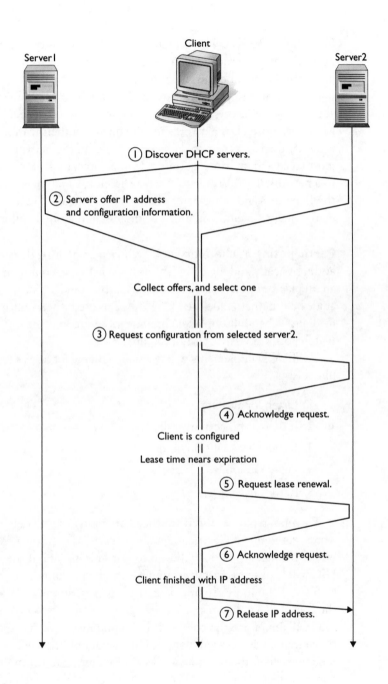

FIGURE 4-3

Acquiring an IP
configuration set

from a DHCP server. The depicted network happens to be a Windows Server–based network, but the process is similar using Linux, Unix, or Mac OS servers.

At this point (once an IP configuration set is received), the device may or may not be able to communicate on the internetwork. Some networks require authentication before any further communications can occur, and others are open. Most enterprise-class networks require further authentication. This authentication is mostly used for access to data and services on the network and is not used for most voice and multimedia traffic. For example, most IP phones do not log on to a network like a Windows Active Directory or a Novell eDirectory. However, the IP phones and other voice clients may be required to authenticate to a call management system, which will be discussed in more detail in later chapters.

Participating in the Network   Once a computing device is "on" the network—both physically and logically—it must abide by the communication standards that are in place. For most modern networks, this standard means that the device must abide by Ethernet guidelines. CSMA/CD is used when multiple devices share the medium. When full-duplex communications are in place from a NIC directly to a switch port, CSMA/CD is not required.

CSMA/CD implements a very simple scheme for access to the network. It flows like this:

1. Listen for activity.
2. If there is no activity, talk.
3. If there is a collision, announce it; otherwise, return to Step 1.
4. Pause randomly.
5. Return to Step 1.

The first step is to "listen" to the medium in order to detect activity. In other words, determine if electrical signals are currently on the medium. If there are signals on the medium, the NIC will wait to transmit its data. If there are no signals, the NIC will "talk" or transmit its frame onto the medium.

Instead of blindly sending data, the NIC will attempt to detect collisions should they occur. If a collision occurs after or during the transmission of the frame, the NIC will announce the collision. This announcement is actually a special jam signal that all devices will detect. This signal tells the devices to initiate their random backoff timers and wait before attempting to transmit on the medium. This announcement (jam) signal is like getting out of your car after an accident and stopping traffic so they know there is a problem. All traffic stops and will continue after some period of time. If no collision is detected, the process begins again until the NIC has no more frames to transmit (remember, there could be many frames in the buffer). Remember that CSMA/CD is only needed when collision domains exist. In a full-duplex connection with a switch, there is no real collision domain, as there is no possibility of collisions, with the exception of broadcast traffic. Since most traffic is not broadcast-oriented on modern networks and such traffic is becoming less common every day, collisions should continue to be reduced or eliminated.

Virtual LANs, or VLANs, are used to change the scope of broadcast domains, as are routers. By default, VLANs cannot even talk to each other. A Layer 3 device like a router or Layer 3 switch would be needed to provide communications between these VLANs. Some engineers will implement a "router on a stick" that references data being sent to a router and coming back in on the same switch port. The router "virtually" routes between two VLANs even though there is only one physical port involved in the communications.

on the

job

When reading documentation, you may see references to "switches" or "the switch;" these references are nearly always to a standard Layer 2 switch. What few references there are to Layer 3 switches usually spell out the full phrase "Layer 3 switch" or "routing switch" to make the intention clear.

Removing from the Network   At the Physical and Data Link layers, there is no real removal from the network unless IEEE 802.1X port-based authentication is being used. This authentication method is covered in Chapter 13 of this book. For now, assume that the only way a device is removed from the network is by disconnecting the cable. Removal from the logical network can be a little cleaner. The computing device may, for example, remove dynamic DNS entries that it created or release the IP configuration set that it acquired at startup.

Wireless

Much like wired network adapters, wireless adapters come in varied forms. In fact, you can buy wireless adapters in all the same form factors that are used for wired NICs. These include ISA, PCI, PCIe, Mini-PCI, USB, CF, and SD, among others. Since these form factors were briefly covered earlier, they will not be revisited here.

Joining the Network   IEEE 802.11 devices have a more complex process for joining the network than do the Ethernet wired devices previously covered. This process of joining the network involves scanning, authentication, and association.

Wireless devices use scanning to locate a wireless network to which they can connect. This scanning can be either active or passive. Active scanning uses what are called probe request frames to locate an AP. The probe request frame may include the SSID of the network it is seeking. In this case, only APs with matching SSIDs will respond with probe response frames. If the probe request simply leaves the SSID blank or null, all APs that hear the probe request will respond unless they are configured in a non-standard way to ignore null probe requests. For security reasons, some vendors do implement this non-standard functionality.

Passive scanning locates networks by listening for special frames called beacon frames. The beacon frame is transmitted by the AP and contains information about the network it manages, such as SSID, supported data rates, and security configuration requirements.

Once a station has located the wireless LAN (WLAN) to which it seeks to connect, which is the first stage of station connectivity, it must go through the authentication and association process, which are the second and third stages of connectivity. This involves being authenticated by the access point or WLAN controller and then determining the PHY, data rate, and other parameters within which the association (connection) must operate. The first step is authentication, and the second is association; each step is covered in sequence in this section. First, we'll look at the IEEE 802.11 concept of the *state machine*.

The *state machine* of the IEEE 802.11 standard can be in one of three states:

- Unauthenticated/unassociated
- Authenticated/unassociated
- Authenticated/associated

All WLAN adapters that claim to meet the IEEE 802.11 standards must implement this state machine. In the initial state, a client station is completely disconnected from the WLAN. It cannot pass frames of any type through the AP to other stations on the WLAN or the wired network. Authentication frames can be sent to the AP. These frames are not sent through the AP, with the exception of a split MAC implementation where a WLAN controller performs the authentication, but are sent to the AP. The distinction is important. Frames must be transmitted to the AP in order to eventually reach the authenticated and associated stage; however, until the final stage is reached, only authentication and association request frames will be processed by the AP.

APs, or WLAN controllers, keep a list known as the association table. Vendors report the stage of the station's state machine differently. Some vendors may report that a client that has not completed the authentication process is unauthenticated, and other vendors may simply not show the client in the association table view.

The second state of the state machine is the authenticated and unassociated state. To move from the first state to the second, the client station must perform some kind of valid authentication. This task is accomplished with authentication frames. Once this second state is reached, the client station can issue association request frames to be processed by the AP; however, other IEEE 802.11 frame types are not allowed. In most APs, the association table will now show "authenticated" for the client station.

Since the interval between reaching the authenticated and unassociated stage and moving on to the authenticated and associated stage is very small (usually a matter of milliseconds), you will not see client stations in that state very often. In most cases, you will either see "unauthenticated" or nothing for the first state and associated for the third state in the AP's association tables. The only exception to this is what is sometimes called "pre-authentication." A station can authenticate with any number of access points, but it can be associated with only one access point at a time. The access point to which the station is associated must be a single entity in order for other devices on the network to be able to reach that station. In some systems, the station is capable of authenticating with multiple access points so that it can roam more quickly when the need arises. This roaming ability can be very important for Voice over WLAN (VoWLAN).

The third and final state is the authenticated and associated state. In order for a station to be in this state, it must have first been authenticated and then associated. The process of moving from state two (authenticated and unassociated) to this state is a simple four-frame transaction. The client station first sends an association request frame to an access point to which it has been authenticated. Second, the access point responds with an acknowledgment frame. Next, the access point sends an association response frame either allowing or disallowing the association. The client sends an acknowledgment frame as the fourth and final step. If the third step resulted in an approval of the association request, the client station has now reached the authenticated and associated state and may communicate on the WLAN or through the WLAN to the wired network if encryption keys match and 802.1X is not enabled.

The association response frame includes a status code element. If the status code is equal to 0, the association request is approved or successful. There are three other status codes that may apply: 12, 17, and 18. A status code of 12 indicates that the association was rejected for some reason outside of the scope of the IEEE 802.11 standard. A status code of 17 indicates that the access point is already serving the maximum number of client stations that it can support. Finally, a status code of 18 indicates that the client station does not support all of the basic data rates required to join the basic service set (BSS).

This last constraint is imposed to ensure that all stations will be able to receive certain frame types that are communicated at the basic data rates. If they cannot receive these frame types, they cannot participate in the BSS, lest they cause interference by not understanding such frames as CTS or by not having the ability to interpret frames at all. The result would be that the station not supporting the basic data rates would cause interference due to an internal misconception that the WM was clear. This overview is a simplification of a complex technical constraint, but it is sufficient for a WLAN administrator to know that a station cannot associate with a BSS if it does not support the basic data rates required. Thankfully, these data rates are specified by administrators, and assuming the station is standards-based and compatible with the PHY being used, this connection shouldn't be a problem with proper configuration settings.

The key point is to realize that you cannot transmit data frames for processing until you've been associated and you cannot transmit association frames for processing until you've been authenticated. Now that you understand the three states in which a station can reside, let's explore the details of how the station can become authenticated and then associated. As you can see, the process of joining a WLAN is much more complicated than that of joining the traditional wired LAN. The process is made even more complex by that fact that most enterprise networks today implement some form of Extensible Authentication Protocol (EAP) authentication.

Participating in the Network   As Chapter 3 briefly explained, WLANs do not use CSMA/CD, but they instead use Carrier Sense Multiple Access/Collision Avoidance (CSMA/CA). This difference is because a wireless device cannot detect a collision on the network medium. Understand that radio frequency (RF) waves theoretically travel an infinite distance. The signal is weakened and eventually disappears into background noise, but it is still there, though it cannot be detected with our current technology. This weakening of the RF wave's amplitude (strength) is known as attenuation, just as it is in wired media. However, attenuation of RF waves in WLANs is less predictable because we do not completely control the objects around and through which the RF waves must pass. This unpredictability means that the RF energy moving out from the AP in one direction may travel 100 feet and still provide a strong signal and the RF energy moving in another direction may travel 300 feet and still provide a strong signal. The point is that we cannot completely control RF signals; we can only limit their detectability.

Remember that the ability to detect an RF signal is more a factor of the receiving antenna than it is the output power of the AP in many situations. The general rule of RF communications is: if I can hear you, you can hear me. In other words, if I have an antenna with sufficient gain to pick up your AP's signal, my client should be able to get frames back to your AP as well. This ability is why highly directional and semidirectional antennas are often used by hackers in war driving—though many of these war drivers do not fully understand why their toolkit works.

So what does all this discussion have to do with CSMA/CA and the inability of wireless devices to detect collisions on the network? Since the RF signal can travel such a great distance, it is entirely possible that a collision will occur closer to the client and farther from the AP (or vice versa) so that the transmitting device is unaware of a collision. In other words, the remote device sees the "other" signal, but the AP does not, since it is too far away. This concept is depicted in Figure 4-4.

For this reason, WLAN devices are configured to retransmit frames when an acknowledgment (ACK) frame is not received in a given amount of time. The transmitting device really doesn't assume that a collision occurred; the designers of 802.11 simply realized that a number of scenarios could cause a frame to be lost in transmission and so they implemented the ACK, and the transmitter knows to look for it. In addition to this error correction scheme, the designers implemented a collision management process that attempts to avoid collisions. This process, known as CSMA/CA, is documented in detail in the following paragraphs.

FIGURE 4-4	
Collisions occurring on a WLAN	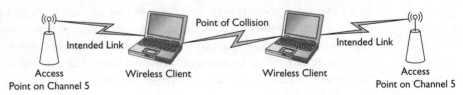

The essence of CSMA/CA is that collisions can happen many places on the medium, at any time during a transmission, and may not be detected by the transmitter at its location. Listening for evidence of a collision while transmitting is thus useless and not a part of the protocol. This architecture is because transmissions cannot be aborted early. Collisions are only inferred as one possible explanation for failure to receive an immediate ACK after transmitting a frame in its entirety. The frame must be retransmitted completely. Under these circumstances there is much value in collision avoidance, and there is much of it in the IEEE 802.11 protocols.

If you've ever had a conversation with another person on the telephone, you've probably experienced a communications collision. When you both started speaking at the same time, neither of you could hear the other effectively. In most conversations when you talk over each other, you will both stop speaking for some amount of time and then one of you will start speaking again. Since the time that both of you choose to wait is slightly different, there is a good chance that one of you will be able to communicate the next time. This wait would be similar to collision detection. The *carrier sense* in CSMA means that the devices will attempt to sense whether the physical medium is available before communicating. The *multiple access* indicates that multiple devices will be accessing the physical medium. In a CD implementation of CSMA, when a collision is detected, both devices go silent for a pseudo-random period of time. Since the time period is different for each device, they are not likely to try communicating at the same time again. This process helps recover from collisions and avoid another collision. In a CSMA/CD implementation, collisions occur because devices can begin communicating at the same time even though they both listened for "silence" on the physical medium. There was indeed silence, but both devices broke the silence at the same moment.

CSMA/CA is used in wireless networks, and it was also used in early Apple LocalTalk networks, which were wired networks that were common to Apple devices. Collision avoidance is achieved by signaling to the other devices that one device is about to communicate. This signaling would be like saying, "Listen, for the next few minutes, because I will be talking" in a telephone conversation. You are avoiding the collision by announcing that you are going to be communicating for some time interval. CSMA/CA is not perfect due to hidden node problems, which result from one node not seeing another node though they both see the AP, but it provides a more efficient usage of a medium like RF than would CSMA/CD.

Carrier sense is the process of checking to see if the medium is in use or busy. If you have multiple telephones in your house and a single line that is shared by all of these telephones, you use a manual form of carrier sense every time you use one of the phones to make a call. When you pick up the phone, you listen to see if someone else is already using the phone. If they are, you may choose to hang up the phone and wait until it becomes available. If you've ever been on the phone when someone else begins dialing without first checking to see if anyone is using the line, you've experienced a form of collision as the tones penetrated your ears and overcame your conversation with noise.

In IEEE 802.11 WLANs, there are two kinds of carrier sense that are performed: virtual carrier sense and physical carrier sense.

Physical carrier sense uses *clear channel assessment (CCA)* to determine if the physical medium is in use. CCA is accomplished by monitoring the medium to determine if the amount of RF energy detected exceeds a particular threshold. Due to the nature of WLAN architectures, there is no requirement for all stations to be able to hear all other stations existing in the same BSS. This design is because the wireless access point forms a kind of hub for the BSS. A station may be able to hear the access point and the access point may be able to hear the other station, but the two stations may not be able to hear each other. This architecture results in what is commonly known as the *hidden node problem*. For this reason, wireless networks must use other forms of carrier sense to deal with medium access control.

The other form of carrier is virtual carrier sense, which uses a network allocation vector (NAV). The NAV is a timer in each station that is used to determine if the station can utilize the medium. If the NAV has a value of 0, the station may contend for the medium. If the NAV has a value greater than 0, the station must wait until the timer counts down to 0 to contend for the medium. Stations configure their NAV timers based on Duration fields in other frames using the medium. For example, if a station detects a frame with a specific duration set in the Duration field, it will set the NAV timer to this duration and will now wait until that time has expired before contending for access.

To be clear, both the physical carrier sense and the virtual carrier sense must show that the medium is available before the station can contend for access. In other words, if the NAV timer reaches 0 and the station uses CCA to detect activity on the medium only to find there is such activity, the station still cannot transmit. In this case, another frame may be pulled from the medium and used to set a new NAV timer value for countdown. While it may seem that this design would prevent a station from ever communicating, the rate of frame transfer is so high that all of these actions usually take place in far less than one second.

After the station has determined that the medium is available, using carrier sensing techniques, it must observe *interframe spacing (IFS)* policies. IFS is a time interval in which frames cannot be transmitted by stations within a BSS. This space between frames ensures that frames do not overlap each other. The time interval differs, depending on the frame type and the applicable IFS type for that frame.

While the IFS implementation in IEEE 802.11 systems can result in the appearance of Quality of Service (QoS), it should not be confused with IEEE 802.11e or any Layer 3 or higher QoS solution. IFS is an 802.11 feature that allows for dependent frames to be processed in a timely manner. For example, a standard 802.11 data frame is transmitted using the DIFS interval, and the Acknowledgment (ACK) to this data frame is sent back using the SIFS interval. Because the ACK uses a SIFS interval, the ACK frame will take priority over any other data frames that are waiting to be transmitted. This way, the original station that transmitted the data frame will receive the ACK frame and not attempt to resend the data frame. In other words, the frame-to–IFS interval relationships that are specified in the IEEE 802.11 standard ensure that frames will be processed in their proper sequence.

I've mentioned some of the IFS types defined by the IEEE 802.11 standard already. These IFS types include the following:

- **SIFS (Short Interframe Spacing)**   The shortest space.
- **PIFS (Point [coordination function] Interframe Spacing)**   Neither the shortest nor the longest space. These spaces are not seen on WLANs, since PCF has not been implemented in any available hardware; however, they are mentioned in the standard.
- **DIFS (Distributed [coordination function] Interframe Spacing)**   Longer than PIFS, but shorter than EIFS.
- **EIFS (Extended Interframe Spacing)**   The longest space. This spacing is used after a corrupted frame is received.

The IFS delay interval is not the end of the wait for devices that are seeking time on the wireless medium. After the IFS delay interval has passed, the device must then initiate a random backoff algorithm and then contend for the wireless medium if the Distributed Coordination Function is in effect. This random backoff algorithm is processed and applied using the *contention window*.

The phrase *contention window* has caused much confusion in the wireless industry, but it is the phrase in use in the IEEE 802.11 standard. This "window" is actually a range of integers from which one is chosen at random to become the backoff timer for the immediate frame queued for transmission. Think of it as being like a contention range instead of a contention window and it will be a little easier for you.

All stations having a frame to transmit choose a random time period within the range specified as the contention window. Next the predefined algorithm multiplies the randomly chosen integer by a *slot time*. The slot time is a fixed-length time interval that is defined for each PHY (physical layer implementation), such as

DSSS, FHSS, or OFDM. For example, FHSS uses a slot time of 50 μs and DSSS uses a slot time of 20 μs. The slot times for each of the currently ratified PHYs are listed here:

- **FHSS**   50 μs (microseconds)
- **DSSS**   20 μs
- **OFDM**   9 μs
- **HR/DSSS**   20 μs
- **ERP – Long Slot Time**   20 μs
- **ERP – Short Slot Time (802.11b compatible)**   9 μs

As you can see, there are definite variations among the different PHYs supported in the IEEE 802.11 standard as amended. Though the IEEE 802.11n amendment is still in the draft stage at this writing, it is expected to use the standard 9 μs slot time used in existing PHYs that support OFDM. You are not required to understand all the specific details about how the random backoff time is generated and utilized in order to become a Convergence+ engineer; however, the following section provides a high-level overview of how the IFS and backoff time come together to provide wireless medium contention management.

Now that you have most of the pieces to the medium contention puzzle, you can begin to put them together in order to understand how a wireless station decides when it should try to communicate on the wireless medium. In order to understand this concept, imagine that a station has a data frame that it needs to transmit on the wireless medium. This data frame will be required to use the DIFS IFS, since it is a standard data frame. Furthermore, imagine that the station uses carrier sense to determine if a frame is currently being transmitted. For discussion's sake, let's assume that the station detected that the frame being transmitted had a Duration/ID field value of 20 μs. The station sets its NAV to count down the 20 μs and waits. The NAV reaches 0 and the station uses carrier sense and detects that the wireless medium is silent. At this time, the station must wait for the DIFS interval to expire, and since the station is using the DSSS PHY, it waits for 50 μs. Next, the station waits for the random backoff time period to expire, and when it does, the station uses carrier sense and detects that the wireless medium is silent. The station begins transmitting the data frame. All of these steps assume the network is using the Distributed Coordination Function, which is the only contention management functionality that has been implemented in widespread hardware at this time.

Ultimately, the carrier sense, IFS, and random backoff times are used in order to decrease the likelihood that any two stations will try to transmit at the same time on the wireless medium. The IFS parameters are also used in order to provide priority to the more time-sensitive frames such as ACK frames and CTS (clear to send) frames. The CCA, IFS, variable contention window, and random backoff times, together, form the core of the Distributed Coordination Function.

Even with all of these efforts, a collision can still occur. In order to deal with these scenarios, *acknowledgment* frames, or ACK frames, are used. An ACK frame is a short frame that uses the SIFS IFS to let the sending device know that the receiving device has indeed received the frame. If the sending device does not receive an ACK frame, it will attempt to retransmit the frame. Since the retransmitted frame will be transmitted using the rules and guidelines we've talked about so far, chances are the next frame—or one of the next few—will make it through.

I've shared this detailed explanation with you for a very simple reason: You must understand how IEEE 802.11 wireless devices communicate in order to implement converged technologies across them. Replay streaming video can be buffered, and so can live streaming video. These buffers can overcome latency problems in a network. However, live video conferencing and voice calls cannot be buffered for more than a portion of a second and not affect the quality of the communication. For this reason, WLANs must be considered carefully and uniquely when implementing converged networks. This topic will be addressed in more detail in Chapters 9 and 11.

Removing from the Network   Much like joining a wireless network, removing or roaming is a complex topic. This roaming ability is really only an issue in wireless networks and not in wired networks. Wired computing devices are usually very stationary due to the fact that they are physically connected to the network. One of the greatest benefits of wireless technology, however, is the mobility. This benefit means the devices will need to have a way to remove themselves from one network and connect to another, and in the case of VoWLAN, they must do it very rapidly and without losing upper-layer connectivity (IP connectivity, for example).

Whenever a wireless device needs to roam from one network to another, it must authenticate and associate with that other network. As was stated earlier, a station can be authenticated with multiple access points, but it can be associated with only one. There are three kinds of frames related to association: association frames, reassociation frames, and disassociation frames.

The process of *association* is very simple. Four frames are transmitted between the client station and the access point station. The first frame is an association request frame, which is followed by an acknowledgment frame from the access point. The third frame is an association response frame, which is followed by an acknowledgment frame from the client station. It is extremely rare for a client station to successfully authenticate and then fail to associate. This rarity is because the client station can usually determine if it is compatible with the BSS by inspecting the Beacon frames or probe response frames sent from the access points. Fast roaming in wireless networks is usually achieved within the client devices. The client device will continually scan for a closer AP (stronger signal) and will automatically roam to that other AP when necessary; however, before the other AP even has a stronger signal, the client device may have already pre-authenticated with it so that it can quickly exchange the four association frames and continue operating.

Reassociation occurs when a client station roams from one access point to another within an extended service set (ESS). In general you could say that there are two types of roaming: seamless and reconnecting. Seamless roaming would be roaming that allows a station to move its association from one BSS to another without losing upper-layer connections. Think of it as like being able to start a large FTP download while associated with one BSS and then walking to another area where you are reassociated with another BSS within the ESS. Seamless roaming allows the FTP download to continue and not fail.

Reconnecting roaming would require a new connection to the FTP server and, unless the server supported failure resume, a restart of the download. This kind of roaming occurs if there is no association hand-off operation that can be performed between the two BSSs, even though they are in the same ESS.

Because the IEEE 802.11 standard does not specify the details of how roaming should occur, it is possible to implement a WLAN using APs from different vendors (or even different model APs from the same vendor) that cannot communicate with each other and will not allow for seamless roaming. If you want to purchase differing hardware and still allow for seamless roaming, either you will have to purchase and test the hardware to ensure that the APs can interoperate or you should ensure that both vendors provide support for the same roaming procedures.

Roaming is only one way that a wireless client may be "removed" from a WLAN. The device may also simply decide to disconnect. This disconnect is done with a deauthentication frame. Deauthentication frames are known as advisory frames. They are so named because they are advising the network of something and the network cannot prevent that event from occurring. In other words, a standard IEEE 802.11–based AP cannot deny a deauthentication frame. This frame would be transmitted to the AP (or other members of the IBSS in an ad hoc network), and the receiving device would simply acknowledge the deauthentication. This frame would also result in a lowering of the state machine's state in the access point's association table.

A deauthentication frame will include the address of the station being deauthenticated and the address of the station with which the deauthenticating station is currently authenticated. The deauthentication frame will have a reason code of 3, which indicates the reason that the deauthenticating station is either leaving or has left the basic or extended service set. You should know that authentication must happen before association can take place. For this reason, a deauthentication frame effectively disassociates and deauthenticates the transmitting client station from the access point.

As you can see, wireless network communications are a bit more complex than wired because of the "open" medium through which wireless devices must communicate. It is important that you understand the inherent overhead of WLANs. Otherwise, you may not properly account for that when estimating the delay that will exist in your network infrastructure as you attempt to move voice packets across it. Voice packets must move very quickly from end to end. If they do not, call quality will suffer and calls may even be dropped due to inactivity and assumed lost connectivity.

CERTIFICATION OBJECTIVE 4.02

Describe Networking Technologies Used in a Converged Network

While this exam objective is covered in the preceding three chapters, it is important to take the topics covered one level deeper. You should understand how a typical operating system implements the TCP/IP protocols and the Ethernet and Wireless protocols. This knowledge will help you implement, secure, and troubleshoot your converged networks.

TCP/IP Implementations

Now that you understand the very basics of the two most fundamental LAN connection options, you'll need to know a little more about how an operating system implements the TCP/IP protocol suite. You'll need to understand the way the TCP/IP protocols are implemented and the tools that are available to configure and troubleshoot TCP/IP communications. In this section, I'll introduce you to the Windows TCP/IP stack in more detail than the others simply because it is more commonly implemented at the end node on enterprise networks. Then I'll review how the other common operating systems implement the protocol suite in summary.

Windows

With the release of Windows NT 3.1 in 1993, the Microsoft Windows platform began to change drastically. Windows 3.*x* and earlier were simply 16-bit graphical shells that ran on top of DOS. Ultimately Windows 95 and its successors were the same, only they were 32-bit shells (in part) that ran on top of a new version of DOS. For a few years, Microsoft developed the "shelled" versions of Windows alongside the NT-based version. We saw the release of Windows 95 in 1995 and Windows 98 came after that. What do we say about them? That they were the last shell-based versions? When Windows 2000 was released in late 1999, this event was the death blow to the "shelled" Windows. From that time forward, Windows NT–based systems have ruled the day.

Today, Windows 2000, Windows XP, Windows Server 2003, Windows Vista, and Windows Server 2008 all have their roots in that original Windows NT 3.1 released about fifteen years ago. The operating system architecture has evolved over the years, but it has remained largely the same. Historically, it was the release of Windows NT 3.5 in 1994 that really began the evolution of TCP/IP in Windows. TCP/IP was there from the beginning, but it was not implemented with the more advanced features that were included in NT 3.5. Version 3.51 repaired a number of bugs in the 3.5 release and was the first version to be considered largely by enterprise environments. I remember working for a company in 1995 that quickly pounced in Windows NT 3.51 as a solution for a particular problem they were having. This organization had more than 6000 nodes on the network in more than 60 locations, so it would certainly be considered an enterprise network. They implemented Windows NT Server 3.51 in early 1996, and it provided a very stable platform for their sales force automation system.

The interesting thing is that the ensuing releases of NT-based systems (Windows NT 4.0, Windows 2000, and Windows XP) didn't change the TCP/IP networking much until the release of Windows Vista in late 2006. Windows Vista made some very important changes that can definitely help or hurt a converged network. Don't get me wrong, there were changes in the TCP/IP tools and higher-layer functions in Windows NT 4.0 through XP, but there were very few changes in the lower-level implementations of the TCP and IP protocols themselves.

With the release of Windows Vista and Windows Server 2008, this model changes. For example, both of these newer systems implement window scaling by default for the TCP protocol. RFC 1323—that's right, it's a standard and not a Microsoft proprietary technology—specifies the use of window scaling for TCP communications. The TCP header uses a 16-bit field to report the receive window size to the sender. The receive window is the amount of data the receiver can buffer before sending an acknowledgment back to the sender. Since the field is 16 bits, it means it can hold any number from 0 to 65,535, representing 0 to 65,535 bytes. In networks with high delay rates, it is advantageous to extend this window's size, and so RFC 1323 specifies a standard for window scaling that allows for window sizes that are larger. This Window Scaling feature is implemented in SYN (synchronization) packets. The new window size can be up to 1GB.

This new feature sounds good and will eventually be of great benefit; however, you must ensure that your routers support RFC 1323 or it can cause very poor performance on your network. If the routers do not support window scaling, you can disable it in Windows with the following command:

```
netsh interface tcp set global autotuninglevel=disabled
```

Keep in mind that this command disables the window scaling feature completely and that you will need to enable it again if you upgrade your routers in the future to support window scaling. Many routers in enterprise networks today already support this feature.

To ensure you understand this window scaling feature of TCP as documented in RFC 1323, consider Table 4-1. The first column, *Scale Factor*, is the value that would be exchanged in the three-way TCP handshake that begins the TCP connection.

TABLE 4-1

TCP Windows Scaling Values

Scale Factor	Scale Value	Initial Window	Window Scaled
0	1	65535 or less	65535 or less
1	2	65535	131,070
2	4	65535	262,140
3	8	65535	524,280
4	16	65535	1,048,560
5	32	65535	2,097,120
6	64	65535	4,194,240
7	128	65535	8,388,480
8	256	65535	16,776,960
9	512	65535	33,553,920
10	1024	65535	67,107,840
11	2048	65535	134,215,680
12	4096	65535	268,431,360
13	8192	65535	536,862,720
14	16384	65535	1,073,725,440

The two end devices, assuming they understand window scaling, would multiply the *Initial Window* value times 2 to the power of the *Scale Factor*. The result is the following formula:

$$ws = 65{,}535 * 2^{\wedge}sf$$

where *ws* is the resulting window size and *sf* is the Scale Factor. As you can see in Table 4-1, the result is window sizes from less than 65,535 bytes to 1GB.

Since I mentioned the three-way handshake, it is useful for you to understand what this term means. The handshake plays an important role in the decision to use UDP for most voice traffic instead of TCP, though it is not the only factor in the decision. Because TCP is a connection-oriented protocol, it requires some method for creating a connection. This method is the three-way handshake. It's really a very basic three-step process:

1. The requestor sends a SYN packet to the recipient requesting a connection. This step is known as an active open.
2. The recipient of the SYN packet responds with a SYN-ACK packet acknowledging the connection.
3. The requestor response to the SYN-ACK with its own ACK packet. This step ensures that the devices can reach each other for future stable communications.

These three packets initiate the connection. A TCP connection is really just a collection of sequenced packets. The first SYN packet sent to the recipient contains the requestor's sequence number, and the SYN-ACK packet includes this number incremented by 1 with the recipient's sequence number as well. The ACK from the requestor to the recipient that finalizes the handshake continues with another increment of the recipient and requestor sequence numbers. These numbers will continue to increment as data is transferred, related to the connection that was initialized with the handshake. This number is used on both ends of the connection to resequence any packets that may arrive out of order. As you can see, there is a lot of overhead introduced in this handshake, and this overhead is usually removed by using UDP instead of TCP for voice communications.

In addition to the window scaling change, Windows Vista is the first Microsoft client operating system to implement IPv6 by default. This protocol may need to be disabled on networks that are not yet utilizing IPv6 to reduce unnecessary delays in communications or traffic on the network. Additional features in Vista, which are aimed at making network browsing easier for the users, may cause congestion on your network, and they may need to be disabled. These are features like Network Discovery or the Link Layer Topology Mapper.

It is always important to understand the network services that are running on your network, and these include both client and server services. Unneeded services result in unnecessary network traffic, and this results in a more congested network. A congested network is a bad network for converged technologies, the key reason being the necessary sequential nature of VoIP packets and multimedia packets. This data cannot be used unless it can be sequenced. This requirement means the data must traverse the network very quickly so that the packets can be resequenced as necessary and placed into a buffer for local conversion back to audio and/or video. To understand how Windows clients function on the network, the following topics will be addressed:

- Windows OS architecture
- Windows networking architecture
- Windows networking services
- Windows networking tools

Once these factors have been investigated for Windows operating systems, other operating systems will be compared and contrasted. Understanding this information will help you work with the many different clients that may be running converged applications on your network.

Windows OS Architecture   The Windows operating system (OS) architecture is depicted in Figure 4-5. This diagram is a simplified representation of the architecture to serve as a reference for describing the various components. The most immediate thing that stands out is the division between user mode and kernel mode.

FIGURE 4-5

Simplified Windows OS architecture

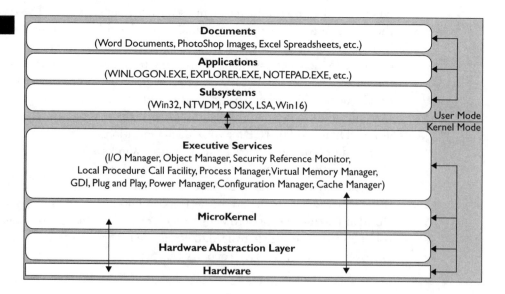

Documents
(Word Documents, PhotoShop Images, Excel Spreadsheets, etc.)

Applications
(WINLOGON.EXE, EXPLORER.EXE, NOTEPAD.EXE, etc.)

Subsystems
(Win32, NTVDM, POSIX, LSA, Win16)

User Mode
Kernel Mode

Executive Services
(I/O Manager, Object Manager, Security Reference Monitor, Local Procedure Call Facility, Process Manager, Virtual Memory Manager, GDI, Plug and Play, Power Manager, Configuration Manager, Cache Manager)

MicroKernel

Hardware Abstraction Layer

Hardware

User mode is the mode in which applications and services run, and kernel mode is home to device drivers and operating system components such as the file system driver, virtual memory, and other important modules.

Each application runs in its own memory space allocated as a portion of a virtual memory pool. The process is allocated 2GB of virtual address space, and it assumes this space is actually allocated. In reality, the operating system allocates memory as a combination of physical RAM and hard drive space used as non-active virtual memory as it is needed. In other words, while a process sees itself as having 2GB of memory, it may only have a few kilobytes or megabytes in actuality. Additionally, with special flags in the process and a special boot switch implemented in the operating system, a process can use up to 3GB of virtual address space on 32-bit versions of Windows and 4GB of virtual address space on 64-bit versions of Windows. When data allocated to virtual address space is actively needed, it must be in RAM; however, when a process does not need the data, it can be swapped out to virtual memory. This process is why the virtual memory file is often called a swap file. Data is swapped in and out of the file as needed.

These applications run in memory, and the documents they work with traditionally load into the applications' memory space. There are exceptions to this rule, but it is the most common behavior. This rule results in a layered approach in the operating system. The documents open in the application's assigned process address space, and the applications communicate with the environmental subsystems. The subsystems communicate with the executive services as needed, and the executive services communicate with the hardware abstraction layer (HAL), kernel, and hardware via device drivers.

The subsystems allow different applications, which may have been designed for operating systems other than Windows, to run on the computer. One such application is a POSIX (portable operating system interface) solution. Of course, 16-bit Windows applications are accommodated through the Win16 subsystem (wowexec.exe), and 32-bit Windows applications run in the Win32 subsystem. These subsystems provide hooks into the operating system for various applications.

The lower level of the architecture, kernel mode, is where the majority of actual networking communications transpire. Of course, Application layer protocols such as FTP, SMTP, HTTP, and Telnet function in user mode, but the TCP, IP, and other lower-level protocols work within kernel mode.

Windows Networking Architecture   The ultimate question that must be answered in this book is, how does an operating system actually communicate on the network? Rather than speak conceptually, this chapter will actually teach you how Windows does just that. Later sections will briefly investigate how other operating systems communicate on networks as well.

The Windows operating system can be said to break the networking communications down according to the OSI model. At Layer 7, Windows runs the network applications such as FTP and HTTP. At Layers 6 and 5, the Networking APIs implement the presentation and session requirements of a networking system. In addition, a *Transport Driver Interface (TDI)* client works as a communication proxy between the user-mode networking APIs and the kernel-mode TDI. At Layers 4 and 3, the TDI transport implements TCP/IP, IPX/SPX, or the needed protocol suite. The *NDIS (Network Driver Interface Specification)* library encapsulates network drivers for NICs and provides the Layer 2 functionality. Of course, the NIC will be responsible for actually transmitting the 1's and 0's on the associated medium. Figure 4-6 depicts this networking architecture, and each of the layers is addressed in more detail in the following sections.

Networking APIs   The Windows operating systems support various Networking APIs that allow legacy and current applications to function on your network. These APIs include Winsock for TCP/IP, NetBIOS APIs for this older protocol, and Remote Procedure Call (RPC), which works over NetBIOS or Winsock. Of course, today's network applications rely more on the Winsock API in Windows, than any other networking API.

TDI Clients   TDI clients run in the kernel mode of the Windows operating systems and provide the needed kernel-mode functionality for networking APIs such as Winsock and NetBIOS. The TDI client is responsible for sending I/O request packets (IRPs) to the lower-level protocol drivers and monitoring for their completion.

FIGURE 4-6

Windows
networking
architecture

L7	• Network Applications Such as FTP and HTTP
L6	• Networking APIs
L5	• Networking APIs • TDI Clients
L4/L3	• TDI Transports
L2	• NDIS Library
L1	• Ethernet or Alternate Physical Layer

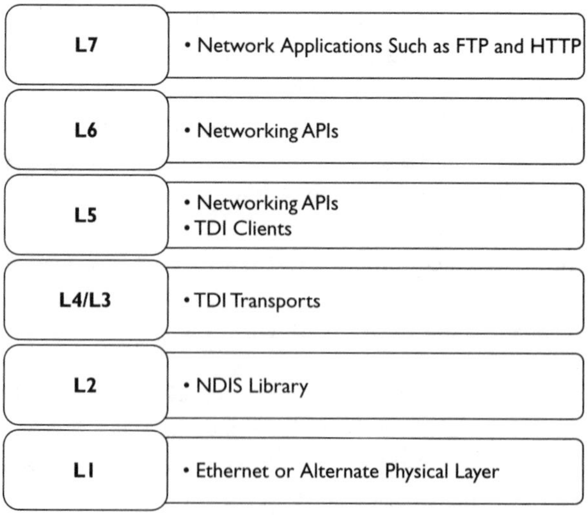

TDI Transports   TDI transports are also known as protocol drivers. These protocol drivers receive the IRPs from TDI clients, which act on the behalf of Networking APIs, and they process the requests contained in the IRPs. As an example, the TCPIP.SYS driver implements the TCP/IP functionality (TCP, IP, ARP, etc.) for Windows systems. TCPIP.SYS is a TDI transport.

NDIS Library   The Windows operating system abstracts the NIC by using the Network Driver Interface Specification (NDIS) that was originally developed in 1989 by Microsoft and 3Com. The NDIS.SYS library translates the local TDI requests into commands the NIC drivers can process, and it translates messages from the NIC drivers into messages the TDI transports can understand.

As you can see, the layered approach does introduce the ability to easily customize or change a given component without necessarily impacting any other components as long as the vertical interfaces remain the same; however, this approach may also introduce some time delays from the moment of communication request to the moment of actual transmission on the wired or wireless medium. This delay is usually so minimal that it does not cause problems, but it must not be overlooked in building networked systems that require low latency and rapid transmission of information.

Windows Networking Services

The Windows operating systems come in two general implementations: client and server. The most popular versions of the Windows clients are Windows 2000 and Windows XP, with some installations of Windows 98 and Windows ME still lingering. The reality is that most corporate networks are running Windows 2000 Professional or Windows XP at this point. Some organizations are beginning to implement Windows Vista now in early 2008, but there is still no massive movement toward this latest version of Windows.

On the server side, most enterprise servers are running either Windows Server 2003 or Windows Server 2003 R2 edition, with some still running Windows 2000 Server at this time. Windows Server 2008 is slated for release in early 2008 and will not likely see adoption until late 2008 or early 2009, and even then, the process will be slow.

The reality is that the networking services differ greatly between the client and server versions of Windows and, in fact, these services are where the greatest differences exist between the two versions. Of course, the server editions of Windows can support more processes and in some cases more RAM, but most decisions seem to come down to the need for networking services. These networking services come with Windows Server and are of great importance to TCP/IP networking:

- DNS
- DHCP
- WINS
- Internet Information Services
- Internet Authentication Services
- SMB 1.0 and 2.0

The Domain Name System (DNS) is very important, as it is the service used on the Internet for name resolution and it is more and more commonly used on private networks as well. On the Internet, it allows us to type a friendly name like www. SYSEDCO.com and be automatically directed to the associated IP address. This correlation means that I can move the web site to a new IP address and my visitors will never have to know, and it also means that it is easier to remember for people who want to come back to my company's web site. On private networks, it is used for host name resolution, but it is also very heavily used for service location. For example, Windows Active Directory Domain Controllers can be located by querying service records in a DNS server.

The Dynamic Host Configuration Protocol (DHCP) is the most widely used protocol for automatic IP address configuration both in Windows networks and otherwise. Most Internet routers that are used in home networks integrate some form of DHCP server, which just goes to show how ubiquitous the network service has become. The Windows Internet Naming Service (WINS) is, thankfully, less and less utilized today. It is a NetBIOS-based name resolution protocol and is a very congestive protocol on networks. Removing NetBIOS communications from converged networks should be a high priority. This removal means upgrading all Windows computers to Windows 2000 and higher, since earlier versions of Windows—including Windows NT Workstation and Server—relied on NetBIOS for the NT Domain name resolution.

Internet Information Services (IIS) is Microsoft's implementation of a web server. Web servers are not only used on the Internet today, but are very commonly used to build intranet applications. My company has assisted more than a dozen organizations in putting their older client-based applications online within their intranet in just the last year (2007) alone. This migration is a very popular trend today. IIS, like other web servers, must be monitored and appropriately patched as necessary in order to maintain security. The nature of this beast is that it is perpetually connected to the network and so it is continually vulnerable to new security threats. If the administrator doesn't keep up with the patches, scenarios like the Code Red and Nimda worms of the past will simply repeat themselves again and again—regardless of the web server you're running.

The Internet Authentication Service (IAS)—not to be confused with the ISA service—is a little-known implementation of a RADIUS server provided with Windows Servers. This service can provide Extensible Authentication Protocol (EAP)–based authentication so that you can implement IEEE 802.1X security if your environment demands it. It was the first step leading to Microsoft's current Network Access Protection offerings.

Finally, the SMB service or protocol is a very important part of Windows. The SMB (Server Message Block) protocol is used in file sharing and printer sharing scenarios in Windows. In other words, when you access a share on a Windows Server from a Windows client, you are using the SMB protocol. For many years, SMB 1.0 has been utilized on Microsoft networks. With the release of Windows 2008 and Windows Vista a new version, SMB 2.0, has been released. This version promises to be less chatty than SMB 1.0; however, until all machines on your network support SMB 2.0, the version that is used will be negotiated at the time of communications.

Windows Networking Tools

Like most operating systems, Windows provides a selection of tools to help you connect to, troubleshoot, and manage the network. Windows provides both GUI (graphical user interface) tools and command-line tools. The GUI tools are used mostly for configuration, like the network settings dialog shown in Figure 4-7.

In addition to the GUI interface, there are many command-line troubleshooting tools. The following list includes the most commonly used commands:

- PING
- IPCONFIG
- FTP
- NETSH
- NETSTAT

Of course, Windows is not the only operating system on our modern networks, so I will provide a brief overview of some of the ways in which other operating systems differ from the Windows systems I've covered here. Remember, this knowledge

FIGURE 4-7

Network settings
in Windows

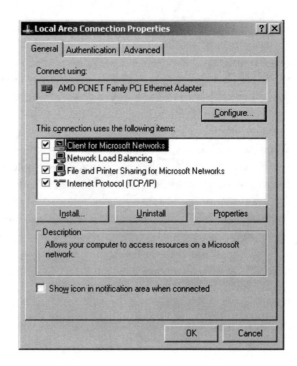

will help you understand how to reduce congestion (removing unneeded services, protocols, and features) and latency in network communications.

Linux, Mac OS, and Mobile Devices

Since the most recent versions of the Mac OS are based on BSD Unix, the networking features of these Linux and Mac operating systems can be covered together. After the unique offerings of the Linux/Unix operating systems are addressed, the mobile operating systems will be briefly covered in this section as well.

Linux operating systems come in many different distributions, including:

- Debian
- SUSE
- Red Hat
- Yellow Dog
- Others

Though there are many distributions, they share a common kernel that may have slight modifications, but does not differ greatly in its the fundamental architecture from one distribution to another. Like Windows, Linux has a kernel space and a user space. Users' applications run in the user space and device drivers run in the kernel space. The users' applications communicate with the device drivers using system calls. This behavior is all very similar to the Windows operating system.

Both the Linux distributions and the newer Mac OS X platforms share this operating system model, as they are based on a similar kernel. The Mac OS is based on the BSD Unix operating system and implements the Mac shell interface instead of such interfaces as Gnome or KDE. Network access is provided through NIC drivers that communicate with Ethernet and Wi–Fi adapters.

The greatest power of Linux/Unix systems is in their TCP/IP support. There is no debate among most IT professionals that the TCP/IP implementation in Unix environments is superior to that in Windows, and this claim should be expected, since Unix has been around for more than three decades; however, the reality is that the TCP/IP tools and features of Linux also seem to be more robust and feature-rich than those we have in Windows. For example, the troubleshooting tools are more powerful and, some would say, more easily scripted, but I suppose the latter argument is really one of familiarity, as I have no difficulty in scripting the Windows command-line tools.

Like Windows, Linux systems require each NIC to have a driver installed that knows how to communicate with that NIC. One big difference is that TCP/IP is in the kernel's memory space in Linux/Unix systems and it is not in Windows systems. Don't confuse the kernel's memory space mentioned here with the kernel mode in Windows. TCP/IP.SYS certainly resides in kernel mode in Windows, but it does not share the memory space with Windows. Many Linux/Unix professionals suggest that this is one of the major reasons the TCP/IP protocol suite is more robust than the Windows implementation.

In the end, the most important thing seems to be that your system can communicate on the network; however, the added importance of efficient communications comes into play on converged networks. The devices must not only be able to communicate, but they must be able to communicate the needed information as efficiently as possible. This leaves bandwidth for other applications and consumes no more network resources than is absolutely necessary.

When it comes to mobile devices, the TCP/IP implementation is very minimalistic. There are usually no troubleshooting tools other than a "connected" or "not connected" indicator of some kind. In order to troubleshoot the communications of these devices, you will likely need third-party software designed specifically for them or you will need to eavesdrop on the wireless communications to see what is actually taking place. The vendors have traditionally released very little information about the inner working of the TCP/IP protocol stack on these devices and instead just inform application developers how to communicate across the implemented stack. In order to discover what features they do or do not implement, one must monitor the communications and look at the actual TCP and IP headers in order to determine how they are communicating. The good news is that you will learn how to use such protocol analyzers later in this book.

CERTIFICATION SUMMARY

This chapter completes the first part of this book. The goal of these first four chapters is to introduce you to the fundamentals of networking so that you can fully understand the later chapters. Since the Convergence+ exam does not have any prerequisites, it is important that you understand these fundamental concepts. This chapter has completed this part of the book by presenting the types of clients that are used on converged networks—wired and wireless—and the way in which network operating systems often implement TCP/IP.

✓ TWO-MINUTE DRILL

Identify the Various Endpoints Used in a Converged Environment

❏ A device driver must be provided for a NIC in a desktop or laptop computer.

❏ Wired devices connect to the physical network by virtue of the real physical connection that they have to the cabling system.

❏ Wireless devices connect to the network by virtue of the association they have with an access point. This association is maintained even through the wireless client is not necessarily connected or communicating with the access point at every moment.

❏ Wired networks using switches do not have collision domains, since the connections to the switches are full-duplex connections.

❏ All networks, both wired and wireless, have broadcast domains.

Describe Networking Technologies Used in a Converged Network

❏ Both Windows and Linux/Unix operating systems divide the operating system into two modes or spaces of operation: kernel mode and user mode.

❏ Applications run in user mode, and device drivers run in kernel mode.

❏ Applications do not have direct access to the hardware, but they access the hardware through device drivers.

❏ The TCP/IP protocol suite is implemented in Windows through the TCP/IP.SYS protocol driver, and it is implemented in Linux/Unix systems as part of the kernel.

SELF TEST

The following questions will help you measure your understanding of the material presented in this chapter. Read all the choices carefully because there might be more than one correct answer. Choose all correct answers for each question.

Identify the Various Endpoints Used in a Converged Environment

1. You are communicating on a wireless network. Which medium access method is employed?
   A. CSMA/CD
   B. CSMA/CA
   C. Direct
   D. Indirect

2. The amber LED on your NIC is blinking very rapidly to the point where it is nearly lit at all times. What does this indicate?
   A. A large number of collisions
   B. Normal network activity
   C. Encrypted data traffic
   D. Unencrypted data traffic

3. When a wireless client connects to a WLAN, what are the three states of the IEEE 802.11 state machine? (Choose all that apply.)
   A. Unauthenticated/unassociated
   B. Authenticated/unassociated
   C. Authenticated/associated
   D. There is no such state machine

Describe Networking Technologies Used in a Converged Network

4. Which one of the following TCP/IP services may run on a network and provide automatic configuration of the IP protocol for devices on the network?
   A. DNS
   B. WINS
   C. DHCP
   D. IIS

5. Which of the following tools are usually found on PDAs and other mobile devices for troubleshooting and analyzing the TCP/IP protocol stack? (Choose all that apply.)

   A. PING

   B. IPCONFIG

   C. NETSTAT

   D. None of the above

6. How can you analyze the TCP/IP protocol implementation on an undocumented device, such as a PDA or mobile device, in order to discover how the protocol is implemented or what features of the protocol exist in the stack?

   A. Protocol analyzer

   B. PDA decompiler

   C. Use a Cisco switch

   D. Use a Cisco access point

SELF TEST ANSWERS

Identify the Various Endpoints Used in a Converged Environment

1. ☑ **B** is correct. Wireless local area networks use CSMA/CA for medium access.
   ☒ **A, C,** and **D** are incorrect. Ethernet networks use CSMA/CD when a collision domain exists and neither wired nor wireless networks use access methods called direct or indirect.

2. ☑ **A** is correct. When the amber light is blinking very rapidly or is lit all the time, it usually indicates a large number of collisions on the network.
   ☒ **B, C,** and **D** are incorrect.

3. ☑ **A, B,** and **C** are correct. The IEEE 802.11 state machine consists of these three states in the order presented in the answers. A machine starts in a state where it is both unauthenticated and unassociated with any access point. The first thing to happen is authentication, so the second state is logically authenticated and unassociated. Finally, the machine is associated as well, which leads to the final state of authenticated and associated.
   ☒ **D** is incorrect.

Describe Networking Technologies Used in a Converged Network

4. ☑ **C** is correct. The Dynamic Host Configuration Protocol (DHCP) is used to dynamically configure the host's IP protocol. This includes assigning an IP address, the subnet mask, the default gateway (router), and other settings like time servers.
   ☒ **A, B,** and **D** are incorrect. DNS is used to resolve domain names to IP addresses, WINS is used to resolve NetBIOS names to IP addresses, and IIS is the web server that comes with the Windows operating system.

5. ☑ **D** is correct. PDAs usually come with no troubleshooting tools for the networking features other than indicators that show a connection either does or does not exist.
   ☒ **A, B,** and **C** are incorrect. These tools are traditionally available in computer operating systems like Windows or Linux.

6. ☑ **A** is correct. A protocol analyzer is an application that can read Ethernet, Token Ring, Wi–Fi, or other frames and extract the data payloads out of them for analysis purposes.
   ☒ **B, C,** and **D** are incorrect.

5

Telephony Fundamentals

> *Mr. Watson, come here, I want you.*
>
> *–Alexander Graham Bell*

Regardless of who gets the credit, there can be no question that the invention of the telephone has changed our culture and our world. Before the telephone, all communications required either signaling methods that demanded human encoding on the transmission end and human decoding on the receiving end, or written messages that had to be transferred by humans from place to place over long distances. The telephone, for the first time, introduced the potential to speak to other humans over tremendous distances. How the telephone and telephone network performs its magic is the primary focus of this chapter.

From the first communication between Bell and Watson to the modern IP telephony networks, there are some concepts that are fundamental and unchanging. One such concept is the behavior of sound and the human voice. If you are to understand how IP telephony works, you must first understand how sound waves work and how they are converted to electrical signals. This conversion will be the first topic of this chapter. Next, you'll need to understand how a telephone takes advantage of our knowledge of sound and electricity to create a communications device. Finally, you'll need to know how these communications devices link together and how they are identified based on numbering plans. When you've completed this chapter, you'll have all of this knowledge and more.

CERTIFICATION OBJECTIVE 5.01

Define Fundamentals of Voice Systems

Whether you are implementing a traditional analog voice system or a modern packet-based voice system, you have to understand the fundamentals of voice communications. This means that you have to understand how the human voice is digitized and transmitted on these systems. You also need to know about the different types of voice systems, including legacy, hybrid, and IP telephony. Finally, you'll need a fundamental understanding of the signaling systems that are used to initiate, process, and terminate communications.

Telephones and Human Speech

Before you can understand the telephone and modern IP telephony, you must understand sound waves and the way they are generated by human beings when they talk. This knowledge will allow you to better understand the way in which telephone systems are implemented. Once you've mastered the basics of sound waves and vocal communications, you'll learn how we convert sound waves into electrical waves and back again. By the end of this section, you'll understand how the telephone cooperates with the laws of physics in order to provide us with a long-distance communications device.

A Brief Introduction to Sound Waves

In order to fully understand voice communications, you must have a basic understanding of wave theory and, specifically, of sound waves. A wave is defined as an oscillation (a back and forth motion or an up and down motion) traveling through space or matter. Another way of defining a wave is to say that it is a disturbance that travels through space or matter without permanently changing the matter or space.

For example, consider a wave in the ocean. Imagine a ball floating on the ocean surface. As the waves pass by, you'll notice that the ball does not travel with the waves. The ball travels with the current, which is another discussion altogether; however, we do see that the waves pass under the ball and that they leave it in place. Sound waves also may be said to pass through space or matter without actually displacing the matter. The sound waves travel through "it" but do not take "it" with them. The reality is that the waves may move the ball in one direction or another over time, but the ball does not latch onto a single wave and pass along with it.

The observance of waves in water was what led to the theory that sound also travels in similar wave forms. Sound waves, like ocean waves, pass through space and matter; however, they make no permanent change to the matter. When you hear a sound, you are hearing the disturbances in the air around you. Your ears are transducers that convert the air pressure changes into signals that your brain can process as sound. A microphone works in much the same way. A microphone, using varying technologies, converts the sound waves into electrical signals and acts as a transducer. A transducer is any device that converts energy from one form to another, such as sound waves into electrical signals. Amazingly, the human ear has an eardrum that processes the sound waves by vibrating small bones in the ear, and these bones convert the vibrations into what could be called electrical impulses—though we call them nerve impulses in the human body. So the ear acts as a transducer in that it converts sound waves into nerve impulses.

The process of converting sound waves into electrical signals is very important. It is what allows us to communicate over great distances. For example, imagine that you live in New York and your first child has been born. Your parents live in California and you want to tell them about it. If you were to attempt this using sound waves without any transduction, you would have to generate tremendous volume and you would cause great suffering to everyone between New York and California. In fact, these sound waves would likely become nothing more than a rumble by the time they reached California anyway. Assuming a 70-degree day with dry weather all over the country, it would take these sound waves approximately 3.5 hours to travel from New York to California. Obviously, this time would create a tremendous delay in a conversation even if it were possible.

What if we convert these sound waves to electrical signals, transmit them from New York to California as electrical signals, and then convert them back to sound waves once they arrive? This method works much better. First, there is no suffering to those in between. Second, the electrical signal can travel from New York to California in about 0.013 seconds. This is faster than the blink of an eye or a speeding bullet. This amazing result is based on the fact that light waves and electromagnetic waves travel at approximately 300,000,000 meters per second. Sound waves travel much slower. By converting the sound wave to an electrical signal, we accomplish very rapid transfer. The question becomes: How do we convert sound waves into electrical signals? To answer that, I'll begin by explaining how the human sound generation system known as the voice actually works.

The Human Instrument

The four major components of the human voice are the larynx, air, vocal tract, and articulators. Figure 5-1 is borrowed from Homer Dudley's 1940 paper, "The Carrier Nature of Speech." This diagram shows the vocal cords or larynx as a backward Pac Man–type entity. Of course, we know that they open and close more like a mouth as they let the air pass through, but they also vibrate. It is at this point that sound begins to develop as the air passes through the vocal cords.

It is important to note that the amplitude of the sound is based on the air pressure pushed from the lungs through the vocal cords. When someone yells, for example, he or she pushes harder from the diaphragm in order to expel more air from the lungs more rapidly. This action results in a sound with a greater amplitude. However, you can yell as loud as you want, but if you keep your mouth closed, the sound will be attenuated by the vocal tract and articulators so that the volume is very low in spite of the greater amplitude at the point of passage through the vocal cords.

FIGURE 5-1

The human
instrument

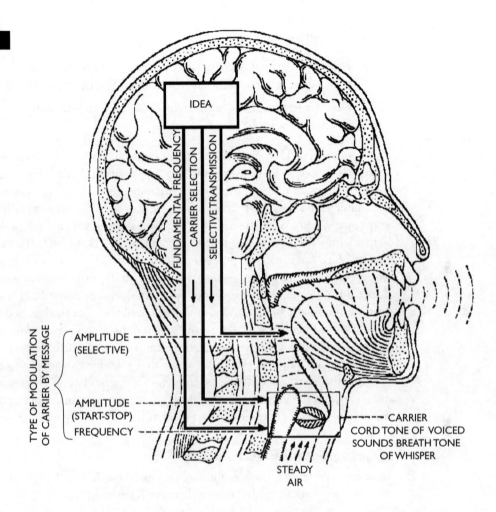

The point is simple: the end sound that is produced by the human voice is
a unique combination of the air pressure on the vocal cords, the shape of the
vocal chords, the shape of the vocal tract leading up to the articulators, and the
individual's manipulation by his or her articulators. For example, to experience the
impact of the articulators (lips, teeth, and tongue), form an "s" sound and lower
your tongue while still expelling the same air pressure from the vocal tract. You'll
notice that the "ssssss" sound quickly becomes an "uhhhh" or "ahhhh" sound. This is
because the resulting sound is greatly impacted by the articulators.

Here's an interesting exercise:

1. Form a "ck" sound while paying close attention to the position or relationship to one another of your articulators (remember, the teeth, lips, and tongue).

2. Now, try to form an "s" sound while placing your articulators in the position for the "ck" sound.

What happened? No matter how hard you tried, you could not create an "s" sound while your articulators were in the "ck" position. But why does all this movement matter to a convergence technology professional? The answer is that you must understand how the human voice produces sound in order to fully understand why voice systems work the way they do. For example, why does the telephone network only look at the frequencies from about 300 Hz to 3400 Hz? Because the human voice can usually be processed and understood by another human's ears, if we capture these sound frequencies. However, limiting the frequencies to this range does reduce the quality of the sound. Sound frequencies below 300 Hz and above 3400 Hz are simply not processed, and this removal causes the low and high sounds to be removed, resulting in the "tinny" sound of telephone communications.

You'll understand this concept better after reading the next section, "Hearing Sound Electronically." For now, just know that sound waves are produced by air pressure, the vocal cords, the vocal tract, and the articulators. The result is a disturbance in the air between the human emitting the sound and the receiving instrument, which may be human or mechanical, as you'll see in the next section.

Hearing Sound Electronically

In order to understand how sounds are processed mechanically, you should be aware of a number of terms related to sound theory. These terms are:

- Frequency
- Amplitude
- Attenuation

The *frequency* of a sound wave is the number of cycles per second that a sound wave vibrates. A higher-pitched sound or tone vibrates at a higher frequency rate, and a lower-pitched sound vibrates at a lower frequency rate. This difference is why the piano keyboard keys have higher frequency rates on higher notes, as depicted in Figure 5-2. You'll notice that a piano tuned to standard pitch has a frequency of 261 Hz for middle-C. This number simply means that the sound generated by that key on the piano vibrates or passes through a wave cycle 261 times each second.

Piano key frequencies

27 Hz 87 Hz 261 Hz 985 Hz 351 Hz

The result of this cycle is that frequency over time equals a sound. While most sounds we hear, including those produced by the human voice, are a complex combination of multiple frequencies, this concept remains true that each of those independent sounds is a vibration at a given rate within a given time frame.

A sound wave of a simple sound (a single frequency sound) can be represented with a traditional sine wave as in Figure 5-3. In this figure, the vertical scale represents the amplitude of the sound wave, and the horizontal scale represents the time. Point "A" represents the first crest of the sound wave, and point "B" represents the second crest. A complete cycle of the wave includes the first crest—the upward arch—and the first trough or the downward arch. Therefore, middle-C on a standard tuned piano would cycle through 261 of these crests and troughs in a single second.

When I grew up in West Virginia, many of my uncles and aunts played musical instruments. As far as I know, they all claimed that they "played them by ear." What did this statement mean? It meant that they reproduced the sounds that they heard in one source on their chosen instrument. For example, my uncle Bob could play most any instrument with a string on it. He could listen to a song on the radio and then play it on his guitar. He did not need sheet music telling him which chord to play when.

According to physics and human biology, what was my uncle Bob actually doing? He was hearing the sound waves, or more specifically the frequencies of the sound waves, and then reproducing similar or compatible sound waves with his guitar, banjo, or mandolin. These musicians often talked about the need to train your ears.

A single-frequency sound wave

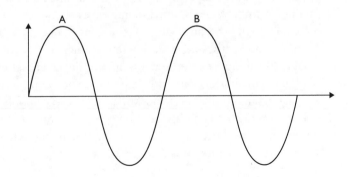

They meant that you had to learn to link a frequency from one instrument—like a trumpet—to another instrument—like a guitar. They could hear when the two sounds were compatible and did not need music notation to tell them where to place their fingers.

In a similar way mechanical devices can be tuned to accept only certain frequencies or to reject certain frequencies. They can also be programmed to compare and contrast frequencies. This latter ability is the core of many active noise-cancellation technologies in microphones and headsets.

While frequency may describe the pitch of the sound, *amplitude* determines the volume of the sound. One of my uncle Bob's favorite pastimes was that of connecting his electric guitar to an amplifier and playing his most cherished tunes. The amplifier took the electrical signal from the guitar and increased the volume as it recreated it through the built-in speaker. The guitar listened for the sound waves with built-in microphones, and these microphones converted the sound waves into electrical signals for the amplifier. This conversion is an example of active amplification. The sound wave is converted to an electrical signal, and the signal is amplified and then converted back to sound waves through the speaker.

Uncle Bob also liked to play his Gibson Hummingbird guitar. This guitar had a hollow body with a hole in the front. The sound generated by the guitar strings entered the hollow body of the guitar through the hole and echoed throughout the body before being thrust back out of the guitar. The result was a passive amplification of the sound wave. If you have access to a hollow-body guitar like this, try placing a thick piece of cardboard in front of the hole behind the strings and then strum the guitar. You'll notice that the volume is greatly diminished, and you'll experience just how powerful passive amplification of sound waves can be.

What this passive amplification really does is redirect the sound energy. Normally, sound waves travel out from the sound-emitting entity in all directions if the air or space around the emitting entity is open. By creating an echo chamber behind the strings, a hollow-body guitar takes many of the backward sound waves and thrusts them back out the front of the guitar. The result is both increased perceived amplitude (though the original sound waves are not really increased in this case) and a melodic sound (because sound at the same frequency is hitting your eardrums more than once and with somewhat different ambience).

Why is this information important? It is important when selecting headsets for VoIP implementations, and it is important when considering whether to actively amplify signals on telephone networks. When you amplify the signal at the source, you are less likely to amplify interference or environmental noises. When you amplify the signal nearer the destination, you are more likely to amplify noises in the results. The major consideration, however, will be in the microphones that you use in the headsets (if headsets are used) for your VoIP communications. Figure 5-4 shows the impact of amplification and attenuation on a simple sound wave without consideration for interference (which would effectively result in a complex sound instead of a simple sound).

When a sound wave is attenuated, the amplitude is decreased. Sound waves are attenuated as they pass through the air, and they can be heavily attenuated as they pass through more solid materials. This attenuation is why closing a door can quiet a room. The door attenuates the sound waves more than the open air. Sound waves do not travel in a vacuum, as they need a medium to disturb in order to travel. The wonderful explosions you hear in outer-space sci-fi movies would never take place, since there would be no medium through which the sound waves could travel. The old question, "If a tree falls in the woods, but no one is there to hear it, does it make a sound?" could be answered, "Yes," but the question, "If a spaceship fires a phaser blast against another spaceship, but no one is there to hear it, does it make a sound?" would be answered, "No!"

FIGURE 5-4

Sound waves with amplitude and attenuation

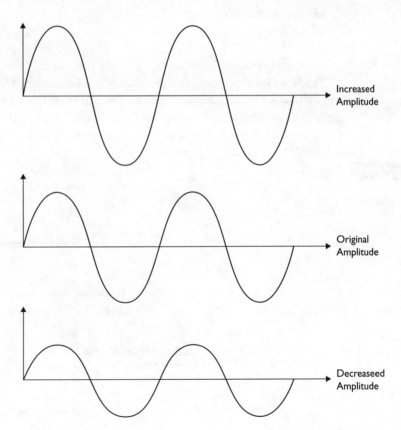

Increased Amplitude

Original Amplitude

Decreaseed Amplitude

Technically, one could argue that the tree only makes a disturbance if no one is there to hear it, but it is a compression and rarefaction wave, regardless of how it is perceived. We are now to the point where we can understand how a sound becomes an electrical signal. If sound waves are disturbances or vibrations that pass through a medium, we should be able to create mechanical devices that can detect these sound waves based on how they impact (disturb) the mechanical device. This detection is exactly what a telephone does. It detects the disturbance and converts it into an electrical signal. Let's see how the telephone works its magic.

The Telephone

The traditional telephone is composed of five major components:

- A microphone
- A speaker
- A ringer
- A switch hook
- A dialing device

These components, with the exception of the ringer, are indicated in Figure 5-5.

The first element is the microphone, or the transmitter. This component converts the sound waves into electrical signals. The way it performs this operation varies depending on the type of microphone used, but I'll explain the traditional carbon-based method that is still used in some telephones today.

FIGURE 5-5

A traditional
telephone

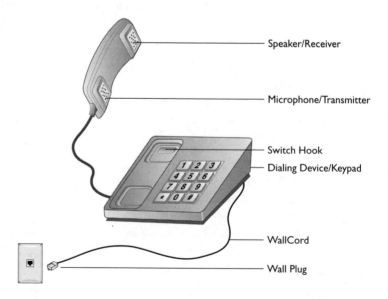

Speaker/Receiver

Microphone/Transmitter

Switch Hook

Dialing Device/Keypad

WallCord

Wall Plug

Inside the handset of the telephone is a diaphragm that is very sensitive to the changes in air pressure caused by sound waves. This diaphragm moves in relation to the sound wave. The movement of the diaphragm compresses and expands the space behind it where carbon particles are located. These carbon particles conduct electricity differently, depending on the compression caused by the diaphragm. The end result is that we can pass electrical current through the carbon particles at a constant input rate, but the output rate will vary depending on the compression caused by the sound waves hitting the diaphragm. When the output is the same as the input, we can trust that there is silence. When the output is different than the input, there is sound.

Since the transmitter (microphone) detected the sound as a vibration against the diaphragm and converted it to electrical signals, the receiver (speaker) must do the opposite. It must take the electrical signal (a varying frequency of electrical energy) and convert it back to sound waves. This task is done by passing the electrical signals to a magnet that vibrates a diaphragm at the rate represented by the electrical signals. This diaphragm creates disturbances in the air instead of being disturbed by the sound waves in the air. The result is that the human brain perceives sound entering through the ears.

If you had two phones connected directly to each other, this would be enough; however, in the real world we connect our phones to a network known as the public switched telephone network (PSTN) and we have to have a way to indicate to that network that we wish to make a call. That point is where the switch hook comes into play. The switch hook indicates to the PSTN that you are about to dial a number with the intention of creating a connection to the device represented by that number.

When your telephone detects an 80-volt alternating current (AC) at 20 Hz coming in on the phone line, it rings using whatever mechanism the phone has implemented as a ringer. Whether it plays music, says your name, or uses the traditional ring is irrelevant. The phone must simply be configured to ring when the 20 Hz signal comes in. While some systems may use different frequencies and voltages for signaling, the concept remains the same. A particular frequency indicates that the phone should ring.

The switch hook basically completes a circuit when you lift the handset, and it breaks a circuit when you replace the handset. If the circuit is closed, the PSTN detects this circuit as an active connection and the central office (CO) returns a dial tone. This connection happens so quickly that the dial tone is there seemingly instantaneously, but there is a microsecond delay while this task occurs. Remember that the electrical signals are traveling at roughly the speed of light, so the communications occur very rapidly. When you replace the handset, the circuit is broken and the CO can end the connection.

Finally, the dialing device or keypad is used to indicate to the PSTN the actual identification of the target phone that is being requested. The dialing devices in most phones today are digital, and they send signals at particular frequencies (tones) to the telephone network in order to indicate the target. This feature is why you can hear the tones when you're dialing the connection.

The modern touch-tone phones use a dual-tone multifrequency (DTMF) dialing method. Each number on the phone generates two simultaneous tones. For example, the number 1 generates both a 1209 Hz tone and a 697 Hz tone concurrently, and the number 6 generates both 1477 Hz and a 770 Hz tones concurrently.

The older pulse dialing systems used rotary dial telephones to create pulses or interruptions of the circuit in order to indicate the target device. This system worked in a simple way, but it was time consuming. You could transmit ten pulses per second, and this time meant that it would take more than five seconds to transmit a ten-digit number. This transmission didn't include the actual dial time, which added even more time to the sequence. The reality is that it took most people longer than ten seconds to dial a long-distance number. Since the other party often did not answer or was on the line, this system became frustrating to many customers. Particularly since there was no inexpensive way to store commonly dialed numbers.

Tone dialing did provide an excellent solution. You can enter the number more quickly, and you can store frequently dialed numbers for speed dial. For this reason, it has become very popular and you rarely see a rotary phone in the United States or Europe these days.

Types of Systems

There are two general categories of telephone systems: public systems and private or local systems. The public system is run by either private organizations or the government, depending on the country in which the system operates, and the private system is run by individual organizations. It is important that you understand both types. Additionally, you should understand the differences among legacy, IP telephony, and hybrid systems that may be used as private systems.

The Public Telephone System

In most cases, the public system is referred to as the public switched telephone system (PSTN), although it is also sometimes called the plain old telephone system (POTS). The major components of the PSTN are the local exchange and the PSTN interoffice network. The local exchange is often analog, and the interoffice network is mostly digital today.

INSIDE THE EXAM

How Does a Telephone Indicate the Target for a Connection?

The answer to this question varies, depending on the type of telephone. A rotary telephone uses pulses, a touch-tone telephone uses frequency combinations, a cell phone uses digital packets, and an IP phone may use IP addresses when connecting to another IP phone. The rotary telephones worked by breaking and completing the circuit a specified number of times for each digit in the target address or target phone number. I won't go into more detail about them here, since they are being used less and less.

The touch tone telephones use the dual-tone multifrequency (DTMF) tones over the analog network. The DTMF tones are combinations of tones or frequencies. The combinations are as follows:

- 1 = 1209 Hz and 697 Hz
- 2 = 1336 Hz and 697 Hz
- 3 = 1477 Hz and 697 Hz
- 4 = 1209 Hz and 770 Hz
- 5 = 1336 Hz and 770 Hz
- 6 = 1477 Hz and 770 Hz
- 7 = 1209 Hz and 852 Hz
- 8 = 1336 Hz and 852 Hz
- 9 = 1477 Hz and 852 Hz
- * = 1209 Hz and 941 Hz
- 0 = 1336 Hz and 941 Hz
- # = 1477 Hz and 941 Hz

As you can see, the top row of digits on a standard dial pad all use the same horizontal frequency of 697 Hz. Additionally, the left column of digits all use the same vertical frequency of 1209 Hz. The intersection of 1209 Hz and 697 Hz is the digit 1, and the intersection of 1209 Hz and 941 Hz is the asterisk.

When you press the button for the asterisk, the tones at frequency 1209 and 941 are both generated simultaneously. The key is that only seven tones are needed to represent all 12 buttons, since the tones are paired with each other in horizontal and vertical fashion. Additionally, the tones that are used rarely if ever occur in nature as pure tones. Remember that the human voice usually operates from 50–100 Hz on the low end to 4000–5000 Hz on the high end, but the sounds emitted are complex combinations of frequencies. They are not pure tones perfectly placed on a unique frequency.

In the end, the CO detects these tones as indicating the phone number of the target node on the PSTN. Once enough digits are received to place a call (possibly seven for local calls in the US, or ten for long distance), the CO switch can use the SS7 network to determine the location of and route to the destination.

Local Exchange   The *local exchange* includes everything from the customer location to the CO. This service includes the cable plant and the cable vault at the CO. The phrase *cable plant* is usually used to reference the cabling or aerial connections, the structures that connect or link them, and any other components involved in connecting customers to the PSTN. The network interface device (NID) is that box on the side of your house that connects to the cable plant. In fact, that box is owned by the local exchange in most cases. Of course, the NID is much more complex and supports more lines in a business setting, but the concept is the same.

The wires from the NID to the cable plant will run either underground or, more commonly, overhead to a telephone pole. The wires from all the houses come together in terminal boxes and then run back to the CO. Along the way, the wires are spliced together as needed using what are called splice cans. These are simply containers that protect the splice points from exposure to weather.

In a commercial installation for a business or organization, the wires come into the customer's building and are usually connected directly to the organization's telephone switch (PBX system), which will be discussed in more detail in the next chapter.

The CO is where all the cable plants come together. The wires come into the CO through the cable vault. The cable vault is where the power systems or batteries are located that allow the local exchange to operate. For example, they provide the power for signaling on the network.

To help you understand how all these components work within the local exchange, consider the flow of a call placed between two telephones connected to the same local exchange. When you pick up the handset and dial the number, the switch determines if that number is in the local exchange or not. In this case, it is in the local exchange, so the switch sends the ring signal to the target telephone. Assuming the target telephone is answered, the switch creates a virtual circuit between the two nodes. At this point, voice communications begin. As long as both nodes keep the circuit open, the switch maintains the circuit. In most modern networks, either node can break the circuit and thereby break the connection.

In addition to intra-exchange calls like the one highlighted in the preceding paragraph, you can also place inter-exchange calls. These calls pass from one local exchange to another. These calls may be handled by multi-exchange switches, or they may require communications between different COs. Either way, the difference is that there will now be a route that your voice communications must travel that is usually more complex than the intra-exchange phone call. The voice signal, which is analog on the local exchange, may be converted to a digital signal in order to traverse the PSTN network. It will then have to be converted back to an analog signal when it arrives at the remove local exchange. This conversion takes extra time, but due to powerful processors and other modern technologies, the latency is usually so low that the delay is not noticed by the communicating parties.

This latency may be greater with a cell tower, since there are more conversions involved. To see this latency in action, you can call yourself using your cell phone and your land line. If you do this test, you'll be able to perceive the delay that occurs from the time you speak something into your cell phone and the time you hear it in your land line. This delay, again, is mostly because of the conversions that must take place along the way and less because of the speed of the actual signals. The signals move at the speed of light, but the conversion and queuing systems are not quite so fast yet.

Interoffice Switching   As was stated previously, most connections from the CO to the subscriber are analog; however, the connections between COs are mostly digital today. These CO-to-CO networks are built on optical fiber lines in a ring structure. Calls are passed from CO to CO through switches that are connected to the fiber ring. These switches are connected via trunk units (TUs) to the ring. When a call is connected from one CO to another, the virtual circuit is created from the subscriber to the CO, from the CO to the ring, from the ring to the destination CO, and finally from the destination CO to the destination subscriber.

The switches at the CO and the TUs use the Signaling System 7 (SS7) system in order to discover which network should be used based on the subscriber's long-distance provider selection and the needed path to the destination subscriber.

INSIDE THE EXAM

How Does a Call Actually Work Across the PSTN?

Let's walk through a telephone call placed across the PSTN from start to finish. The first step in placing a call is the creation of an off hook condition. This means that the telephone handset is lifted. When you lift the handset, the circuit of the local loop with the CO is closed and current flows through your local loop. The CO switch detects this current almost immediately and returns a dial tone (this tone is an electrical signal that is converted to sound waves by the receiver or speaker in your telephone) on your line.

In actuality, the CO switch may perform line tests before sending back the dial tone in order to ensure that the line is in a condition that will support a voice call. This test happens very quickly, though, and you will rarely if ever pick up a phone that is operational and connected to the local loop and not receive a dial tone immediately.

Now that you have a dial tone, you can begin entering the phone number for the node you wish to call. As you press the buttons, the DTMF signals are sent across the local

INSIDE THE EXAM

loop back to the CO switch. Keep in mind that these signals are sound waves in your earpiece, but they are electrical signals sent to the switch. The switch disables the dial tone as soon as it detects the first signal indicating that you are dialing a number. This action is why the dial tone goes away when you begin the actual dialing process.

Modern CO switches are very intelligent and may interrupt the dialing process if they detect an error. For example, if I am dialing from area code 937 to area code 614 and I must dial a 1 before the ten-digit phone number, the switch may detect that I've dialed 614-555-12 and immediately understand that I've dialed incorrectly. It can send a recorded message to my phone indicating that I need to dial a 1 before a long-distance number. If I've dialed the number correctly, however, the next step for the CO switch is to determine how it will create the connection to the target node on the PSTN.

From here, the call proceeds differently depending on whether the target phone is connected to the same switch or a different switch. If it is connected to the same switch, the switch simply connects the call and reports to an administrative module that the call has been placed in case billing must

be performed. If the call is targeting a phone connected to another switch at another CO, the call is routed through a tandem switch to the appropriate switch at the other CO, which connects the call to the end node.

Finally, if the call is targeting a phone in a completely different region, such as another state or country, the call is routed up to higher-level COs until eventually a route is determined for the connection. This routing all happens in a matter of a few seconds at most, and once the circuit is established, the voice communications can travel across it very quickly (remember, the signals travel at roughly the speed of light).

Before any voice communications travel across the circuit, the target phone must be answered. This task means that the target human (the person who owns or answers that phone) must be notified that a call is coming in. In order to notify the target of a call request, the CO sends a ring signal to the phone, and it sends a ring sound back to the calling receiver. Assuming the person on the target end picks up the phone, the off hook condition exists and voice communications can ensue. Of course, the traditional "hello" message is optional from the target to the initiator, but it is quite courteous.

Private Telephone Systems

When an organizations desires to implement a private telephone system, they will implement a *private branch exchange (PBX)*. PBX systems can be hardware devices that operate as stand-alone equipment. They can be software applications that run on computers and control hardware, and they can be IP-based PBX systems. I'll cover each of these in more detail in Chapter 6.

When an organization implements a PBX, they must connect to the PSTN in some way. These connection points to the PSTN or CO are called *trunks*. The trunks are typically either analog FXO lines or digital T1 or ISDN lines and can be configured to supporting incoming, outgoing, or two-way calls. The number of incoming trunks needed will depend on the number of lines that must be supported. For example, a telemarketing company would need many more outbound lines than a customer service department; however, the customer service department would likely need more incoming lines than the telemarketing company.

The good news is that a single inbound or two-way line can usually support multiple actual internal phones. This support is possible because users are not on their phones all the time (with the possible exception of the telemarketing company mentioned previously). For example, you may have 1000 users at a location that have telephones in their offices. However, you may determine that at any given time only 125 of them are using the telephone. You may further determine that a peak call volume would be around 400 active connections. This ratio means that you do not need the bandwidth capabilities between your PBX and the CO to support 1000 concurrent calls. You only need to support 400 concurrent calls.

It's also important to keep in mind that you will not have to have bandwidth to the CO for internal calls. Calls between internal users (those among the 1000 in this case) can be routed internally by the PBX. In Chapter 6, you'll learn more of the details about how these PBX systems actually work.

Voice Transmission

In order to transmit voice across the PSTN, there must be methods for setting up calls, routing calls, and managing calls. These methods are known as *signaling*. There are two primary signaling areas you need to be aware of: the local loop and the digital PSTN network. The local loop is the connection between you and the CO. The digital PSTN network uses the SS7 signaling system for communications. I'll provide you with an overview of both here.

Local Loop Signaling

Earlier in this chapter you learned about the basic process used to create a phone call. You learned about the closing of the circuit to create an off hook condition. This circuit is the first signal used to establish a phone call. The closing of the circuit signals the CO switch that you are about to attempt a call. The switch, in turn, sends back the dial tone, which could be called a signal, since it does indicate to you that you can begin dialing the target phone number.

In addition to this step, you'll remember that the dial pad or keypad sends DTMF signals back to the CO switch in order to connect to the target phone. The switch may send the ring signal to the target, if the target is connected to the same switch. The switch may also send the ringing sound back to the dialing phone. These are all signals that happen in the local loop.

SS7

When you need to connect to a telephone outside of your switch and certainly when you need to connect to one outside of your region, the PSTN uses the Signaling System 7 (SS7) switching standard that was developed in 1981 by the ITU-T, formerly known as the CCITT. SS7 is an out-of-band signaling architecture. This architecture means that the signals are sent in a different frequency range than the voice data. It can also mean that the signals are sent on a different wire than the voice data. In-band signaling, like the local loop signaling, is actually sent using the limited available bandwidth in the channel. In the case of the SS7 system, a separate line is used or an entirely separate network is used.

e x a m

ⓦ **a t c h**
Remember that SS7 is used between network elements in the PSTN and not between your telephone and the CO. Remember that SS7 uses out-of-band signaling in that it sends the signals on one channel and routes the call on another. Finally, don't forget that SS7 is digital and packet-based.

In North America, a unique architecture has been developed for SS7 signals. This architecture is known as the North American Signaling Architecture. The North American Signaling Architecture stipulates a separate signaling network. The network is built out of the following three components and is connected through signaling links:

- Signal switching points (SSPs)
- Signal transfer points (STPs)
- Signal control points (SCPs)

FIGURE 5-6

Call routing
through the
SS7 network

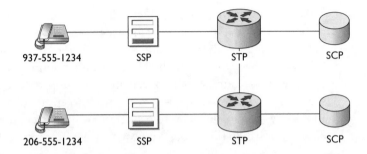

| 937-555-1234 | SSP | STP | SCP |

| 206-555-1234 | SSP | STP | SCP |

The SSPs are responsible for the communications between the CO and the SS7 network. The STPs are the packer routers that allow SSPs to communicate with each other. Finally, the SCPs are basically databases of call processing features and capabilities. As with the Internet, in order for the network to function, there must be redundancy. For this reason, the hardware that acts as the STPs and SCPs is redundant and therefore fault tolerant.

In order to understand how these different SS7 components interoperate, consider the diagram in Figure 5-6. As you can see, local loops are connected in to the SSP, which is a CO switch capable of talking to the STPs in the SS7 network. When the user in area code 937 attempts to call the user in area code 206, the call is routed through the SS7 network.

PSTN Calling Features

In North America and most other parts of the World, the telephone networks offer many calling features or line features. These features include call waiting, conference calling, custom ringing, caller ID, and blocking.

Call Waiting

The *call waiting* feature that is available in most areas now provides you with the ability to be notified when a call comes in while you are on the line. If you are currently connected to someone's phone and are involved in a phone conversation and another call comes in destined for your line, the CO switch sends a tone on your line that you can hear. This tone indicates that another call is coming in. If you want to take the call, you can press a particular button or take the specified action that sends the signal to the CO switch to allow the call in. Most of these services allow you to switch to the incoming call and then switch back to the original call.

When call waiting is enabled on a line, it can usually be turned off before dialing a phone number. This task is usually accomplished in the U.S. with the *70 prefix. When you pick up the phone and receive a dial tone, if you dial *70, it tells the CO switch that you want to disable call waiting during the call that you are about to make. This feature is useful if you are about to participate in a conference call or some other communication where you want to ensure that you are not interrupted. In the days of using modems to connect to the Internet, it was very helpful in ensuring you didn't lose your connection when someone called you.

Three-Way Calling and Conference Calling

A popular feature in small businesses and home offices is *conference calling*. This feature was called three-way calling early on because it allowed you to create a conference among three different telephone connections; however, the technology evolved to allow for more than three connections, and many telephone companies provide conference calling as an optional or included part of the package you get when you become a subscriber. Three-way and conference calls effectively create a private circuit network among the involved subscribers and broadcast the audio signals from all subscribers to all subscribers.

Custom Rings

In order to allow subscribers to have multiple phone numbers without the need for multiple lines, many service providers offer *custom rings* or distinctive rings. This feature causes the phone to ring differently depending on the incoming number dialed. For example, you could have both 555-1234 and 555-4321 as valid numbers for a single land line. When someone calls your land line by using the number 555-1234, the phone may ring twice quickly and then pause before ringing twice quickly again and so on. When someone calls the 555-4321 number, the phone may simply ring once and then pause as a normal phone ring does. You will know which number was dialed from the ring. This difference allows you to answer the phone as one logical entity when it rings one way and as another logical entity when it rings another way.

Caller ID

You are able to identify the phone number and usually the name of the calling party when you subscribe to the *caller ID* feature. If you've been around long enough to have used a 1200-baud modem, then you've experienced the technology that allows caller ID to work. Telephone line modems use an encoding mechanism known as frequency shift keying (FSK) where one frequency represents a 1 and another frequency represents a 0. This allows the modems to transmit any binary or digital data they desire. The caller ID technology uses exactly the same FSK algorithm as the old 1200-baud modems. This message that is encoded with FSK is sent between the first and second rings. This placement is why you will usually notice that the caller ID information does not show up until shortly before the second ring or sometimes it appears to show up at the same time as the second ring. In order for caller ID to work across different carriers, the carriers must be connected by SS7. Luckily, the vast majority of the telephone service providers are connected this way. Caller ID is also known as Calling Number Delivery (CND).

Blocking

There are really two kinds of blocking that are important for you to understand. The first is *caller ID blocking,* and the second is *number blocking.* The first protects your location identity when you are the caller, and the second protects you from calls that originate at a specific location or number.

Caller ID blocking, like disabling call waiting, is achieved by dialing a special code assigned by your telephone service provider before dialing the target phone number. This code is usually *67. Effectively, caller ID blocking simply does not allow the phone number to be sent by the modem technology that was described in the preceding section on caller ID.

With the number blocking feature, you can specify that you do not want to receive calls from particular phone numbers. When individuals at the specified numbers attempt to call you, your phone does not ring and instead they receive a recorded message in most cases. In addition to this service, many telephone service providers now offer a *privacy manager* service. This service will give you the option to disallow any unknown, anonymous, or blocked calls. Instead, the caller will receive a message telling them to unblock their call if they want to get through. It may also give them the opportunity to enter a special access code to get through anyway.

CERTIFICATION OBJECTIVE 5.02

Describe Components of Number and Dialing Plans

With a little thought, it becomes clear that there must be some structure to the numbering plans used on the PSTN. If not, numbers would be reused and the PSTN would be unable to route calls appropriately.

North American Numbering Plan

The North American Numbering Plan Administration (NANPA) is responsible for managing the phone numbers available in North America. I'll briefly cover their numbering plan as an example of how one might work. The North American Numbering Plan (NANP) consists of Numbering Plan Areas (NPAs), Central Office Codes (COCs), and Caller Identification Codes (CICs).

The NPAs, more commonly known as area codes, are three digits in length. There are specific area codes that are not allowed to be used for general purposes, and they are as follows:

- **N11** The N11 area code or prefix is set aside for specific purposes. Certainly the most well-known implementation in the U.S. is the 911 code. In the N11 code, N can be any number from 2 to 9. The FCC only recognizes 211, 311, 511, 711, 811, and 911, but 411 has been implemented as directory information and 611 is usually implemented as a repair request line. It should be noted that 611 is specific to the carrier that owns the circuit you are using to connect to the PSTN and will connect you to their service center.

- **N9X** The N9X code is reserved for future expansion. The N can be any number from 2 to 9 and the X can be any number from 0 to 9. Of course, the 9 is only a 9. This is why you never see area codes like 497 or 598.

- **37X and 96X** These two blocks of ten NPAs are set aside in case they are needed for some undefined reason in the future. The X, again, can be any number from 0 to 9.

COCs take on the format of NXX. This means that a COC can have any number from 2 to 9 in the first position and then any number from 0 to 9 in the second and third positions. This means that 218 is a potentially valid COC, but 187 is not. The COC is also often called the local exchange number.

Finally, the CICs are the last four digits of the phone number and can contain any numbers from 0 to 9. Therefore, the format for the NANP numbering plan is NPA-NXX-XXXX, where NPA is a valid code based on the constraints specified earlier.

In some parts of North American, you may be required to enter the NPA even for local calls. You will usually not be required to include the country code in these scenarios. The country code (CC) for the U.S. and Canada is 1, and the CC for Mexico is 011. In other areas of the country, the NPA may be added for you automatically. This feature is known as *digit translation*. Digit translation alters dialed numbers according to predefined rules and is mostly used to convert internal numbers on private networks to compatible external numbers on the PSTN.

Toll Fraud

Toll fraud occurs when someone successfully places a billable call and accomplishes charging it to your account instead of theirs. You could also refer to it as service theft. In the early days of the phone networks, there were tone generators that could be used to perform tool fraud. Today's networks are too complicated to fall prey to such simple methods, but attackers are continually seeking ways to get services for free.

Emergency Services

According to federal law in the United States, the 911 NPA is set aside for use by local emergency services. The PSTN must redirect all 911 calls to the nearest emergency services center. This emergency services center is known as a public safety answering point. The PSTN will determine the location of the caller by identifying the number from which the call is placed. Once the caller is identified, the 911 call can be routed to the nearest public safety answering point. When on a cell phone, the cell provider uses either GPS chips or triangulation based on cell towers to locate a 911 caller and then direct them to the nearest emergency center.

Not all countries use 911 as the emergency services number. Other countries may even use 911 as a valid area code. Following is a list of emergency numbers in some other countries:

- **Australia** 000
- **European Union** 112
- **Britain** 999
- **Zimbabwe** 995 (police) or 999 (all emergencies)
- **China** 110 (police), 120 (medical), 122 (traffic accident), or 119 (fire)

CERTIFICATION SUMMARY

This chapter began peeling back the layers of telephony. You learned about the basic components of the PSTN, including the CO and the SS7 network. You also learned that a PBX can implement the same fundamental capabilities of the PSTN for private networks. In Chapter 6, you'll learn more details about the functionality of a PBX.

You also learned about the numbering plans that are used in order to ensure uniqueness in phone numbers and that allow for communications across national boundaries. You learned about Numbering Plan Areas, Central Office Codes, and Caller Identification Codes.

This chapter also introduced you to some of the basic features offered by the PSTN. Most of these features are also available in PBX systems, as you'll learn in the next chapter. Finally, this chapter began by introducing you to the basic concepts of how voice communications work.

✓ TWO-MINUTE DRILL

Define Fundamentals of Voice Systems

❏ A sound wave is an air pressure disturbance that can be detected by mechanical devices.

❏ Sound waves are converted to electrical signals for transmission across the PSTN.

❏ The public switched telephone network (PSTN) is composed of the local loop, the central office, and the SS7 network.

❏ The SS7 network allows for out-of-band signaling and is a packet-based network.

Describe Components of Number and Dialing Plans

❏ What we call an area code is called a Numbering Plan Area (NPA) in the North American Numbering Plan (NANP).

❏ The local exchange is referred to as the Central Office Code (COC).

❏ The individual phone numbers are called the Caller Identification Codes (CICs).

❏ The format of the NANP is NPA-NXX-XXXX.

SELF TEST

The following questions will help you measure your understanding of the material presented in this chapter. Read all the choices carefully because there might be more than one correct answer. Choose all correct answers for each question.

Define Fundamentals of Voice Systems

1. You have created an off hook condition. What action have you most likely taken?
   A. Disconnected the telephone cord from the wall plug.
   B. Lifted the handset.
   C. Dialed 911.
   D. Dialed 411.

2. When the remote telephone you are dialing rings, the SS7 network sends the actual sound of the ringing phone back to your telephone earpiece. True or False?
   A. True
   B. False

3. When your phone rings, you notice that the caller's ID is not showing on the display. Which of the following are likely causes? (Choose all that apply.)
   A. The call is coming from an unsupported area.
   B. The caller blocked caller ID.
   C. The caller ID modem cannot connect.
   D. The SS7 network does not support caller ID.

Describe Components of Number and Dialing Plans

4. Which of the following are actual portions of the telephone number as defined in the North American Numbering Plan?
   A. Number Plan Area
   B. Caller Identification Code
   C. Caller ID
   D. Central Office Code

5. Which one of the following could not be a valid telephone number under the North American Numbering Plan?

    A. 387-537-1214

    B. 964-432-0010

    C. 304-863-9351

    D. 937-265-8707

SELF TEST ANSWERS

Define Fundamentals of Voice Systems

1. ☑ **B** is correct. When you lift the handset you cause an off hook condition. This means that the circuit is closed and the CO will assume you are about to make a call.
   ☒ **A, C,** and **D** are incorrect.

2. ☑ **B** is correct. It is not true that the actual ring of the remote phone is sent back to your phone. Instead, your CO switch sends a ringing sound back to your phone.
   ☒ **A** is incorrect.

3. ☑ **A** and **B** are correct. The caller ID may not show up if the caller's CO is not connected to your CO with the proper technology. Also, the caller ID may not show up if the caller blocks caller ID.
   ☒ **C** and **D** are incorrect.

Describe Components of Number and Dialing Plans

4. ☑ **A, B,** and **D** are correct. These three are the core components of the North American Numbering Plan.
   ☒ **C** is incorrect.

5. ☑ **B** is correct. The problem with this number is in the NPA. The NPA 96X is set aside for any future needs where a sequence of ten NPAs may be needed.
   ☒ **A, C,** and **D** are incorrect. These are all valid numbers.

6

Telephony Hardware

> *The Linux philosophy is 'Laugh in the face of danger'. Oops. Wrong One. 'Do it yourself'.*
> *Yes, that's it.*
>
> *—Linus Torvalds*

W hile this is a brief chapter, it is very important. This chapter introduces you first to the traditional PBX systems we've used in enterprise networks for years and then to the modern hybrid and fully IP PBX solutions. Understanding the different components that make up a PBX system is a first step to understanding modern converged networks. The PBX solutions allow you to, in the words of Linus, "Do it yourself."

PBX Systems

The private branch exchange (PBX) is, in large part, to telephony what Network Address Translation is to IP networking. While they are not exact equivalents, the PBX systems do allow you to run a private network of telephones with features similar to those of the PSTN or POTS networks. You will need to understand the basic features and components of TDM, IP, and Hybrid PBX systems for the Convergence+ examination.

In addition to the PBX systems, another telephone system type known as a key system is available. This telephone solution is not covered in the Convergence+ objectives and will not be covered here in detail. A key system uses telephones with multiple buttons, or "keys," and lights that indicate which lines are in use. Someone who wants to place a call just presses a button to select a line and begin dialing. Each line will connect the user directly with the telephone company's central office for a dial tone. The Hybrid PBX solution referenced in Convergence+ refers to a TDM and IP combined PBX and not to a key system/PBX combination. Table 6-1 provides a comparison of PBX systems and key systems.

TABLE 6-1	PBX Compared to Key Systems

PBX Systems	Key Systems
Digital or analog telephones	Digital or analog telephones
Used in larger organizations (>50 users)	Used in smaller organizations (<50 users)
Works like a CO switch	Shared lines
Key code for outside line (like 9)	Uses the line button for outside lines
Provides advanced user features (call forwarding, conferencing, contact center)	Provides basic user features (call hold, Music on Hold, intercom)

TDM

Time-division multiplexing (TDM) refers to the signaling and transmission method used in the phone system. Traditional PBX units use TDM for the transmission of voice and signals and interconnect with the PSTN using the TDM solution. These TDM-only PBX systems are often called legacy PBX systems today; however, there are still many thousands of them in production environments, though the sales of new legacy PBX systems have dropped significantly over the past decade.

TDM requires the use of specialized switches that route calls and are designed specifically for telephony purposes. TDM is a process that provides the appearance of running multiple signals on a single channel by segmenting the different signals and sending the segments in short bursts. By giving some time to each of the signals (time division), TDM can provide for both the transfer of the voice communications and the signaling related to those communications across one line (multiplexing).

At one end of the link is a multiplexer, which is the device that joins the signals before sending them as segments across the TDM link. In addition to signaling, the TDM link may contain voice communications from multiple conversations. The signal that is sent across the link is sometimes called a composite signal because it is a composition of multiple original signals. On the receiving end of the link, a demultiplexer will separate the signals into their original individual signals based on timing. In reality, each end of the link will have a multiplexer and a demultiplexer to provide two-way communications.

TDM links are usually implemented as T-lines or E-lines and allow for multiple calls to be placed across the link simultaneously. Traditional or legacy PBX systems use these T- and E-lines to link to the PSTN and use TDM for communications across the links. Internally, direct dedicated lines usually run from the endpoints (telephones) to the central PBX, and the TDM communications occur across the trunk lines that are either connected to the PSTN for telephone calls outside the organization's network or are connected to other PBX systems within the organization for internal telephone calls.

TDM has also been called *time-domain multiplexing* because it multiplexes within the time domain, whereas frequency-division multiplexing, also called frequency-domain multiplexing, multiplexes in the frequency domain.

You'll remember, from the last chapter, that the central office (CO) has a connection to the larger network of telephony devices around the world. The TDM PBX or legacy PBX is used to connect to the traditional telephone network. It is connected through a CO trunk, and the PBX system acts much like the CO for your company while connecting to the CO for outside communications. Each phone or endpoint device is connected to the PBX and has its own ID or extension. In most cases, these internal phones will be digital phones using 1's and 0's for communications with the PBX.

However, if one of the phones dials the number indicating the need for an outside line (frequently the 9 digit), the PBX will grab an available line and return the dial tone to the user again or the user will simply be expected to continue dialing. If one user within the organization wants to communicate with another user within the organization, the connection can be made through the PBX and will not require an outside line.

You can also connect PBX systems in different offices using what are called *tie lines*. By connecting PBX systems with tie lines, you enable a user in one city to call a user in another city by simply dialing that user's internal PBX-assigned extension. These tie lines usually route through the CO and may interconnect through different providers, but they provide dedicated channels for direct communications between the PBX systems at the various locations.

IP Only

Many modern PBX systems are IP only. These systems convert all voice signals to IP packets and route them through optimized IP networks. They usually provide the same traditional features such as call waiting and voice mail that the TDM PBX systems have, and they usually provide enhanced features as well. Many service providers now support pure IP routing of voice packets, enabling the entire enterprise infrastructure to use IP telephony while only converting calls that must be routed across the PSTN to traditional voice calls. The codecs and QoS demands of IP PBX systems will be discussed in Chapters 7 and 9.

Hybrid

By combining TDM and IP, you can create a hybrid PBX system. The vast majority of hardware devices on the market today are hybrid PBX systems. It is becoming increasingly difficult to locate a pure TDM PBX, or at least one that cannot be expanded to support IP telephony. Companies like Avaya and Nortel have protected customer investments by providing upgrades to hybrid systems through new line cards. This upgrade option prevents the need for forklift implementations (a forklift upgrade or implementation occurs when the entire system must be replaced).

Hybrid PBX systems usually support connecting to the PSTN through both TDM trunks and lines dedicated to running IP telephony. These IP telephony lines may be used to connect different sites of the same enterprise. Ultimately, a hybrid PBX system is a traditional TDM PBX system with a Voice over IP gateway that provides for the IP telephony support, or the system uses a special processor card instead of the TDM processor. The special processor card allows the phones to run as traditional digital sets, but the brains or intelligence of the PBX is now based on IP instead of traditional telephony.

Common PBX Features

There are features that are common to most PBX systems, and then there are a few features, or ways in which the features are implemented, that help to differentiate one vendor's offering from another's. The following list includes the most common features offered by PBX solution providers.

- Auto attendant
- Automated directory assistance
- Automatic ring back

- Call accounting
- Call forwarding
- Call through
- Call transfer
- Call waiting
- Call return/call back
- Conference calling
- Extension dialing
- Follow-me
- Music on Hold (MOH)
- Voice mail
- Voice recognition

Auto Attendant

The auto attendant feature provides the capability to have a single number that outside callers dial and then allow those callers to choose from a predefined selection of options. The caller may choose to enter the extension of the target individual and be connected directly to that extension. If the PBX system supports directory services, the caller may choose to look up an individual's or department's extension in the directory. There may also be other options, such as listening to prerecorded messages and so on.

Automated Directory Assistance

When a caller doesn't know a specific extension number, he or she can be given access to a company directory. This service is usually a dial-by-name directory where the customer begins entering the letters in the target's name until a match is found or until a sufficient limitation is imposed that results in fewer potential targets than some upper threshold. For example, if you have five people with the last name of Carpenter and six people with the last name of Carrel, this would result in eleven people matching the first three letters of "Car" as they are keyed in. As soon as the letter p is added to the sequence, the total number has been limited to five people. We can now assign each of the five Carpenters to the keypad numbers 1 through 5 and ask the caller to select the intended target from that list. The caller may also be informed of the actual extension, once the name is selected, for future reference.

A few PBX systems will handle this task automatically. When a user sets up the telephone for the first time, he or she is asked to key in his name and record an audio file that contains his or her name. This way, when individuals call in and attempt to spell a name, the user's name will be in the list. The caller will actually hear the user stating his name instead of some electronically generated voice. It is much more pleasant, and the sound of his or her voice is probably known to the caller.

Automatic Ring Back

Automatic ring back is the answer to that common dilemma of trying to reach an individual who is on his or her line. Instead of trying to call again and again, you can simply choose to have the PBX system ring you back when the line become free. This service saves a tremendous amount of time and greatly increases user productivity.

Call Accounting

It is very important to many departments that they have the ability to track the actual time spent on the telephone. Sales managers like to see how many calls their salespeople make. Customer service managers like to see how many calls their customer service representatives take. Every organization needs to be able to perform cost management by watching the calls that are placed and ensuring that the telephone system is not being abused. For example, I remember working in one organization in the 1990s where an individual, who was from another country, was placing long-distance calls back home that were costing the company more than $300 every month. This example is just one reason that call accounting is very important. It can also be used to audit the service provider to ensure that you are not being overcharged for the services to which you are subscribed.

Call Forwarding

Call forwarding, just like in our home telephone systems, is used to forward a call that comes into a particular extension to another extension. For example, you may need to work outside of your office for some period of time. You may have a land line in your office, but you may also have a mobile VoWLAN phone. You could have the system forward all calls placed to your land line over to your VoWLAN phone with this feature. It is also very common to forward a phone to a conference center when you are in meetings for long periods of time and you are expecting an important call to come in during that time.

Call Through

Some organizations provide the call through feature to traveling sales people and others whom they want to be able to place calls by first dialing into the PBX. A toll-free number will frequently be used for this purpose. So a traveling employee can call in to the 1-800 number and then enter an authorization code that gives her an outside line. Now she may call any number that is allowed by the PBX. The end result is that she is calling through the PBX.

Call Transfer

I'm sure you've experienced receiving that call that was not intended for you. In the early days of PBX systems, you often had to ask the caller to call back in order to reach the proper extension; however, most modern PBX systems will allow you to transfer the call (though, in my experience, the average user is not very good at doing it, at least not if the times I've been "hung up on" are any indication).

Call Waiting

This PBX feature is much like the call waiting feature you have at home. When you are on the line, the telephone will provide you with a specified tone in order to indicate that another call is coming in. In addition, most phone systems will also show the telephone number and any other caller ID information on the telephone's LCD so that you can make a better decision as to whether you should answer the call or not.

Call Return/Call Back

This feature allows you to redial the number that last called your extension. This may work both internally and externally, or it may only work internally, depending on the implementation method. This is very useful if you've just received a message from someone who asked you to call back, but did not leave his or her number. Many phone systems will also allow you to scroll back through the list of previous calls, but this is usually a feature of the telephone endpoint and not the PBX system itself. This is why you have this feature in your caller ID on your home phone. The CO is not keeping this list for you. The list is stored in the memory of your telephone. This is also why the list will disappear if your telephone loses power.

Conference Calling

The reality is that we can rarely accomplish big things in business with just two people involved. Sometimes it's easier to get everyone involved in a problem or a solution on the telephone at the same time. This is particularly true when they are located in different physical areas. Conference calling has been the answer for the last couple of decades, and most modern PBX systems will support this feature. Depending on the PBX system, you may only be able to include extensions within your PBX system's control, or you may be able to include a PSTN number in the conference call. The latter is true for most modern PBX solutions.

Extension Dialing

PBX systems will allow users that are connected to the system to dial other users connected to the system by entering a simple four- or five-digit number. This both keeps the call within the PBX system and makes for more efficient dialing.

Follow-Me

The follow-me feature of phone systems is a little different than call forwarding. Whereas call forwarding forwards the phone call to one other specified number, follow-me includes a list of numbers, and the PBX continues to ring the numbers down the list until either it exhausts the numbers or you answer one of them. While you could accomplish a similar feat by forwarding your phone to another phone, which is forwarded to another, which is forwarded to another, you would not be accomplishing quite the same thing. In the latter scenario, the only phone that would actually ring—in most implementations—is the very last phone in the forwarding chain. With follow-me, all the phones in the list will ring and you can answer where you are.

Music on Hold (MOH)

Of course, what is a PBX system worth if it can't play music and other messages while on hold? Some organizations use MOH to simply act as a notice to the caller that he or she is indeed still connected. Other organizations use it very strategically. For example, I called an ISP back in the last 1990s and still recall the on-hold message. It went something like this, "We appreciate your business and want to help you get the most out of your Internet connection. For this reason, we thought you'd rather get some tips about how to get more out of the Internet instead of hearing music that you do not find particularly appealing. Your call will be answered within the next five minutes, but here's your first tip while you wait . . . ." I thought this was brilliant, and more and more organizations are realizing that this is indeed something that they should take advantage of. Of course, you'll have to be careful.

I know that I don't want a company to try to sell me something while I'm on hold with a complaint about what I've already purchased from them. I suppose it's about balance in the end, but this is just an example of what MOH can be used for other than simply playing the music.

Voice Mail

No list of PBX features would be complete without voice mail. Other than e-mail, and voice mail has been around longer, voice mail is probably one of the most important business tools in use today. I'm not suggesting that it's always used well, but many of us have come to rely on it for many of our daily routines. Modern PBX systems not only provide you with voice mail, but many of them also allow you to have the voice mail messages sent to your e-mail as either links to audio files or embedded audio files. Internal voice mail messages from your peers can usually be responded to by simply pressing the right key to respond and then speaking your response into the telephone. This keeps you from having to write down extensions or other information when it's not needed.

Of course, a PBX with a voice mail system is a PBX with a hard drive. This audio data has to be stored somewhere, and this means that you're dealing with a more complex PBX solution. In some cases, PBX systems link to external voice mail applications rather than providing their own voice mail solution, but I don't know of a modern PBX solution that does not in some way provide for voice mail services.

Voice Recognition

Some PBX systems either come with or allow for expansions to include voice recognition. This provides the feature you may have encountered where the automated attendant allows you to "press or say one," just as a simple example. Needless to say, these systems are not infallible, but they have really come a long way. Ten or fifteen years ago, you would never have been able to implement flexible voice recognition technology on the scale that it has been implemented today. The ability to understand a single human's voice was hard enough, but today the recognition systems do pretty well at recognizing hundreds of humans when they speak particular words or phrases. I'm still impressed that I can read my entire credit card and the company can process it—not that I would, since someone may be listening nearby, but it's still impressive.

Special PBX Features

In addition to the common features, there are special demands placed on PBX systems for unique scenarios. For example, a call center may need automatic call distribution (ACD) that maps a single incoming number to any number of internal representatives. Such a system will likely need a queue as well in order to reduce the total number of employees needed to handle the call volume. I remember my first job in Information Technology was that of a help desk technical support staff member. There were probably twenty of us that answered incoming calls, and we could see the number of calls waiting in the queue on our telephones' LCDs. These features have obviously been with us for some time now in the traditional PBX solutions.

Hardware Components of a PBX

If you're going to work with PBX systems, it's going to be very important that you understand the hardware and software components that form a PBX. In addition to the phone lines, there are specialized components that are used in these systems. A PBX will have line cards and trunk cards in it. The line cards provide connectivity to your phones, and the trunk cards are used to connect to the CO. Hybrid PBX solutions will also have network interface cards for connectivity to the data network.

The control complex of the PBX is the logic processing section of the system. This is where the real controlling power of the PBX is located. The features, such as call waiting and call forwarding and more, are provided by this intelligent component. In addition, the PBX will have a switch backplane that is used to make the connections between the telephone endpoints. Keep in mind that you will likely be required to use telephones from the same vendor that manufactured your PBX in order for them to communicate with the PBX appropriately. The newer IP PBX solutions may be an exception to this, since they communicate using many standardized protocols.

Of course, the PBX will have to have a power supply and cooling fans, since it is basically a computing device, whether it is a TDM PBX, an IP-only PBX, or a hybrid PBX.

In summary, you could say that a PBX provides all the components of a CO on a smaller scale. The end result is that you're able to implement your telephone system as a stand-alone system, but it will also be able to connect to the CO for PSTN connectivity.

CERTIFICATION SUMMARY

This chapter was very brief and for a good reason: you will not have to know a tremendous amount about traditional PBX systems in order to implement the modern IP PBX solutions. Chapters 7 and 9 will be heavily focused on running IP telephony, and you will learn about the specific codecs and Quality of Service mechanisms required in those chapters. The point of this chapter was simply to expose you to the basics of enterprise telephony and the features that have been provided through PBX systems for the past few decades. You should now be able to know what TDM, IP-only, and hybrid PBX systems are and how they differ. You should be aware of the common features of PBX solutions of all kinds, and you should be prepared to begin your discovery of IP telephone and streaming video and conferencing technologies, which make up the next four chapters in this book.

✓ TWO-MINUTE DRILL

PBX Systems

❑ Traditional or legacy PBX systems were TDM PBX systems.

❑ IP-only PBX systems implement a purely IP telephony internal communications network while connecting to the PSTN for outside calling.

❑ Hybrid PBX systems include both TDM and IP telephone capabilities.

❑ Line cards are used to provide connectivity to the endpoints or telephones.

❑ Trunk cards are used to provide connectivity to the trunk lines connected to the CO.

❑ Tie trunks or tie lines connect PBX systems together, and CO lines or CO trunks connect a PBX to the PSTN for outside call routing.

SELF TEST

The following questions will help you measure your understanding of the material presented in this chapter. Read all the choices carefully because there might be more than one correct answer. Choose all correct answers for each question.

PBX Systems

1. You want to connect three PBX units together that are at separate locations. One is in Columbus, OH. The second is in Miami, FL, and the third is in Lexington, KY. What kind of trunk lines will you need to create the connections required?

   A. CO trunks

   B. PBX trunks

   C. Tie trunks

   D. Private trunks

2. You need a PBX solution that will allow you to use a digital telephone and twelve IP phones. What kind of PBX do you need?

   A. TDM

   B. IP-only

   C. Hybrid

   D. Trunk

3. You want to have the PBX system automatically try to ring you at three different numbers in order to ensure that an important call gets through to you regardless of which location you are in. Which PBX feature do you need in order to perform this?

   A. Call forwarding

   B. User homing

   C. Follow-me

   D. Caller ID

4. You need to install more telephones in your environment. The PBX system you are using can be expanded by adding more cards for more telephones. What kind of cards will you need to add in order to provide connectivity for more telephones?

   A. Trunk cards

   B. PCI cards

   C. Network interface cards

   D. Line cards

SELF TEST ANSWERS

PBX Systems

1. ☑ C is correct. Tie trunks are used to connect PBX units together when they are at separate locations.
☒ A, B, and D are incorrect.

2. ☑ C is correct. A hybrid PBX will allow you to connect a digital phone as well as an IP phone, so you will need to implement this kind of PBX solution.
☒ A, B, and D are incorrect.

3. ☑ C is correct. The follow-me feature provides this functionality. The call forwarding feature will only ring one number and would not notify you if you were in the area of the other two specified numbers. Only follow-me will ring you at all three locations.
☒ A, B, and D are incorrect.

4. ☑ D is correct. Line cards must be added in order to provide more lines.
☒ A, B, and C are incorrect.

7

Data and Voice Converged

> *It's easy to get the players; it's getting them to play together that's the tough part.*
>
> –Casey Stengel

U p to this point, you've learned how data networks work and the fundamentals of how telephone networks work. Now it's time to start making the players "play together." In order to accomplish this task, you must understand the hardware components that act as the interface between the data networks that are carrying voice as data and the telephone networks that carry voice as traditional telephony signals. This difference means you'll need to learn about gateways and gatekeepers, and they are both covered in this chapter.

It will also be important for you to understand the edge devices or endpoint devices that are used on a converged network. These devices include specialty soft phones and various communications devices. You'll also need to understand how we track these endpoints when they are implemented as mobile devices. This knowledge will include such topics as network roaming and presence.

In earlier chapters, you learned about voice encoding where the telephone converts sound waves into electrical signals. In this chapter, you will learn about specific encoding and decoding protocols that are used in voice over IP networks. Specifically, you will learn about H.323 and SIP, which build the IP telephony framework and G.7xx coders. Additionally, you'll learn about traffic shaping concepts and the different messaging and collaboration applications that are used on converged networks. It's clear that this chapter will present a broad array of information, and I will begin by discussing how the voice network will overlay the data network that is traditionally implemented. I'll also explain a few additional components that must be added to these traditional networks in order to allow the Voice over IP solution to work effectively.

Voice Layered over Data Networks

You've learned about data networks and telephone networks, but how do we get telephone conversations to travel over data networks? The answer is a simple one: covert the audible telephone conversation into digital data packets. This conversion is accomplished with encoding standards, and the voice data packets are then transferred using standard or proprietary voice communication protocols. It's really no different than any other data communications process as to its technical nature; however, voice data packets do come with demands that are not seen in traditional data packets.

For example, if you are sending a file to a server using FTP, it doesn't matter if a few packets arrive out of order or if there happens to be a delay between the arrival of one packet and the arrival of the next packet that is greater than a particular threshold. Of course, if there is an extended delay it will slow the communications down, but the data will eventually arrive at the destination. Voice traffic will not tolerate such occurrences. If there is an extended delay, the call will be dropped or quality will suffer. If you think about it for a moment, you'll understand why this low tolerance is true.

There are humans at both ends of a Voice over IP (VoIP) communications link. They will both talk and listen, and they have expectations that have been set by the analog telephone network. If they do not hear any sound for some variable length of time, they will assume that the call has been dropped or the person on the other end has disconnected. If the sound quality is inferior, particularly to the point where they cannot understand one another, they may give up on the conversation. There are simply expectations of quality that must be met with VoIP data that have not traditionally been required of other data types. In fact, we often refer to "carrier grade" or "carrier quality" VoIP communications. This term means that we have accomplished a quality of sound and communication speed that is, at least, equivalent to the traditional PSTN.

Since we are transmitting the VoIP packets over the same physical network as our traditional data packets (think e-mail, database access, file transfer, printing, etc.), we can say that we are layering voice over the data network. We are using the same network devices, cables, and software that are used for our traditional data to transfer our voice data. This layering places a new demand on the network. The demand is that the data network must be able to differentiate between different packet types and give priority to voice data so that the quality expectations of the VoIP users are met. The needed technology is known as Quality of Service and will be discussed extensively in Chapter 9.

Since voice traffic must move at a rapid speed across the network and since it would not provide a benefit to resend the traffic if it is corrupted or lost in transmission, we use UDP to send most VoIP data packets. You'll remember from previous chapters that UDP is a connectionless protocol, unlike TCP. TCP has far too much overhead to transmit voice packets as rapidly as they must be transmitted.

You may wonder why there is no benefit from resending corrupted or lost voice packets. The reason is simple. Think about how long it takes you to say the word *don't*. If you're like me, it will take you far less than a second. Now, imagine you're having a conversation on a VoIP phone and you say the following sentences, "Don't push the button. Pull the lever." Further imagine that the word *don't* was lost in transmission and the system decided to resend it. Because of the sequencing problem, the user on the other end hears the following, "Push the button. Don't pull the lever."

This reordering could theoretically happen because the phrase *push the button* made it through while the word *don't* didn't make it through. When the word *don't* was retransmitted, it was placed before the phrase *pull the lever*. The result is a complete opposite of the intended message. Do you see why retransmitting lost audio packets would be useless and possibly damaging?

Instead, the listener would just not hear the word don't; however, the reality is that it all gets a bit more complex. More than likely the listener would hear something like, "D---t pu—the ---ton. Pu-- --- --ver." All the dashes represent either sounds that are unintelligible or complete silence. The point is that the network doesn't usually drop exact words, but rather portions of audio much less than a complete word, resulting in what we usually call a "bad connection."

CERTIFICATION OBJECTIVE 7.01

Identify the Functions of Hardware Components as Used on a Converged Network

In previous chapters, you reviewed the basic hardware components that are used to build a traditional data network or traditional enterprise telephony networks. In this section, you'll learn about the two basic building blocks of VoIP networks: gateways and gatekeepers. PBX systems, which manage the internal enterprise telephony network and may include a gateway component for connection to the PSTN, were covered in Chapter 6.

Gateways

When one type of network must be connected to another type, a gateway is used. Gateways come in many different forms, but the three primary gateways of concern to us are:

- Media gateways
- TDM/IP gateways
- SIP gateways

Gateways enable the convergence of voice and data by connecting different network types such as voice networks and data networks.

Media Gateways

Media gateways are most usually placed between the PSTN and your private IP telephony network. According to the Megaco protocol (also known as H.248), a media gateway is managed by a *media gateway controller* (MGC). Figure 7-1 shows where a media gateway would fit into a telephone infrastructure. H.323 gateways are used in H.323 network implementations

TDM/IP Gateways

While the line is often blurred between media gateways and TDM/IP gateways, you might be assisted by thinking of a TDM/IP gateway as a subset or feature of a media gateway. You are basically working with two different concepts in telephony gateways: the conversion of signaling protocols and the conversion of the actual voice data/packets. TDM/IP gateways are responsible for converting IP packets to TDM transfers and TDM transfers to IP packets. TDM is used on the PSTN and IP on the local VoIP network.

SIP Gateways

The Session Initiation Protocol (SIP) is widely used for VoIP communications, and a SIP gateway is used to allow SIP users to place calls to traditional land lines or other VoIP users that may be connected to H.323 networks or other proprietary networks.

Gatekeepers

In order to understand gatekeepers, you'll need to understand the basics of the H.323 protocol. I'll review those basics momentarily; however, I do want to point out that a gatekeeper is an optional element and it is one that can provide extra features and security to an H.323 network.

FIGURE 7-1

Media gateway placement

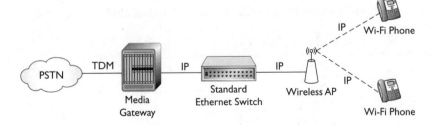

What Is H.323?

H.323, like SIP, is a signaling protocol for VoIP networks. H.323 is responsible for call setup, maintenance, and teardown. H.323 is not a stand-alone entity in that it requires the foundation of an IP network under it. This requirement means that H.323 depends on a solid IP, DNS, DHCP, and other IP services infrastructure. H.323 will use the Real-Time Transport Protocol (RTP) and the Real-Time Transport Control Protocol (RTCP) for communications and control of communications, respectively.

The H.323 protocol suite is a standards-based solution for IP telephony that defines four functional units within the H.323 network:

- Endpoints
- Gateways
- Gatekeepers
- Multipoint control units

The H.323 endpoint is the origination or destination for telephony traffic. An endpoint can place or receive a call. The H.323 gateways allow interconnection with traditional or different telephony networks. The gateways translate between the H.323 protocols and the protocols of the connected network. An H.323 gatekeeper, which is optional, handles the IP address resolution and the access to the H.323 network. The gatekeeper will provide address translation between alias addresses and actual IP addresses. This translation allows a user to call the "Help Desk" instead of having to call 10.56.78.132. In addition, the gatekeepers control endpoint access to network services and resources. While gatekeepers are optional, they must be used by endpoints when present.

Finally, multipoint control units (MCUs) allow for conference calls that include three or more parties. H.323 provides for both centralized and decentralized conferencing. All H.323 endpoints participating in the conference call must be able to establish a connection with the MCU.

H.323 Protocols

H.323 uses many protocols to accomplish its objective of implementing a secure and reliable VoIP network. These protocols include:

- H.225/Q.931
- H.225.0/RAS
- H.245
- RTP and RTCP
- Codecs (G.711, G.723.1, G.729)

The codecs are discussed in more detail later in this chapter, and RTP/RTCP have been mentioned previously; however, the H.2xx protocols in this list require some further analysis here.

H.225/Q.931 is used by H.323 to define the signaling for call setup and teardown. This use involves the establishing of the source and destination IP addresses, the ports utilized, and relevant country codes, and specific H.245 port information. Don't confuse this protocol with H.225.0/RAS, which actually specifies the messages that are used between an endpoint and a gatekeeper. If you have implemented the optional gatekeeper, H.225.0 specifies the messages that are used for registration, admission, and status management.

H.245 is used to discover and negotiate the capabilities of endpoints. It is also used to determine the master/slave relationship in the H.323 call and the logical channel information for media streams. You can think of the logical channel established by H.245 for voice transfer as being similar to the TCP connections made between two hosts communicating across TCP. There is no guarantee that the voice data will travel a specified path from endpoint to endpoint, but the session is established and this is the logical channel.

H.323 is an ITU recommendation. For more information about H.323, view the Wikipedia article found here: http://en.wikipedia.org/wiki/H.323

Controlling Access to the Telephony Network

This information brings us back to the original question: what is a gatekeeper? As you've seen, the gatekeeper is used to allow access to the H.323 network for endpoints. It is also useful to know that the gatekeeper provides connectivity for MCUs as well. Gatekeepers manage zones, and a zone is simply a collection of endpoints, MCUs, and gateways that are registered with the gatekeeper. In the end, gatekeepers provide the following required functionality:

- **Address translation** This process is used to translate H.323 IDs, such as HelpDesk@company.net, and/or standard telephone numbers to IP addresses.
- **Admission control** Using Registration, Admission, and Status (RAS) messages, the gatekeeper provides access to the H.323 network.
- **Bandwidth control** Bandwidth management is provided through three RAS messages. These messages are Bandwidth Request (BRQ), Bandwidth Confirm (BCF), and Bandwidth Reject (BRJ).

In addition to these required functions, a gatekeeper may provide call authorization, bandwidth threshold limits (rejecting connections that do not meet minimum bandwidth requirements), and other proprietary features provided by the vendors.

SIP

H.323 is not the only solution used for IP telephony. Another popular protocol is the Session Initiation Protocol (SIP). SIP uses the Real-Time Streaming Protocol (RTSP), Session Description Protocol (SDP), and Session Announcement Protocol (SAP), all defined by the Internet Engineering Task Force (IETF). SIP is defined in RFC 2543, though this RFC has been revised a number of times. The newest at the time of this writing is RFC 3261.

Like H.323, SIP acts as a signaling protocol for the setup, teardown, and management of call sessions. RTP is also used in SIP to transport the actual voice packets. SIP devices can communicate directly with each other, or they can communicate through a SIP proxy. SIP addresses take on the form of a Uniform Resource Indicator (URI), which is very similar to an e-mail address. In fact, they often look just like e-mail addresses. For example, *tom@sipnetwork1045.com* could be a valid SIP address.

For more information about the SIP solution, see the Wikipedia article at: http://en.wikipedia .org/wiki/Session_Initiation_Protocol.

e x a m

ⓦ a t c h   *ENUM unifies traditional telephony with modern VoIP solutions by translating the standard telephone numbers into universal identifiers that can be used by VoIP. The term ENUM comes from E.164 Number Mapping. ENUM is defined in RFC 3761. Ultimately, ENUM provides the mechanism to translate standard phone numbers (E.164) into SIP or H.323 IP addresses.*

CERTIFICATION OBJECTIVE 7.02

Identify the Various Endpoints Used in a Converged Network

A voice and data infrastructure isn't much use without endpoints to take advantage of it. When implementing a converged network, many new endpoints are introduced that have not been traditionally used on typical data networks. Most of these devices are used for telephone conversations, but they may be hybrid devices with both voice and data capabilities. The endpoints you'll need to be aware of for the Convergence+ exam include:

- PC-based endpoints
- Voice terminals

- SIP phones
- Wi-Fi phones

PC-Based Endpoints

PC-based endpoints are just what they sound like: personal computers that communicate on the network. However, there are really two primary types of endpoints that fit into this category. There are traditional personal computers (desktops and laptops), and there are handheld computers (PDAs).

PC-Based Softphones

If you want to use your desktop computer or laptop to communicate on an IP telephony network, you will need a software application known as a softphone. A *softphone* is an application that supports various microphones so that you can talk across the IP network and that plays the audio coming from the other party through your PC speakers (or through the headphones or headset). PC-based softphones provide many advantages:

- Installation is simple, with no additional hardware required. This configuration assumes that you have a microphone and speakers.
- Enhanced features are supported by the powerful processing capabilities of the computer.
- Softphones often support video calls with the addition of a webcam.

There are also negatives. For example, softphones that run on desktop and laptop computers are clearly not as portable as Wi-Fi phones, SIP phones, PDA softphones, or H.323 mobile phones. Additionally, PDA softphones rarely support video calls, since the processing power is insufficient to handle the video and audio processing simultaneously. Figure 7-2 shows an example of a PC-based softphone application.

FIGURE 7-2

PC-based
softphone

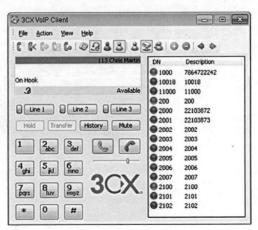

Dedicated Endpoints

While PC-based softphones are non-dedicated devices in that the computers or PDAs are used to perform functions other than call processing, there are many devices dedicated to acting as a VoIP phone or endpoint. These devices include:

- Voice terminals
- SIP phones
- Wi-Fi phones

Voice Terminals

A *voice terminal* is a specialized device that allows you to connect a regular analog telephone to a VoIP system. These are also called *analog telephone adapters*, or ATAs. The voice terminal communicates with the VoIP server using the configured protocol, which may be H.323, SIP, MGCP, or some other protocol supported by the voice terminal. Figure 7-3 shows an example voice terminal from D-Link.

SIP Phones

A phone specially designed to communicate on a SIP network is called a *SIP phone*. This SIP phone may be a hard phone (a device dedicated to telephony), or it could be a classification of softphone software. A stationary SIP phone will have an RJ45 connector for connectivity to the Ethernet network across which it will operate. Figure 7-4 shows a D-Link SIP phone that is intended for stationary use. This particular phone includes the following features:

- Redial
- Menu
- Mute

FIGURE 7-3

D-Link VoIP terminal

FIGURE 7-4

D-Link SIP phone

- Transfer
- Hold
- Voice mail
- 3-way conference
- Speaker
- Address book (200 records)
- Special dial (10 records)
- Review address records
- Speaker/ringer volume adjust

Wi-Fi Phones

Wi-Fi phones are simply dedicated telephony devices that communicate across 802.11 network instead of Ethernet networks. These phones support SIP, H.323, and other standards as well. Some wireless phones are hybrid in nature, supporting both cellular communications and Wi-FI IP telephony so that they can use whichever network is available. Figure 7-5 shows a D-Link Wi-Fi phone. You can see that they look almost identical to traditional cell phones.

FIGURE 7-5

D-Link Wi-Fi
phone

CERTIFICATION OBJECTIVE 7.03

Identify Methods of Encoding, Decoding, and Compression

Digital voice communications require encoding and decoding. This encoding and decoding takes time, and there is additional time accrued due to the actual transfer of the data from one endpoint to another endpoint. While encoding and decoding allow for the transmission of voice as data, compression allows for faster transfer of this data through the utilization of less bandwidth. This section outlines all three factors: encoding, decoding, and compression.

Analog to Digital

To convert analog sound waves (specifically voice waves in this case) into digitized bit streams, a coder/decoder (codec) is used. In order to convert the smooth variations in analog information to the abrupt changes in digital information, sampling rates are applied. The *sampling rate* is the number of times per second that the analog information is analyzed to convert it to digital information.

A typical sampling rate is 8000 times per second. This notation would be written as 8 KHz (kilohertz) because it is a sampling rate or sampling frequency of 8000. The Nyquist Theorem suggests that a sampling rate of at least two times the highest frequency component of the analog information set results in an acceptable and accurate representation in a digital form.

When a higher sampling rate is used, the audio quality is improved on playback. However, a higher sampling rate means that more data is generated. When more data is generated, more bandwidth is consumed to transfer the data. Therefore, an acceptable loss in audio quality must be tolerated in order to use the bandwidth most effectively. This loss is why an individual's voice sounds different on a telephone than it does in person. We only sample enough and in the right frequency range to transfer intelligible communications with acceptable quality.

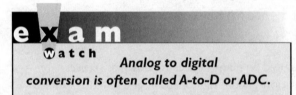

Analog to digital conversion is often called A-to-D or ADC.

Standards for Encoding

There are many standards for encoding audio in telephony networks. These include H.261, H.264, and the G7xx encoding algorithms.

H.261 and H.264

The International Telecommunications Union (ITU) developed the H.261 codec in 1990 for the transmission of video over ISDN lines. H.264, also known as MPEG-4, is a video encoding standard also developed by the ITU. H.264 is used for video conferencing, HDTV, and VoIP. The big difference between H.261 and H.264 is that H.261 does not work well over Frame Relay or TCP/IP networks and H.264 does work well in these environments. You'll learn more about video over IP in Chapter 8.

G711, G729a, G723.1, and G722

When it comes to audio codecs, the G7xx series of standards are the most commonly utilized. Audio capability is the minimum standard required of an H.323 endpoint (they may support video as well), and so any H.323 endpoint must support the G.711 codec at a minimum. The G.7xx series of codecs are outlined as follows:

- **G.711**   Supports uncompressed audio encoding at 64 Kbps. This codec is required by H.323 in that any device claiming H.323 compliance must support it. You are not required to use it, but it must be supported.
- **G.722**   Encodes at 64, 56, and 48 Kbps.
- **G.723/723.1**   Encodes at 5.4 and 6.3 Kbps.
- **G.729/729A**   Encodes at 8 Kbps.

A form of bandwidth reduction used by many codecs is known as *silence suppression*. In most voice conversations, about half of either end's audio is silent. The G.729 codec will remove this silence before encoding the audio. G.711 doesn't remove this silence and makes for much larger bandwidth consumption rates. In addition, codecs may use advanced compression algorithms that look for redundant bits and other factors that allow for even greater compression of the voice data. Ultimately, the compression algorithms must balance the need for reduced bandwidth consumption and quick transfers of the audio data. All the compression must happen in a matter of microseconds for each voice packet.

μ-Law and a-Law

Voice companding refers to the process of compressing the audio on the transmitting end and expanding the audio on the receiving end. Compressing and expanding equals companding. I know, it sounds like something my six year old would create, but it's an important part of IP telephony.

G.711 defines two different companding algorithms. Each is based on mathematical formulas or laws, and each is named for this law. The first is *μ-Law* (pronounced *mu-law*), and the second is *a-law*.

Here, μ-Law is a mathematical algorithm used for pulse code modulation (PCM). The algorithm represents analog amplitude values as compact digital codes. Three binary bits represent one of eight different ranges of amplitude values. Four more bits are used to represent an amplitude value within the range specified by the first three bits. The μ-Law algorithm is used in T1 lines in the Unites States and J1 lines in Japan.

A-Law is also a mathematical algorithm. The specifications are the same as for μ-Law in that the first three bits represent the range and the last four bits represent an amplitude in that range. A-Law companding is used primarily on E1 lines in Europe.

CERTIFICATION OBJECTIVE 7.04

Identify Different Types of Messaging Applications

Two primary types of applications have emerged in modern computer networks that relate to communications: messaging and collaboration. Here, I'll investigate messaging applications. Messaging applications fall into two general categories. The first is audio messaging, which has been available for decades, and the second is text messaging, which is much newer. The latest addition is unified messaging, and this latter type of messaging will also be discussed here.

Voicemail, Video, and Text Messaging

Audio messages, also known as *voicemail*, have been available for decades. Telephone answering machines have been used in small implementations for years, and they have evolved from analog tape-based machines to modern digital recording units. In PBX systems and VoIP systems, the audio is recorded in a compressed digital format so that it can be played back at a later time.

Voicemail is a very important application in many organizations. Customer service departments must be able to allow overflow callers to leave messages, and all employees need a way to receive important messages when they are unavailable. Voicemail is so common today that many people have three or four places where they receive it. I have voicemail on my business phone, my home phone, my cell phone, and a client's telephony network for whom I provide support. Many people have even more voicemail boxes than me. It's obviously an important part of modern business communications.

While less common than voicemail, some organizations now employ video messaging. There are certainly times when it is beneficial to leave a video message instead of an audio message. This benefit is particularly true when you can show the recipient something useful in the video; however, the applications are still few and so most organizations are not providing video messaging.

Text messaging is very popular among the youth today (I'm afraid I can't place myself in that category anymore), and it's also useful in business. A quick note to a colleague that indicates the reason that you will be late for a meeting is faster than a phone conversation and prevents frustration on the part of the colleague.

How many times have you avoided calling someone because you feared he would pressure you to do what he wanted? Text messaging helps in alleviating this very common and very human problem.

In addition to modern text messaging, many VoIP systems integrate well with traditional SMTP e-mail. These systems allow you to send an e-mail from your VoIP phone that may include an audio file as an attachment or that simply contains a text message. This attachment is a very useful feature when integrating with users who do not have text messaging features on their phones.

Unified Messaging

The concept of unified messaging involves the combination of voicemail and e-mail. In this case, voicemail messages are automatically forwarded to e-mail boxes for target users. One of the most obvious benefits of this capability is that you can receive your voicemail messages from anywhere as long as you have Internet access. These systems will usually compress the audio in MP3 format today so that the files are as small as possible. Unified messaging may also include support for fax messages. This feature provides the user with the ability to receive faxes through his or her e-mail application. eFax is an Internet service that provides such a feature, and the other features of unified messaging have all been provided through Vonage as well. VoIP systems that are implemented in enterprises may also provide the same capabilities, and these capabilities should be considered when investigating a VoIP solution.

CERTIFICATION OBJECTIVE 7.05

Identify Different Types of Collaboration Applications

Collaboration applications have become increasingly important with the modern teleworker movement. Many employees work from home, and in large organizations even on-premises workers are often at remote locations. Some inexpensive solution is needed that allows these workers to collaborate in real time. The two most common solutions are audio and video conferencing.

Audio and Video Conferencing

Audio conferencing allows multiple endpoints to aggregate their audio so that each endpoint hears the audio transmitted by every other endpoint in the conference. This aggregation is nothing new and has been available in traditional PBX and PSTN networks for decades. What is new is the ability to form a conference call across the Internet with no additional fees other than those required to connect to the Internet.

There are different ways to create an audio conference call:

- **Ad hoc**   This conferencing methodology allows individuals to connect to the conference as they are available.
- **Scheduled**   A conference call that is scheduled on a weekly or some periodic basis and is structured more like a formal meeting.
- **Opportunistic**   A conference call that is set up the moment all needed attendees are available. This kind of call can be very useful when people must collaborate who are rarely available at the same time. Rather than trying to discover a time when all parties are available, the system can automatically detect the unified availability.

Video conferencing is similar to audio conferencing in that multiple parties can be involved. You'll need video camera hardware in addition to audio hardware, and the bandwidth requirements will be much higher. Video conferencing will be discussed in greater detail in Chapter 8.

Data Sharing

One important element of collaboration is data sharing. Modern solutions, such as Microsoft's SharePoint service, Cisco's MeetingPlace, or WebEx, allow you to form discussion groups, share data documents, and even leave audio and video messages in some implementations. H.323 also offers data sharing as an optional capability. When available, data can be shared with whiteboard or file transfers, and even full-screen application sharing can be used in conference training sessions. The recommendations that describe data sharing are known as the T-series. They include T.122–127 and T.134.

CERTIFICATION OBJECTIVE 7.06

Identify Components of Mobility

Mobility is all about flexibility in communications. Users want to be able to move around while having a conversation. Today, they want to move around in buildings, between buildings, and across state and country lines. We owe much of this expectation to the proliferation of cell phones. How do we allow users to communicate across our VoIP network and also across the cellular network using the same device? How do we find out if the user is available or not? How does the network determine where the user is and how to best place the call? The answers to these questions lie in understanding presence, cellular integration, and other mobility features.

Presence

Presence, in the world of telephony, is about communicating the user's desire and ability to communicate, and the medium through which that communication may occur. If a user is connected to the Internet, presence may dictate that the user wants to be reached through the medium of IP telephony. If the user is driving down the road in his car, presence may dictate that the user wants to be reached through his cell phone. The point of presence is to allow the user to be located and contacted wherever the user is physically using the preferred method of the user. Additionally, the user can indicate the he or she does not want to be contacted using any method. This ability is the core concept of presence.

How does presence work? It can only work if the calling user knows the presence server that contains the target user's contact profile. This case would occur when a user within a company attempts to contact another user in the company, and this area is where presence is most useful today. In the future, with large-scale centralized contact servers, it may be possible to have a centralized database for contacting any user who wishes to be contacted, but there is no such solution at this time. While this global presence scenario may seem a bit intrusive to some, it could prove acceptable as long as users are allowed to opt in or opt out.

In addition to logical location (connected to the Internet or the cell network, for example), presence may also provide the actual physical location of the user.

This tracking is accomplished with GPS or cell tower triangulation with cell phones and may be accomplished with RFID tracking or wireless access point triangulation in 802.11 networks. Other physical tracking mechanisms may also be linked into the presence system. These tracking mechanisms include swiping smart cards and logging on to computers with physical MAC addresses and physical locations on the network.

Cellular Integration Services

Cellular integration allows the features of an IP-PBX to extend to the users' cell phones. For example, the user can allow the cell phone to ring using the same number that is used to call her desk phone. Cell phones may join in conference calls and be used to send text messages across the PBX system.

Other Mobility Features

Many modern devices have both an 802.11 radio and a cellular network radio. This dual-radio feature allows the device to place calls across SIP or H.323 VoIP networks when connected to the 802.11 network and to place calls across the cellular network when not connected to the 802.11 network. The result can be a tremendous cost savings for a large enterprise with dozens or hundreds of users using such a device. Instead of having every call billed through the cellular provider, only those calls that are placed when the users are out of range of the enterprise WLAN will be toll calls.

CERTIFICATION SUMMARY

This chapter introduced you to the terminology and technologies involved in integrating VoIP with traditional telephony. You learned about gateways and gatekeepers and the role they play in an IP telephone network. You were then introduced to the basic endpoints that are used on VoIP networks and the applications that they support. The various encoding, decoding, and compression methods were introduced, and the concept of mobility was explained. The next chapter will provide a similar overview of video over IP solutions before we move on to learn about how it all fits together in Quality of Service terms in Chapter 9.

TWO-MINUTE DRILL

Identify the Functions of Hardware Components as Used in a Converged Network

❑ A gateway provides an interface between two different network types.

❑ Gatekeepers provide access to the VoIP network in an H.323 implementation.

❑ Gatekeepers are optional, but if they are used, the endpoints must communicate with and through them.

Identify the Various Endpoints Used in a Converged Network

❑ PC-based softphones are composed of software and hardware. The software is an application that communicates on the VoIP network, and the hardware includes microphones and speakers.

❑ Wi-Fi phones are VoIP phones that connect to the network using 802.11 instead of 802.3.

Identify Methods of Encoding, Decoding, and Compression

❑ Analog-to-digital encoding is known as A-to-D or ADC.

❑ G.711 is an encoding scheme that uses no compression.

❑ The μ-Law companding algorithm is used in T1 in the U.S. and J1 lines in Japan, and the A-Law companding algorithm is used in E1 lines in Europe.

❑ Companding is compressing and expanding.

Identify Different Types of Messaging Applications

❑ The most common types of messaging applications are voicemail, text messaging, and e-mail.

Identity Different Types of Collaboration Applications

❑ Collaboration applications include audio conferencing, video conferencing, and data sharing.

❑ Data sharing is an optional part of the H.323 specification.

Identify Components of Mobility

❑ Mobility includes the ability for a user to move around and for other users to be able to locate him or her.

❑ Presence means that the user can be located and the user can specify the medium through which he or she wishes to be contacted.

❑ Cellular integration is a feature of many IP-PBX systems that allows the user to specify that his or her cell phone should ring using the same number as his or her VoIP phone.

SELF TEST

The following questions will help you measure your understanding of the material presented in this chapter. Read all the choices carefully because there might be more than one correct answer. Choose all correct answers for each question.

Identify the Functions of Hardware Components as Used in a Converged Network

1. You are implementing a VoIP solution, and the client has stipulated that a device must be installed that will register and manage VoIP endpoints. What is this device called?

   A. Gatekeeper

   B. Gateway

   C. Firewall

   D. MCU

2. Which of the following are common gateways used on VoIP networks for the purpose of translating one form of telephony communication to another?

   A. SIP gateway

   B. TDM/IP gateway

   C. Internet gateway

   D. Media gateway

Identify the Various Endpoints Used in a Converged Network

3. You place a call while walking down the hallway in your company. You are not connected to an outside service provider. What kind of phone are you most likely using among the following choices?

   A. Cell phone

   B. Desktop PC softphone

   C. Wi-Fi phone

   D. Hard phone

Identify Methods of Encoding, Decoding, and Compression

4. The VoIP phone is stripping out silent portions before sending the voice data. What is this process a part of?

A. Encoding

B. Decoding

C. Compression

D. Decompression

Identify Different Types of Messaging Applications

5. You need a solution that includes the ability to have voicemail messages sent to users' e-mail boxes. What is this technology called?

A. E-voicemail

B. Electronic messaging

C. Digital voicemail

D. Unified messaging

Identity Different Types of Collaboration Applications

6. Users state that they need the ability to distribute documents among their teams. What form of collaboration provides this functionality?

A. Data sharing

B. Unified messaging

C. Audio conferencing

D. Video conferencing

Identify Components of Mobility

7. A user states that she must have the ability to specify the method she wants to use when people call her and needs the ability to specify that she should not be disturbed. What is the general technology that provides this capability?

A. Mobility

B. Presence

C. Roaming

D. Cellular service

SELF TEST ANSWERS

1. ☑ **A** is correct. A gatekeeper is responsible for registration, access, and status.
   ☒ **B, C,** and **D** are incorrect.

2. ☑ **A, B,** and **D** are correct. A media gateway, a TDM/IP gateway, or a SIP gateway would be used to translate one form of telephony communications to another.
   ☒ **C** is incorrect. An Internet gateway is used to connect a private network to the Internet, but it is not used to translate one form of telephone communications to another.

3. ☑ **C** is correct. Only the Wi-Fi phone could be utilized within these constraints. The hard phone would have to be connected to an Ethernet cable, and the cell phone would require a service provider other than the organization.
   ☒ **A, B,** and **D** are incorrect.

4. ☑ **C** is correct. Part of compression is removing silent audio in many algorithms.
   ☒ **A, B,** and **D** are incorrect.

5. ☑ **D** is correct. The ability to send voicemail to e-mail boxes is part of unified messaging.
   ☒ **A, B,** and **C** are incorrect.

6. ☑ **A** is correct. Data sharing is needed for these users.
   ☒ **B, C,** and **D** are incorrect.

7. ☑ **B** is correct. Presence is the needed technology.
   ☒ **A, C,** and **D** are incorrect.

8

Multimedia and Video

> *You may have brilliant ideas, the kind that could revolutionize the world, but unless you can express them effectively, they will have no force, no power to enter people's minds in a deep and lasting way.*
>
> —Robert Greene

The introduction of video and multimedia has improved learning processes in many arenas. Grade school students find it easier to understand another country by watching video of the daily lives of the people in that country. High school students understand physics concepts more easily by viewing animated processes on their computer screens. Employees of Fortune 500 companies understand how to use Excel and PowerPoint better by watching the software demonstrations on a computer-based training DVD than they ever have by reading instructions in a book alone. The point is simple: video and multimedia can have a positive impact, and they are likely to grow in use even more now that they can be streamed across our computer networks.

I remember working with "new media" in the 1990s and thinking that it was going to be big. I really had no idea just how big it would be. Just this morning, I visited my favorite news site on the Internet and added 15 to 20 video clips to "My Playlist" and then began watching my own customized newscast. I remember my father sitting in his favorite chair and complaining about every other news piece and how it wasn't important to anyone. Today, he can build the newscast that he wants in less than five minutes and watch just what he wants with very little in the way of commercial interruption.

We can do all of these things because of video over IP and streaming media. This chapter will present the basics of video over IP and such solutions as video conferencing and streaming media. You will learn about the different video compression technologies that are available and gain an important understanding of the impact that streaming video will have on your network. These video technologies can help your organization better communicate the brilliant ideas that can take the organization to the next level.

CERTIFICATION OBJECTIVE 8.01

Identify Methods Used for Rich Media Transmission

There are as many rich media transmission methods as there are rich media encoding algorithms; however, we need only to investigate the primary categories of web conferencing, audio and video streaming, and webcasting. Webcasting, or any kind of casting, can be in one of three categories: broadcasting (which is extremely rare in webcasting), multicasting, or unicasting. All of these concepts will be investigated in this section.

First, I would like to provide a generic definition of rich media. *Rich media* can be defined as interactive or non-interactive media that may include the use of text, graphics, sound, and video or animations, or may not include one or more of the specified elements. Some specific definitions indicate that rich media is always interactive, but many texts have used the phrase to reference media that is non-interactive. Additionally, rich media does not demand the use of all components, but it may include only text and graphics with interactivity providing the "rich" factor in the media.

In this chapter, I will focus specifically on the video subset of rich media applications. It is beyond the scope of this book and the Convergence+ certification to understand the full complexity of multimedia development and delivery environments such as those used by modern e-learning systems. The focus here will be on the applications used and the technology supporting these applications for video delivery. Creation of video content is also not a primary focus of this chapter.

Video Conferencing

Video conferencing initially required expensive video cameras and cabling as well as specialized subscriptions to ISDN lines, or some other subscriber line, for the transfer of the video between locations. Today, thanks to video over IP, the cost of implementing a video conferencing system has been reduced dramatically. Assuming you have sufficient Internet bandwidth, you can implement a video conferencing system that connects multiple sites for just a few thousand dollars in additional equipment.

You'll need at least three items to make this video conference system work. The first is a collection of cameras and microphones (or cameras with built-in microphones).

The second is a video monitor or projector to display the video feed from remote locations and often the local video as well. Finally, you may decide to implement a centralized controller or server that can be used to set up the video conferencing calls, and schedule availability to ensure that bandwidth is not overutilized in a way that would be detrimental to business-critical applications that may be running across the same network. In addition to the media controllers for video, you will need QoS. Without a definitive QoS plan, any video project is doomed to fail.

If you decide to share a single Internet connection between standard user data communications and a video conferencing solution, you have to be very aware of the impact that the video conferencing solution will have on the user's data transfer. You will most likely give preference to the user's data transfers and sacrifice video quality instead. This prioritization is because the e-mails and other Internet communications may be more important to day-to-day operations. For this reason, many organizations will purchase two Internet connections. One Internet connection will be used for traditional user utilization and another for video conferencing. Video conferencing over IP can also travel over your private WAN and avoid the Internet altogether, and this solution is usually the preferred one.

Audio and Video Streaming

The Internet has made audio and video streaming very popular. Web sites like YouTube.com, OnePlace.com, and various podcasting, screencasting, and video blogs have added to this popularity. Audio and video sources may provide streaming, or they may provide download-only access. An additional alternative source for video and audio is something called progressive download.

File-based distribution, or download-only access, is the least useful for real-time applications. The user must download the entire video or audio file before he can play the file. This time to download means that even a five- or ten-minute clip may not be available for play until after a three- or four-minute delay.

Progressive download is a feature supported by many multimedia formats. Progressive download works by allowing the playback of the video or audio to begin once a sufficient portion of the file is downloaded. Using progressive download removes the need for streaming servers; the files can simply be stored on a standard file server. Formats that support progressive download include

- Microsoft Advanced Streaming Format (ASF)
- QuickTime (MOV)
- Flash Video (FLV)
- RealMedia (RM) and RealAudio (RA)

Progressive download has a few negative aspects as well. For example, the entire file is ultimately downloaded and stored on the user's computer. With the file stored on the computer, the user could potentially copy and share the file and possibly breach intellectual property laws; however, many new encryption and protection mechanisms allow the user to download the file and play it, but not copy it to another machine for playback.

An additional limit of progressive download is the lacking support for live events. Unlike streaming, progressive download only supports previously recorded events, since the entire file must be on the server for the user to begin the download. Additionally, the user may be able to move backward through the content, but she will not be able to move forward, since there is no way to move ahead in the download.

Streaming delivery is the real real-time distribution model. Streaming resolves the negatives with progressive download and file-based delivery methods because you can implement live video feeds and the user can skip forward during playback of previously recorded events. The users can begin viewing or listening to the content almost immediately, since the stream can be reduced in quality to match the user's bandwidth. The streamed content is not saved to the user's computer, and theft is made much more difficult. Effectively, a user would have to record the audio or video in real time in order to share it with anyone else. Of course, this recording does happen; however, due to the time and effort involved, it is not as common as the theft of file-based delivery content.

The downside of streaming is that a streaming server will be required. The Windows Media Server and Helix server are two of the most common. The Windows Media Server is provided by Microsoft for streaming of Windows Media, and the Helix server is provided by RealNetworks for RealAudio and RealVideo streaming. Additional technologies are coming to the market very rapidly. The streaming takes place over a Real Time Protocol (RTP) connection or a Microsoft Media Server (MMS) protocol connection in most cases. Some organizations choose to block streaming protocols at the Internet firewall in order to prevent users from consuming too much bandwidth with streaming audio and video.

In order for a client device to be able to stream an audio or video feed, it must be able to keep up with the demand. For example, the CPU must be fast enough to decode the stream and display the results at an acceptable rate. In addition to CPU, the memory and system bus may need to meet certain specifications. Among the many streaming clients are Winamp, Real Player, Windows Media Player, KMPlayer, and the QuickTime player.

Webcasting

Webcasting is a rather modern concept that includes screencasting, podcasting, and other concepts used on web sites and blogs. Webcasting generally falls into one of two transmission methods: unicasting or multicasting. Whichever method is chosen, the delivery of multimedia content definitely takes the Internet to a whole different level when it comes to communications.

There are many popular uses of webcasting, including:

- E-learning and seminars
- Corporate meetings
- Presentations
- Concerts
- Radio broadcasts
- TV broadcasts
- Marketing seminars

Unicasting and Multicasting

Unicasting uses a separate connection for each individual accessing the content. Both file-based and progressive download delivery methods are unicasting transmission technologies. The web server will track each request and the packets traveling to and from the server using that request's connection. If dozens of users want to access the same video file, the server must send the video file dozens of times. This redundancy of data transfer impacts both the server and the network. The server is impacted because it must do the work of sending the same data to multiple locations in separate connections. The network is impacted because it must handle the bandwidth utilization of all connections even though they are transferring the exact same data. In scenarios like this one, multicasting may be more useful if you have a scheduled live event.

Multicasting is the transmission of a single data stream to multiple concurrent users. You can think of multicasting as being similar to TV and radio broadcasts. You may be listening to your favorite radio station at the same time as ten thousand other listeners. The radio station does not send out the same audio signal 10,001 times; instead, it sends out one audio signal (modulated on radio frequency waves) one time and the 10,001 listeners are all tuned into this signal.

Multicasting is intelligent enough to send the data only to end nodes that are listening for the data. This method is different than broadcasting, where the data is sent to every node on the network, whether they need the data or not. Therefore, multicasting reduces overall network utilization.

In addition to scheduled webinars (seminars on the web) and other scheduled webcasting events, multicasting is useful anytime a single stream is to be transmitted without the ability to start listening to or viewing the stream at an earlier point in time than "now." For example, Internet TV, Internet radio, or audible stock tickers fall into this category. True Internet radio stations do not allow the listener to listen to the radio stream from a point in time earlier than "now," but a listener can join in and start capturing the stream at any time. (Note: The listener may be able to "rewind" within the time span the listener has been connected. This capability is not the same as tuning in initially at an earlier time than "now.")

Ensure you do not confuse multicasting with on-demand streaming. On-demand streaming uses streaming technologies, but does not provide the bandwidth utilization advantages of multicasting. This difference is because a user may begin streaming content at any time from any point in the stream and other users may very well be at different points in the stream at the same time. The server must send the needed data to allow users to generate the proper audio and video for the point in the stream where they are.

CERTIFICATION OBJECTIVE 8.02

Identify Methods of Encoding, Decoding, and Compression

Like audio, video must be encoded and decoded, and in order to reduce bandwidth utilization, it must be compressed. It is important for you to understand the different video formats (NTSC, PAL, etc.) as well as the compression techniques available.

NTSC, PAL, and SECAM

The TV video that you're used to watching is not as it first appears. From a normal viewing distance, it looks as though you are viewing movement on the screen or at least as though you are viewing entire frames (still pictures) flashing on and off the screen at a rapid rate. The reality is that a TV picture is composed of many lines.

The lines are made up of individual dots or pixels. Interlaced video draws every other line on the screen in one pass and comes back through to paint the remaining lines in-between. Non-interlaced systems paint frame by frame. The actual specifics used for video are called video standards, among which there are four primary standards:

- NTSC
- PAL
- SECAM
- HDTV

The *National Television Standards Committee (NTSC)* was established in 1940 by the FCC and was established with the intention of developing an analog television standard. The NTSC standard specifies 585 lines. Four hundred eighty-four of the lines are used for the actual picture, and the remaining lines are used for synchronization, closed captions, vertical blanking, and any other needed signals. You will often see the frame rate rounded to 30 frames per second (fps), but NTSC actually specifies a frame rate of 29.97 fps. This corresponds to a 59.94 Hertz (Hz) refresh rate. The NTSC standard is used in the United States, Japan, South Korea, the Philippines, Canada, and some other locations.

PAL, or *Phase Alternating Line,* is a common video standard used in Asia and Europe. The original specification was for 625 lines at 50 Hz, but modern PAL systems often operate at 100 Hz in order to remove flicker effects. Many of the 625 lines are used for signaling just as in NTSC. In this case, 576 lines are used for the picture itself. With this greater number of lines devoted to the picture, PAL offers a higher-quality (resolution) picture than NTSC, though the difference is minimal in perception. PAL is heavily used in Europe, South America, Australia, Asia, and Africa.

The *Séquentiel couleur à mémoire (SECAM)* is an analog video standard that was initially developed in France (you might have guessed that from the name). SECAM specified an 819-line signal in the beginning; however, this was later reduced to 625 lines in order to help accommodate some level of compatibility and conversion to and from the PAL standard. Again, 576 lines are used for the video picture and the remaining lines are used for signaling. Like PAL, SECAM operates at 50 Hz. SECAM is used in France, in parts of Africa, and in the former USSR.

High-Definition TV (HDTV) is the new video standard that supports much higher resolution and, therefore, quality. Comparing HDTV to NTSC, you will find that NTSC has a pixel height of 720 and a pixel width of 490; however, the lowest-resolution HDTV, which is sometimes called 720 HD or HD 720,

	Standard	Height	Width	Ratio
TABLE 8-1	NTSC	720	480	4:3
	PAL	768	576	4:3
Video Standards	SECAM	768	576	4:3
	HDTV-720	1280	720	16:9
	HDTV-1080	1920	1080	16:9

supports a pixel height of 1280 and a pixel width of 720. HDTV standards are defined by the International Telecommunications Union (ITU-R) as having 1080 active interlace or progressive scan lines or 720 progressive scan lines. The aspect ratio is 16:9 (16 by 9), which means that there are 16 pixels horizontally for every 9 pixels vertically. Compare this to traditional TV which is 4:3 (4 by 3), or 12 pixels horizontally for every 9 pixels vertically. As you can see, HDTV uses a widescreen aspect ratio. Additionally, the audio transferred for HDTV can be encoded as Dolby Digital 5.1 for surround sound. Table 8-1 provides an overview of the specifications for these various video standards.

H.261 and H.264

In the preceding chapter, you learned that H.323 is a protocol used for VoIP implementation that is also used for video over IP. H.323 is the umbrella that covers multiple other standards for sending and receiving audio and video across packet-based (in our case, specifically, IP-based) networks. H.320 is a standard for sending multimedia content over ISDN lines. The H.320 substandards include H.221, H.230, and H.242. ISDN lines may use G.711 and G.723 just like IP-based networks. Both ISDN (H.320) and packet-based (H.323) networks support the H.261 and H.264 video codecs.

H.261 is a video codec that operates between 40 Kbps and 2 Mbps. It can adjust the quality of the video to match data transfer speeds within the 40 Kbps to 2 Mbps range. H.261 was originally designed to be used over ISDN lines and is now used in H.323 multimedia systems.

The evolutionary step from H.261 led us to *H.263*. H.263 is based on the lessons learned through implementations of H.261, MPEG-1, and MPEG-2. H.263 is provided in H.323 and H.320, as well as SIP implementations. Ultimately, if your video system supports H.263, you will find that it provides better-quality video than does H.261 at an equal bitrate. There is a version 2 of the H.263 standard known as H.263v2. Version 2 is sometimes called H.263+.

The current ultimate evolution is H.264. This codec provides a high level of data compression and is actually MPEG-4 Part 10. H.264 is exactly the same as the MPEG-4 standard in the implementation of Part 10.

MPEG-x

The MPEG standards are closely related to the H.26x standards. For example, MPEG-2 is the equivalent of H.262, and MPEG-4 Part 10 is exactly the same as H.264. If one standards body has developed a video compression standard that works well, why should another standards body start from scratch? They shouldn't. This is why many of the ITU and ISO standards overlap.

CERTIFICATION OBJECTIVE 8.03

Identify Benefits of Using Different Video Standards and the Impact on the Performance of the Network

Many factors can influence the performance of a video over IP system. These include the bandwidth available on the network, the video codecs implemented, and the number of users using the system. We'll look at each of these factors in this section. Additionally, we'll investigate the H.323 and H.320 standards in more detail with the intent of understanding the performance features or capabilities of these standards. We'll also look at the different Common Intermediate Format standards and their impact on performance.

Bandwidth Requirements

The phrase *bandwidth requirements* is usually used to refer to the needed data throughput in a given scenario. Notice the stress is on data throughput and not data rate. The difference between the two is simple: data throughput indicates the amount of useful data that can be sent across the network in a given interval of time, and the data rate indicates the total available bandwidth, including both useful data and management overhead. This management overhead can include acknowledgment frames with the TCP protocol and collision detection algorithms within broadcast domains. It reduces the total amount of throughput available for useful data. I'm using the phrase *useful data* to reference the data that the user perceives as valuable. This useful data would be the actual video frames in a video stream and not the authorization or synchronization data. The influencing factors that must be considered when determining bandwidth requirements include:

- Number of concurrent users
- Applications in use
- Percentage of active time
- Peak activity measurements

When calculating these factors, you will usually begin by determining the applications that will be used and the data rate demanded by each instance of the applications when they are communicating. From here, you will move on to calculate the percentage of time that the applications are actually busy on the network. You're now ready to consider the number of concurrent users that will be using the applications. Considering the percentage of active time, you should be able to calculate a peak load that will be used to determine the needed bandwidth in order to handle that load.

If your calculations reveal that your current infrastructure cannot handle the peak loads, you may have to segment your network into smaller groups of users or implement a technology that increases the overall bandwidth available. Strategically using multicasting instead of broadcasting can also reduce bandwidth consumption. Finally, you could consider implementing relay servers to retransmit video or audio streams to users on either side of the servers. For example, you could have a single server sending a stream to five servers, which in turn send the stream to 30 users each. In the end, you serve 150 users, but they all experience the service as if there were only 30 users viewing the stream, based on the redistribution model.

Codecs and Performance

The chosen codec has a tremendous impact on performance. Some codecs use very high compression levels and others use no compression. When using no compression, you will find the need for bandwidth is excessive and will usually prohibit the use of video over IP technologies. On the other hand, if you compress the video too much, the quality will become so poor that many users will choose not to use the system. The key is to strike a balance between compression and quality that works within your bandwidth constraints.

There are two impacting components in any video over IP solution: audio and video. The audio codec and the video codec, along with compression levels and packet lengths, will determine the overall bandwidth needed. Table 8-2 lists audio codecs that are common in VoIP and the bandwidth needs for a single connection.

TABLE 8-2

Audio Codecs
and Bandwidth

Audio Codec	Bandwidth	Actual Estimated Network Bandwidth Consumed
G.711	64 Kbps	87 Kbps
G.729	8 Kbps	32 Kbps
G.726	32 Kbps	55 Kbps

The *Actual Estimated Network Bandwidth Consumed* column contains a rough estimate of the bandwidth consumed due to network management overhead.

When you lower the bandwidth, as in Table 8-2, you lower the audio quality. This difference is a fact of audio codecs. If you compare CD-quality audio to AM radio–quality audio, there is a very noticeable difference; however, this lowering of quality may be accepted as a trade-off with video needs. You may find that your users prefer to have superior video quality and inferior audio quality in many scenarios. Alternatively, users may be more tolerant of reduced video quality for the sake of improved audio quality. The good news is that most systems will allow you to select the codecs for each of these (audio and video) so that you can tune the system to the users' expectations.

Video codecs also come with ranges of compression options. You can think of video compression as similar to image compression. If you open a bitmap image (a photograph, for example) and save it as a JPEG, you will usually have the opportunity to select the level of compression you want to apply. If you highly compress the image, you'll begin to notice tremendous loss of quality; however, you can usually use JPEG compress of about 70 percent without doing tremendous damage to the image (from a perceptive point of view). In similar ways, videos can be compressed so that they reduce the needed bandwidth for transmission. The later section "H.323 and H.320" will relay more information on video compression in video conferencing and video over IP solutions.

on the
job

As audio and video codecs become more capable (able to compress the data more and consume less bandwidth while providing great quality) or as bandwidth becomes more available, you will likely see more conferencing occurring from the desktop instead of in meeting rooms. Users can have their weekly staff meetings without ever leaving their offices. This decentralization of conferencing resources will increase productivity on the one hand and introduce new distractions on the other. Additionally, networking requirements begin to increase quite quickly when deploying video conferencing to the user's workspace.

CIF, SIF, QCIF, FCIF, and HD

Video signals come in different resolutions. The Common Intermediate Format (CIF) describes these different resolutions. Table 8-3 outlines these resolutions and their specifications.

SIF (the name stands for the *Standard Interchange Format*) is a resolution similar to that of a VCR. It uses a resolution of 352 × 240 for NTSC and 352 × 288 for PAL. SIF is defined in MPEG-1. HD resolutions are defined in the H.264 specification. Additionally, high-definition (HD) resolutions are often considered anything above standard definition (SD).

e x a m

🐶 **a t c h**
SD includes NTSC, PAL, and SECAM. HDTV is four times the resolution of standard television, and HD uses square pixels with an aspect ratio **of 16:9 as opposed to SD pixels with a ratio of 4:3. The resolution of HDTV is either 1080 or 720 lines.**

H.323 and H.320

The H.323 and H.320 standards, as was stated previously, encompass many different standards. Depending on the subprotocol used, performance will vary. For example, utilizing H.264 with a high resolution will require far more bandwidth than using H.263 with a low or moderate resolution.

H.323 and H.320 support the H.261, H.263, and H.264 video specifications. The H.261 specification was the first really practical video standard; however, it has been made obsolete by newer standards like H.263 and H.264. H.261 was designed to operate on ISDN lines. It supports from 1 to 30 ISDN lines running at 64 Kbps.

TABLE 8-3	Format	Aspect Ratio	Size
Video Format Types	CIF (also known as full CIF or FCIF)	4:3	352 × 288
	QCIF	4:3	176 × 144
	SQCIF	16:11	128 × 96
	4CIF	4:3	704 × 576

The result is a data rate from 40 Kbps to 2 Mbps. H.261 supports both CIF and QCIF resolutions and luma resolutions of 176 × 144 and 88 × 72. H.261 was the basis of a number of codecs or standards, including the following:

- MPEG-1
- MPEG-2/H.262
- H.263
- H.264

INSIDE THE EXAM

More on the H.2xx Standards

The popular video codec standards include H.261, H.263, and H.264. However, H.263 is now available in multiple revisions. We have the original H.263 and the H.263+. H.263+, also known as H.263v2, was enhanced to have better interoperation with the RTP protocol. The enhancements provide improved operations when the bit stream is packetized (running on a packet-switched network). Additionally, H.263+ reduces video delay and allows each video segment to be decoded independently.

H.263 was designed for video conferencing and for use on video telephones. The bandwidth used is between 20 Kbps and 24 Kbps instead of from 40 Kbps to 2 Mbps as with H.261. This improved use of bandwidth makes H.263 much more usable on today's networks. H.263, like many other voice and video over IP solutions, uses the Real-Time Transport Protocol (RTP) for stream control and management.

The H.263 codec supposed multiple resolutions and can implement CIF, QCIF, SQCIF, 4CIF, and 16CIF. SQCIF provides half the resolution of CIF and 4CIF provides four times the resolution of CIF. As you might have guessed, 16CIF provides sixteen times the resolution of CIF.

H.264 is a portion of the MPEG-4 standard. It is referenced as MPEG-4 Part 10. The purpose of H.264 was to implement a video standard that used less than half the bandwidth required by H.263 or MPEG-2. H.264 was designed with packet networks in mind from the beginning; therefore, modifications are not needed to make it play nice on an IP network.

Packet Size and Performance

Multimedia data packets come in varying sizes. We generally speak of a long packet or a short packet, and each packet type has its benefits and drawbacks. Since management information is related to each frame or packet, longer packets incur less overhead. This conclusion is simple math. If you have a total amount of data that you wish to transmit and you transmit that data with short packets, you will incur more overhead. That exact same amount of data sent with long packets incurs less overhead, since there are fewer packets. To clearly understand this concept, consider the fictional packet sizes in this table:

Total Data	Packet Size	Management per Packet	Total Packets	Total Management
100k	10k	1k	10	10k
100k	2k	1k	50	50k

As the table shows, a packet size of 10k results in 10k of management overhead, assuming a 1k measurement for management data per useful data packet. However, a 2k packet size increases the management overhead by five times (interestingly, the same factor by which it reduced the packet size). The resulting principle is that long packets usually incur less management overhead than short packets.

Long packets also reduce the packet processing load. The devices that forward the packets on the network must de-encapsulate the packets from the frames and re-encapsulate them to send them on to their destination. The result of using long packets is that there are fewer iterations of the de-encapsulation and re-encapsulation process per total data payload. Think back to the table that listed management overhead for a long packet as opposed to a short packet. The same factor applies here. The long packet configuration would demand only 10 iterations of the packet processing, and the short packet would require 50 iterations.

Another possible benefit of long packets is that you achieve greater network loading. More time is spent sending data as opposed to determining where to send data. This greater loading capability means that the network bandwidth may be more evenly consumed and the network performance may be more predictable.

The negative aspects of long packets become the positive benefits of short packets. Long packets do create more problems when packets are lost. Since there is so much data in an individual packet, the loss of just one packet can greatly impact the quality of a video or audio stream. Shorter packets are less harmful in such situations. In fact, some scenarios where a packet is lost will hardly be noticeable to the user.

When longer packets are used, latency is increased. Latency is a measure of the difference in time from when a packet is transmitted to when the same packet arrives. Most streaming video and audio technologies, as well as VoIP in general, have upper boundaries that must be imposed on latency variables. One way to reduce latency is to shorten the packet size. Smaller packets move through the network faster due to the fact that there are simply smaller bits to transmit and Quality of Service processing can "fit" the smaller packets in more easily.

Finally, short packets have less need for fragmentation. Upper-layer (OSI) data must be chunked or fragmented if it is too large for the lower TCP, UDP, or IP layers. Fragmentation incurs extra processing for both the sender and the receiver and can cause unnecessary extra delay. A simple method for removing much of the fragmentation is to set your application to send data chunks that result in smaller packets and, therefore, less fragmentation.

T-120

The *T.120* recommendation was created by the ITU and specifies protocols that enable real-time communications. This recommendation is used by Microsoft's NetMeeting products and Lotus SameTime, among other products. These applications support application sharing, video conferencing, and audio over IP with the T.120 series of recommendations. The provided recommendations are outlined in Table 8-4. The term multipoint indicates that the T.120 system allows for application sharing, text messaging, and conferencing among more than two users.

TABLE 8-4   T.120 Standards

Standard	Capability
T.120	Protocols for multimedia conferencing
T.122	Multipoint communication service (used for service definition)
T.123	Data protocol stacks that are network-specific and provide for multimedia conferencing
T.124	Generic Conference Control
T.127	Binary file transfer specification
T.128	Application sharing (previously called T.share)
T.134	Text chatting
T.135	Transactions between the users and the reservation system within T.120 conferencing
T.136	Remote control protocol
T.137	Virtual meeting room administration (services and protocols are specified)

When you implement a T.120-based system, it is important to consider the impact of the advanced features such as remote control, binary file sharing, and text chatting. All of these add extra overhead above what is imposed by standard video/audio conferencing solutions. For example, a binary file transfer can be many megabytes in size, and numerous text messages can add a large sum of consumed bandwidth as well. Ensure that you have sufficient bandwidth to handle the extra demands of the solution.

CERTIFICATION SUMMARY

This chapter covered the basics of video solutions that operate on IP networks. The quality of service factors discussed in the next chapter will impact the video technologies discussed here as well as the voice technologies discussed in earlier chapters. Key concepts to take away from this chapter include the facts that video conferencing systems can reduce costs but also introduce additional network resource needs, that video standards provide interoperability, and that different codecs result in different demands on the network infrastructure.

✓ TWO-MINUTE DRILL

Identify Methods Used for Rich Media Transmission

❑ Webcasting includes on-demand webcasts and live webcasts.

❑ A live webcast is streamed across the Internet in real time.

❑ Streaming media transmission provides reduced bandwidth and reduced likelihood of content theft.

❑ File-based media transmission may require the entire file to be downloaded before it can be played.

❑ Progressive download is a file-based technology that allows the file to be played back while it is downloading.

Identify Methods of Encoding, Decoding, and Compression

❑ The NTSC standard is used in the United States, Japan, South Korea, the Philippines, Canada, and some other locations.

❑ PAL is heavily used in Europe, South America, Australia, Asia, and Africa.

❑ SECAM is used in France, in parts of Africa, and in the former USSR.

❑ H.261 is a video codec that operates between 40 Kbps and 2 Mbps.

❑ H.263 is a video codec that operates between 20 Kbps and 24 Kbps.

❑ H.264 is a video codec that was created to run over packet-switched networks using the Real-Time Streaming Protocol (RTP).

Identify Benefits of Using Different Video Standards and the Impact on the Performance of the Network

❑ To determine the needed bandwidth, you must determine the encoding technology used, the number of concurrent users, and the way the users will use the technology.

❑ H.264 was designed with support for packet-based networks and may reduce bandwidth over H.263 with the right implementation parameters.

❑ An implementation of H.264 with a high-resolution video format will consume more bandwidth than an implementation of H.263 with a low-resolution video format.

SELF TEST

The following questions will help you measure your understanding of the material presented in this chapter. Read all the choices carefully because there might be more than one correct answer. Choose all correct answers for each question.

Identify Methods Used for Rich Media Transmission

1. Which technology allows you to begin downloading a file and start playback after buffering, but will not allow you to start playback at a specified point in the file?

   A. Streaming

   B. Casting

   C. Progressive download

   D. FTP

2. You want to stream a live presentation across the Internet. There will be more than 70 users watching the stream. It is a business update presentation given by your CEO. What technology will you use for this need?

   A. File-based video

   B. Progressive download

   C. Webcasting

   D. Broadcast TV

Identify Methods of Encoding, Decoding, and Compression

3. You are working with a television station in Japan and another in South Korea. What video standard is likely to be used with the video stored in the archives of these stations?

   A. NTSC

   B. PAL

   C. SECAM

   D. DivX

4. Which video codec standard has a data rate of between 20 Kbps and 24 Kbps?

   A. H.264

   B. H.263

   C. H.261

   D. H.262

Identify Benefits of Using Different Video Standards and the Impact on the Performance of the Network

5. Which of the H.2xx standards was created from its inception to support packet-based networks?

   A. H.261

   B. H.262

   C. H.264

   D. H.297

6. You are implementing a video conferencing over IP solution. You need to ensure that you have sufficient bandwidth. You know that an upgrade to the infrastructure will be needed, since you still have 10 Mbps links in many areas. What factors do you need to consider when upgrading the infrastructure to support video over IP? (Choose all that apply.)

   A. Number of concurrent users

   B. Bandwidth requirement of applications

   C. Management overhead

   D. 10 Mbps compression

SELF TEST ANSWERS

Identify Methods used for Rich Media Transmission

1. ☑ C is correct. Progressive download allows you to start the downloading of a file and begin playback after some amount of buffering.

☒ **A, B,** and **D** are incorrect.

2. ☑ C is correct. Webcasting will provide the needed capability. The key here is that you want to stream the session across the Internet, which would rule out TV broadcasting.

☒ **A, B,** and **D** are incorrect.

Identify Methods of Encoding, Decoding, and Compression

3. ☑ A is correct. Japan and South Korea use the NTSC standard like the United States.

☒ **B, C,** and **D** are incorrect.

4. ☑ B is correct. H.263 uses a video stream data rate of between 20 Kbps and 24 Kbps.

☒ **A, C,** and **D** are incorrect.

Identify Benefits of Using Different Video Standards and the Impact on the Performance of the Network

5. ☑ C is correct. The H.264 standard was created from its inception to support packet-based networks. H.263 also supports packet-based networks, and H.263+ or H.263v2 enhanced this support.

☒ **A, B,** and **D** are incorrect.

6. ☑ **A, B,** and C are correct. In a situation like this, you must ensure that you consider the number of concurrent users, the bandwidth demands of the applications, the utilization methods, the management overhead, and peak load times.

☒ **D** is incorrect.

9

Quality of Service

> *The quality of expectations determines the quality of our action.*
>
> —Andre Godin

Quality of Service (QoS), in the domain of computer networking, can be defined as "the tools and technologies that manage network traffic in order to provide the performance demanded by the users of the network." As Andre Godin stated, though with different original intentions, the quality of expectations determines the quality of our action. I will pour a different meaning into this statement. Here is a paraphrase of the quote, "The accuracy of your requirements determines the benefits of your efforts." If the quality requirements are not accurate, the QoS solutions that you implement are not likely to prove beneficial to your users. In fact, a poorly designed QoS solution can cause more trouble than it solves. Therefore, successful QoS implementation begins with quality requirements development.

Once you have a clear picture of the quality requirements in your network, you can begin to consider the different QoS tools that may assist you in meeting those requirements. These tools will be targeted at reducing latency, jitter, and delay, and they may accomplish these tasks through the use of prioritization, queuing, and dedicated connections. You can think of QoS tools as a collection of technologies that allow you to use any physical network more efficiently. QoS tools cannot make the network faster, but they can enforce or suggest better use of the network. This fact will become increasingly clear as you read through this chapter.

Before you can fully appreciate QoS, you must understand the more general concept of network performance analysis. Therefore, we will begin by discovering the terminology, challenges, and potential solutions related to network performance problems.

CERTIFICATION OBJECTIVE 9.01

Analyze Network Performance

When you implement *video or voice over IP (VoIP)* on an existing network, it is essential that you consider the impact of such an action. Implementing VoIP on an in-place data network is a lot like adding go karts to an interstate highway.

The go karts do not individually take up as much space as a traditional car, but they still consume a lot of highway space due to free space required between the go karts and other vehicles. If you think about it, you would need to leave just as much free space for a go kart as you would a full-sized automobile. You can think of this free space as being like management overhead. If the same protocols are used to transmit voice packets and data packets, the overhead will be similar unless the protocol has specific characteristics that allow it to detect a voice packet and handle it differently. This is one reason VoIP communications use specific protocols aimed at sending voice data instead of traditional protocols (FTP, HTTP, etc.) aimed at sending standard data.

The most important point of this analogy is that adding a new type of data to the network does more than consume bandwidth. This new type of data will likely have different characteristics and therefore impact the network differently. For example, does the data vary in size greatly from packet to packet, or is it more consistent? Traditional data tends to vary (think about e-mail and HTTP traffic), whereas voice data is more consistent. Therefore, you may have an easier time calculating the bandwidth needed to handle a particular number of calls than you will a particular number of web site connections. That's the good news. The bad news is that streaming video, text messaging, and other technologies can change the landscape quickly.

This section will provide you with an overview of the tasks involved in network performance management. You'll learn the definitions of important terms related to network performance and how to create a baseline of your network. Finally, you'll discover the impact and needs of converged technologies, such as VoIP and video over IP, on your network.

Network Performance Terminology

Any topic that is discussed must include a shared vocabulary. For this reason, I'll begin the coverage of network performance on converged networks by defining some common terms that are utilized in the network performance and QoS domains. These terms include:

- Bandwidth or data rate
- Throughput
- Delay
- Jitter
- Loss or packet loss
- Talker overlap
- Availability

Bandwidth or Data Rate

Bandwidth can mean different things in different contexts. Sometimes the difference is subtle, and sometimes it is large. In this context of converged technologies and standard networking, *bandwidth* is a reference to *data rate*. If we say that a particular network connection has 10 megabits per second (Mbps) of bandwidth, we're saying that data can be passed across the connection at a rate of 10 million bits per second. This measurement is the bandwidth or the data rate. Stated differently, 10 Mbps represents the rate of data flow. This data flow is inclusive of management overhead and the actual data that can be transmitted across the connection.

In the world of radio frequency (RF) engineering (which includes wireless local area networks), bandwidth is a reference to the range of frequencies within which a connection operates. For example, an IEEE 802.11g channel uses 22 MHz of bandwidth. An IEEE 802.11b channel also uses 22 MHz of bandwidth. Now, this is where the term bandwidth can get confusing. When referencing the frequencies that are used, both IEEE 802.11g (ERP-OFDM) and IEEE 802.11b (HR-DSSS) use 22 MHz of bandwidth; however, the bandwidth provided by 802.11g is 54 Mbps and the bandwidth provided by 802.11b is only 11 Mbps. This seeming conflict is because the first use of the term bandwidth references the frequency range and the second use of the term bandwidth references the data rate. For this reason, much of the wireless networking literature uses the phrase *data rate* to reference the total flow of data that can pass across the connection and the term *bandwidth* to reference the range of frequencies used to establish the connection. In the end, the wired networking world generally applies the word bandwidth to mean the total data flow capability of a link and the wireless networking world generally applies the phrase data rate to mean the same thing. However, it is also acceptable to use the phrase data rate in the wired networking domain.

Just in case you're wondering how 802.11b and 802.11g can each use 22 MHz of frequency bandwidth and yet provide different data rates, let me give you the short explanation. 802.11b uses the direct sequence spread spectrum (DSSS) signaling architecture and 802.11g uses the orthogonal frequency division multiplexing (OFDM) signaling architecture. Specifically, 802.11b uses high-rate (HR)-DSSS and 802.11g uses extended-rate PHY (ERP)-OFDM. The point is that the two modulation systems provide for differing data transfer mechanisms. The result is that the HR-DSSS modulation system cannot transfer as much data as the ERP-OFDM system using the same frequency bandwidth. You can think of it in much the same way as different problem-solving techniques we use as humans. If my problem-solving technique involves trying every possible solution to see if it works and your technique involves a pre-trial algorithm that weeds out 82 percent of the possible solutions in 5 percent of the time it takes to try the same 82 percent of the solutions, you will most likely find the solution to the problem in much less time than me. Did you have more "brainwidth" (this term is my own and I'm sure you'll not find it in the dictionary) than me? Not necessarily (though that may be true). The reality is that you simply used your brainwidth better than I used mine. That's also the big difference between 802.11b and 802.11g. 802.11g uses the bandwidth better, which results in a higher data rate.

Throughput

Now that you understand bandwidth, you can properly understand throughput. *Throughput* can be defined as the actual amount of user data (sometimes called useful data) that can be passed through a connection in a given time period. For example, a wireless connection with a 54 Mbps data rate may have a throughput value of only 22 Mbps. Bandwidths or data rates force an upper threshold on throughput. If the bandwidth of a wired connection is 100 Mbps, you cannot have a throughput rate of 10 Gbps. It's simply not possible. In fact, the throughput will always be a factor of bandwidth, overhead, and efficiency. The last of these three, efficiency, is often overlooked.

Network efficiency is more than traffic shaping, though that is an important component that will be covered later in this chapter. Network efficiency is impacted by devices in the path of a connection, and these devices include routers, switches, hubs, firewalls, and even the network interface cards (NICs) in client devices. I've seen many situations where the device drivers for a NIC were poorly written, resulting in underutilization of the network's potential. Stated differently, the network card can't talk as fast as it is rated. We see this situation with wired and wireless NICs all the time, and it is one of the major factors that should be included in vendor selection.

In addition to the NICs, the infrastructure devices can have a tremendous impact on network efficiency. Routers need to have plenty of memory, sufficient processors, and well-written code. Switches need to operate as close to line speed as possible. Firewalls can benefit greatly by increased RAM and processing power as well, and hubs simply shouldn't be used in modern networks where performance is paramount.

Security solutions, other than firewalls, can also have a tremendous impact on throughput. Virtual private networks (VPNs) employ encryption, and the use of encryption impacts the throughput for the connection. As the data comes into the VPN device, it must be decrypted before being forwarded on to the destination.

This decryption increases delay and reduces throughput for the connection that is employing the VPN technology. In the same way, when a client sends data to a VPN device that must be passed through to the remote end, the VPN device must encrypt the data, and again, this task incurs overhead. VPN technologies can be thought of as imposing overhead on the connection.

Other overhead factors include IP headers and checksums, TCP headers, and Application layer and Session layer information. For example, the TCP header is 20 bytes long. This header means that every packet of data that your application transmits will have an extra 20 bytes appended, if the application uses TCP. If the application uses UDP, the header size will be only 8 bytes. As you can see, a UDP packet (or datagram) contains 12 bytes less overhead than TCP. This smaller overhead is why UDP is traditionally used for real-time applications. Though UDP does not have delivery verification mechanisms, its reduction in overhead makes it more appealing.

You're probably beginning to see how the following formula works:

$$\text{Throughput} = \text{Bandwidth} - (\text{Bandwidth} \times \text{Overhead}) \times \text{Efficiency}$$

If bandwidth is 10 Mbps and overhead is 20 percent and efficiency is 80 percent, you end up with about 6.4 Mbps of throughput. The important thing to take away from this example is not an exact measurement or even the memorization of this formula, but a realization that you will never achieve the data rate as throughput, due to overhead and less-than-perfect efficiency. The reality is that we often determine throughput by monitoring performance rather than calculating expectations. The work is usually performed by using this monitoring method because very few vendors will admit that their NICs do not run at 100 percent efficiency. You will have to test this capability for yourself.

Delay

Delay is the measurement of time required to move data from one point to another across a network. For useful purposes, delay is usually measured from endpoint to endpoint. In a VoIP implementation, it would be a measurement of the delay in transmitting the VoIP packet from the sending phone to the receiving phone—whether that phone is a hardware-based or software-based phone. This delay is also known as *propagation delay*, as it is a measurement of the time that it takes for the data signal to propagate from one endpoint to the other.

Delay is impacted by a number of factors. These factors include bandwidth, efficiency, and utilization. Bandwidth impacts delay because it limits the amount of total data that may transit the network. When more data can move across the network in a given window of time, delay usually decreases for each transfer.

Efficiency has a tremendous impact on delay because a slow router or switch can cause major increases in delay even though very high-bandwidth cabling and signaling technologies are used. Ultimately, I'm saying that all gigabit switches and routers are not created equal. For example, I have a router that is classified as an enterprise router by the vendor that sells it, and this router does not perform nearly as well as an inexpensive Netgear router that was manufactured during the same time according to the same signaling standards (100 Mbit Ethernet). You cannot assume that all devices by all manufacturers perform the same just because they support the same standards. This dilemma is why testing is so important to those of us who actually have to implement the technology.

Of course, utilization impacts delay through the simple math of large numbers. Think of it as being like a highway. If you have a specific route that you travel to work and you've traveled that route for a few years, you've probably noticed that congestion is getting worse (at least, if you're in a larger city). Why is this congestion happening? It's the math of large numbers. There are more cars trying to get to the same relative area today than there were a few years ago. Assuming that the lane-width (the number of lanes) has not changed, the cars must either move more efficiently or take longer to get to their destinations. The point is simple: as more cars travel the same route you travel to get to work, your delay increases. There are really three possible solutions that we've seen implemented:

- Install more generic lanes.
- Flag one or more lanes as high-occupancy lanes.
- Teach people to drive better.

Installing more generic lanes will increase the lane-width (similar to network bandwidth) and reduce your delay. However, this solution is a very expensive one and can slow down traffic while the extra lanes are being installed. If we apply this concept to the data network, we face the same negative factors. It can be very expensive to rip out a 100 Mbps infrastructure and install a 1 Gbps infrastructure, if the cabling that is already installed cannot handle the faster speeds. Additionally, there may be major disruptions in service during the upgrade. These factors must be considered.

The second option, of flagging a lane for high-occupancy vehicles (HOV), will not have the same impact as adding more lanes, but it will be far less expensive and should not result in as many delays during initial implementation. Installing HOV lanes is analogous to implementing QoS on a network. QoS cannot increase the available bandwidth, but it can indicate that specific traffic types have a higher priority.

The third and final option is probably the most difficult in real life. Teaching people to drive better is a coded way of saying that we want to teach people to work together so that all drivers arrive at their destination in a window of time that is relatively acceptable to all drivers. In other words, one driver is not gaining an advantage while delaying other drivers even more. Ultimately, we're talking about implementing better collaborative driving algorithms. Thankfully, while this task is very difficult with human drivers on the highway, it is fairly simple to accomplish with data that is driven across our networks.

Reflect back on the explanation of why 802.11g provides a higher throughput and data rate than 802.11b even though they use the same 22 MHz of frequency bandwidth. You could say that both 802.11g and 802.11b use 22 lanes for data transfer, but 802.11g gets nearly five times more data through. How is this difference in throughput possible? The answer is that 802.11g drives better.

I'm sure you're beginning to see how these different terms relate to one another. Bandwidth or data rate provides the upper boundary of data throughput, and the actual throughput is a factor of bandwidth, overhead, and efficiency. Delay is impacted by bandwidth and efficiency. Utilization of the network also has a large impact on delay. In the end, these three (bandwidth, throughput, and delay) are interrelated and you can seldom impact one without impacting one of the others. For example, if you want to decrease delay, the actions you take will likely also increase throughput—even if bandwidth is not changed; however, if you increase bandwidth, both throughput and delay will usually be impacted as well.

e x a m

W a t c h *The ITU-T recommends in Recommendation G.114 that the round trip time (RTT) or round trip delay not exceed 300 ms in a telephony network. A quick calculation reveals that this RTT cannot be met unless the unidirectional delay is 150 ms or less as an average.*

VoIP and Delay VoIP implementations must address delay from multiple perspectives. Effectively, delay is broken into components, and the delay is considered at this component level. For this reason, you'll want to be aware of delay from three perspectives:

- Encoding delay
- Packetizing delay
- Network delay

The concepts I've described up to this point are network and packetizing delay, with the greatest emphasis on network delay. In a VoIP system, we must remember that the real delay that is measured is the delay that occurs between when one user speaks the words into the phone and when the other user hears the words from her speakers (desktop speakers, handset speakers, headphones, etc.). For this reason, we must also consider encoding delay. The time it takes to encode a G.711 audio stream is different than the time it takes to encode a G.729 audio stream. However, the G.729 audio stream sends the same sound waves through the VoIP network with smaller packet sizes. Therefore, we see an increased encoding delay for G.729 and a reduced network delay. The reality is that the reduced network delay is usually far greater than the increased encoding delay, and so G.729 is going to be a better VoIP protocol from a delay perspective.

Packetizing delay, which I've only hinted at before now, is the delay incurred by putting the voice data into UDP and IP packets. You can and usually do have more than one voice frame in a single IP packet (remember the HOV lane analogy?). When you have more voice frames in a single IP packet, packetizing delay increases; however, network delay decreases due to reduced bandwidth consumption, with fewer UDP and IP headers traveling the network for the same amount of audio data. When you have fewer voice frames in a single IP packet, packetizing delay decreases, but network delay may increase with more UDP and IP headers required.

on the job

The difference between delay and round-trip time (RTT) is simple, though many seem to confuse the two. Remember, delay is the time it takes for data to go from source to destination across a network. RTT is the time it takes for a request to get from the requestor to the requestee added to the time it takes to process the request added to the time it takes for the response to get from the requestee to the requestor. RTT may be impacted by the number of requests flooding into the requestee at a given moment, while delay will not be, since it only accounts for the time it takes to travel from request to requestor or vice versa.

Jitter

While delay is a measurement of the time it takes for data to travel from endpoint to endpoint, *jitter* is the variance that occurs in this delay. There will always be jitter in any network that is not an exclusive, dedicated network between two endpoints. For this reason, jitter must be considered and addressed. I've defined jitter as the variation in delay; however, others frequently define jitter as the variation in the time between packets in a stream arriving at a receiving endpoint.

We are basically saying the same thing. I focus on the delay, and the alternate definition focuses on the actual time span between arriving stream packets. If there is a variation in the time span between packets in a stream, it is because there is a variation in delay. If there is a variation in delay, it will result in a variation in the time span between arriving packets in a stream.

Why is jitter a problem? Jitter is a problem in VoIP networks because it can cause calls to drop (if the variation is too great) and it can lead to poor quality. Have you ever been listening to someone on the telephone while waiting for your opportunity to talk? If so, you've probably had the experience of interrupting someone accidentally. You thought they were finished talking because of an unusually long pause, but they were actually just catching their breath or trying to think of the right word. The same thing can happen on a VoIP network because of jitter. The jitter could cause a break in the audio stream that results in an artificial pause. The outcome is that you begin talking over the other party. To avoid this circumstance and also to provide consistency in the audio streams, many systems implement something called a *jitter buffer*.

Jitter buffers work at the receiving endpoint in most cases. Rather than sending the voice frames immediately up for processing, the jitter buffer holds the incoming voice frames for a very short time and then begins sending them up for processing at regular intervals resulting in a smooth stream of audio projected through the speaker. The hold time or delay is very short, usually well under 200 ms. More often than not, the jitter buffer will be 100 ms or less.

Figure 9-1 illustrates the concept of a jitter buffer. Notice that the incoming packets have variable delays. The delay between the first packet coming into the buffer and the second packet is 15 ms, while the delay between the second and third packets is 30 ms. Within the jitter buffer, the packets are stored and sent out at 20 ms intervals.

It's very important to understand the difference between a static jitter buffer and a dynamic jitter buffer. Static jitter buffers are of a fixed size. This fixed size means that the buffer is vulnerable to underruns—meaning that there are too few packets in the buffer to service the needed upstream rate to the telephone or software. Using a dynamic jitter buffer may resolve this issue. A dynamic jitter buffer evaluates the incoming delays and adjusts the jitter buffer size accordingly. While the dynamic

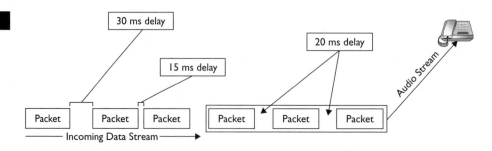

FIGURE 9-1

Jitter buffer

jitter buffer may utilize more processing power on the receiving device, it is usually preferred due to its ability to function best in a network with variable utilization levels.

Loss or Packet Loss

Packet loss or simply *loss* is the term used to describe packets that are missing in a stream. At the receiving end, lost packets can be detected through the analysis of sequence numbers. When packet loss occurs, there are different techniques that can be used to address it. These techniques include:

- **Ignoring packet loss**   This solution is actually what most VoIP implementations do. Rather than trying to accommodate for lost packets or retransmit them, the lost packets are simply ignored up to a specified threshold.
- **Interpolating lost packets**   Interpolation is the process of estimating unknown values based on known values and patterns. Interpolation can be very processor intensive and can increase delay. Depending on the codec used, interpolation may not be an option.
- **Retransmitting packets**   Retransmission is very common with standard data transfers; however, it is not very useful with VoIP streams, since the retransmitted data would likely arrive too far out of synchronization. The receiver would have to send a request for the sender to retransmit the data, since UDP is used. This solution means that the sender would have to buffer transmitted packets for some period of time, and it also means that the delay between the sender and receiver would have to be phenomenally low. For these reasons, retransmission is seldom used with lost packets in VoIP systems.
- **Transmitting redundant packets initially**   An alternative to retransmission is to transmit multiple copies of the same data all the time. The problem with this solution is that it will unnecessarily utilize network bandwidth when there are no dropped or lost packets.
- **Using codecs with higher compression**   Implementing codecs that compress the audio to smaller sizes, and therefore smaller packets, may reduce packet loss to an acceptable level. For example, implementing G.729 instead of G.711 can have a big impact on reducing packet loss.
- **Upgrading the network**   Of course, when all else fails, you can upgrade the network infrastructure to be able to handle VoIP better. Sometimes this solution means installing a faster standard, and sometimes it just means installing specialized equipment that can better handle VoIP.

Availability

The ultimate requirement of any network is that it be available when it is needed. If it performs well from 9 A.M. to 10 A.M., but heaviest utilization time is from 10 A.M. to 12 P.M. and the network performs poorly during these hours, the network is not sufficient. The network needs to perform well and be available when it is needed. *Availability* is the term used to reference the requirements related to network up time. How much of the time must the network be up? What does it mean to say that the network is up? Does it mean only that the network has to be there regardless of delay, jitter, and packet loss? More than likely, when you say that the network must be up a particular portion of the time, you mean that the network must be there and must perform as required. This definition is the core concept of availability.

CERTIFICATION OBJECTIVE 9.02

Demonstrate Application of Traffic Engineering Concepts

How do you ensure availability? In order to make sure the network can perform as required and perform in this way when required, you must know how your network is currently performing and how it will be impacted once you've implemented voice or video over IP. These two tasks will be outlined next: capacity baselining and measuring the impact of convergence.

Network Capacity Baselining

Network capacity baselining is a process that results in the documentation of the network's performance. The process involves measuring the capacity of the network and the standard operating efficiency. The key to remember is that baselining is not estimating. Baselining involves the actual measurement of the network's performance using network analyzer software and hardware. Estimating is a process that uses computations involving bandwidth, concurrent users, and overhead averages. There are pros and cons of each method as outlined in Table 9-1.

Capacity baselining is very accurate, since you are actually measuring or reporting on real network performance. Of course, this measurement assumes that the network is in place or that you can implement a scaled version of the network and extrapolate estimates from this smaller test network. The latter option of performing capacity

TABLE 9-1	Capacity Baselining	Capacity Estimating
	Very accurate to within 5–10 percent	Accurate to within 20–40 percent
Capacity Baselining Versus Capacity Estimating	Time-consuming process	Fast process
	Resource-consuming process	Minimal resource consumption
	Greater cost	Less cost

baselining against a scaled network is a trade-off between the two extremes and can be useful, since it will be more accurate than simple capacity estimations and cheaper than capacity baselines. Capacity baselining is also very time consuming, since you must gather all of the data. There is an exception to this time consumption, and that exception is seen when you have an infrastructure that is self-reporting. A self-reporting infrastructure is an infrastructure that generates capacity baselining reports (among others) automatically. While saving you time, a self-reporting infrastructure is likely to be more expensive and more resource intensive (the report processes require consumption of resources that would otherwise be available for users). Yes, there is always a trade-off.

on the **job**

It is important to remember that the term baseline is used in many knowledge domains, such as project management and even page layout and design for graphics designers. The term means slightly different things in these different knowledge domains. For example, a baseline—in project management—is a view of the project (schedule, budget, quality, etc.) at a specific point and time that can be compared with later states in order to calculate changes in the health of the project. In the case of network analysis, it is a measurement of the network's performance at a given state that can be used to determine the impact of additional applications or services on the network.

Defining a Baseline

Now that you understand what a baseline is and the fundamental difference between a baseline and an estimation, you can analyze the different tools and methods used to build a network capacity baseline. These tools include:

- Network analyzers
- Operating system reports
- Specialty applications

Network Analyzers   Many vendors have placed both hardware and software network analyzers on the market. The software-based network analyzers run on PCs or laptops with the appropriate NIC installed. These network analyzers may support both wired and wireless analysis, or they may be dedicated to only one of these network types. The wired network analyzer software packages are usually compatible with a very large number of NICs; however, the wireless network analyzer software packages usually have a very limited list of supported wireless NICs. If you need to perform wireless network baselines, make sure you have the right wireless NIC or can buy one before you select a wireless network analyzer.

The network analyzers can be used to capture information such as the network utilization and the type of traffic that is consuming the network resources. Depending on the application, you may be able to report on the most heavily used protocols or applications on the network. Network analyzers can usually decode the packets and allow you to see the actual payload (contents of the network frames) as well.

Operating System Reports   Many network operating systems (NOSs) provide built-in reporting methods related to network performance. These reports may be generated and stored on the system's hard drive, or they may be retained temporarily in memory and only be reported if they are queried. An example of this method is the Performance tool in Windows operating systems. The Performance tool is actually a Microsoft Management Console (MMC) snap-in configuration that includes an ActiveX control used to read performance statistics and an interface for configuring logging and alerts. Figure 9-2 shows the Performance tool on a Windows Vista computer with the network counters running.

Specialty Applications   You can also purchase specialty applications that are designed to gather network performance statistics and report on anomalies and results. Figure 9-3 shows the nGenius software solution that reports on network performance and behavior.

---

e x a m

ⓦatch

It is important to remember that you can calculate and store multiple baselines. A baseline is a picture of the network's performance at a point in time, and you can compare the values in one baseline with the values in another. You can also compare values of many baselines over time. This latter process is sometimes called trending.

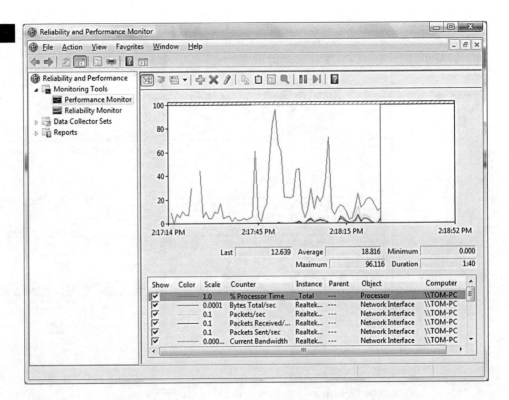

FIGURE 9-2

Performance
monitoring in
Windows Vista

Impact of Converged Applications on the Network

Once you've established the network capacity baseline, you can simulate or estimate
the impact of adding converged applications to the network. For example, you can
answer questions like these:

- Will the network be able to support VoIP communications?
- Will the network be able to support on-demand video?
- Will the network be able to support streaming live video?
- Do I need to upgrade the network in order to support converged technologies?

Supporting VoIP on the Network

Supporting VoIP communications requires a network with low latency and
sufficient bandwidth. If your network baseline shows that the network is currently
at 60–70 percent utilization, it is not likely that the network will support VoIP
effectively without some major changes. You'll learn more about the impact of

FIGURE 9-3

nGenius network
performance
management

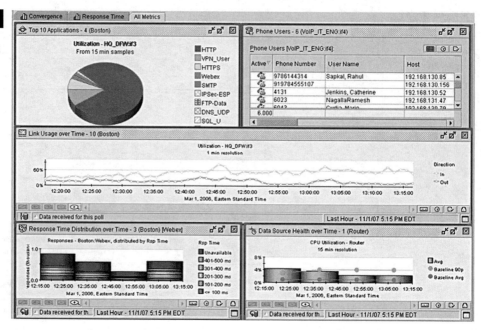

adding VoIP to a network in the later section titled "Performance Requirements" in the Defining QoS portion of this chapter. For now, it is sufficient to know that you have to have the bandwidth to support VoIP, otherwise, your users will be frustrated with the resulting system. You may have to upgrade the network in order to accommodate VoIP.

Supporting Video over IP on the Network

Video over IP is far more impacting on the network than VoIP. This additional impact is because video streams consume much more bandwidth than audio streams. Consider this example: a video stream almost always includes an audio stream. Stated differently, the audio stream that is required of VoIP is still there when you implement video over IP. Therefore, the demands on the network are much higher. There is some good news, however. In most cases when VoIP and video over IP are implemented, the majority of users will use VoIP and a small portion of the users will take advantage of video over IP at any given time. The ratio is often seen to be 10:1 or better, meaning that ten users are using VoIP for every one user using video over IP at a given time. The result is that the impact of video over IP may not be as damaging as the first glance makes it appear.

It is important that we dose this good news with a reality check. Video over IP is always intensive in comparison to VoIP on a one-to-one basis. Additionally, video over IP introduces the difficult challenge of dealing with spikes in network utilization. For example, during a video conference that involves six people, a very large amount of network bandwidth may be consumed. This conference may only last ten minutes, but the other users on the network still need to their work done and the conference attendees need acceptable video conferencing performance. How do you solve this problem? The answer is really twofold: ensure you have sufficient bandwidth and use QoS to manage the traffic effectively when the video conferencing is taking place.

Upgrading the Network to Support Converged Technologies

There are ultimately three strategies that you can employ to get the performance you need for VoIP out of your network. First, you can increase the bandwidth. Second, you can decrease utilization. Finally, you can implement QoS. The first two strategies will be covered here, and the implementation of QoS will be covered in a later objective covered by this chapter titled "Define QoS, Describe Implementation Techniques, and Show the Importance of QoS."

Increasing Bandwidth   A more accurate statement of this strategy might be increasing available bandwidth. This more accurate statement reveals that increasing bandwidth is not always about upgrading to a different and faster networking standard. I alluded to this earlier, but the reality is that two different 100 Mbit switches can perform very differently. One may function in such a way that it would be better called a 70 Mbit switch, while the other may perform closer to the standard specification. For this reason, you can often increase available bandwidth by using better infrastructure equipment to communicate across your in-place cabling. This strategy will keep you from having to replace all the cabling in order to support a faster standard.

Sometimes you have to upgrade the entire infrastructure or portions of it in order to get the performance you need. Certainly, if you are currently running a 10 Mbps infrastructure, it's time to upgrade. Implementing 100 Mbps down to the endpoints with a 1 Gbps core is still very common today; however, it is not uncommon to see gigabit to the endpoints. A network that supported VoIP five years ago may not be able to do so today. The new features may impose too large of an impact on the network.

Decreasing Utilization An alternative to upgrading the network is to downgrade the utilization. I've seen networks that were running services and applications that they no longer needed. Some modern networks are still running NetBIOS protocols even though they no longer need them. I'm not suggesting that no network needs the NetBIOS over TCP/IP support that is in Windows; however, I am suggesting that many networks do not need it and they are still running it. This practice is a bad idea and only results in unnecessary overutilization. Removing protocols like NetBIOS and services like WINS (Windows Internet Naming Service) can help reduce utilization.

Implementing and enforcing network utilization policies can also help decrease utilization. For example, you may choose to enforce a policy that disallows streaming radio from the Internet down to your client computers. Many network administrators tell their users that streaming radio and video as well as real-time weather applets and the like are not allowed. Other network administrators implement policies, firewalls, and endpoint security solutions that prevent users from using such applications even if they wanted to. I'm definitely in favor of the restriction enforcement, but some companies will not adopt such a strict policy. You may find it easier to convince users to stop using such applications if they begin to experience problems with their VoIP connections. Sometimes it takes a little negativity to move people beyond their habits.

CERTIFICATION OBJECTIVE 9.03

Define QoS, Describe Implementation Techniques, and Show the Importance of QoS

The definition of QoS, as provided in the opening paragraph of this chapter, needs to be mined further. That definition was, "the tools and technologies that manage network traffic in order to provide the performance demanded by the users of the network." In order to perform this mining expedition, we'll look at three topics in detail:

- Performance requirements and levels of traffic prioritization (performance demanded by the users)
- QoS types (tools and technologies)
- Application of QoS types (managing network traffic)

Performance Requirements

Performance requirements are important with an information systems project, and a VoIP or video over IP implementation is no exception. In order to understand the performance requirements, you'll need to understand just how VoIP impacts a network. This impact will be a factor of the packets generated by VoIP applications and the requirements of those packets (for example, they should be delivered quickly and sequentially).

VoIP Packet Characteristics

VoIP packets vary in their characteristics, depending on the codec and VoIP protocols used. The most important impacting characteristics are the codecs and the sizing used. Different codecs work at different bit rates. These bit rates, in addition to compression techniques, affect the final size of the VoIP packets. Table 9-2 lists the common voice codecs and their bit rates. You'll also notice that the table lists the MOS, or mean opinion score. The MOS is a numerical rating system that reflects the quality of a codec or VoIP system. Notice how the MOS goes down as you move from G.711 to G.728, but it moves suddenly upward when you reach G.729. This is because G.729 uses a lower bit rate, but a much better compression technique.

In general, an MOS of 4.0 or better is considered *toll quality*. This phrase, toll quality, simply means that the VoIP system provides a quality equal to or better than the PSTN. It does not indicate that it is the best quality possible. According to the ITU-T, in recommendation P.800, the following MOS value interpretations can be used:

- 1 = Bad
- 2 = Poor
- 3 = Fair
- 4 = Good
- 5 = Excellent

TABLE 9-2	Codec	Bit Rate	MOS
	G.711	64 kbps	4.3
Codecs, Bit Rates, and MOS	G.726	32 kbps	4.0
	G.728	16 kbps	3.9
	G.729	8 kbps	4.0

Reflecting back on Table 9-2, you'll recall that the G.711 codec has an MOS of 4.3. This MOS is the highest value in the table and for good reason: the G.711 codec doesn't compress the VoIP data, but it does limit the frequency range to a range that captures the majority of the sound waves from the majority of human voices. The other codecs limit the frequency range in a similar way, but they also compress the data. MOS values are calculated by using surveys of human participants. A group of usually 30 or more individuals will listen to a telephone conversation and then rate the quality of the connection on the five-point scale used by the MOS system. A mean score is calculated based on the results of these user surveys. Remember from your school days that the mean average is the one calculated by adding all the values together and then dividing by the total number of inputs. For example, assume that you were performing an MOS test and you had 37 people listening to a connection that uses your new codec. Your goal is to calculate an MOS for the codec. Table 9-3 represents the breakdown of the participant's ratings. Based on these ratings, you determine that the MOS value for your codec is approximately 3.65.

As you can probably imagine, some vendors may desire to inflate the scores for their chosen codecs and deflate the scores for the codecs chosen by their competitors. The reality is that one independent test may result in a score of 4.2 while another will result in a score of 4.0. This variation can be the result of unseen variables such as participants with hearing difficulties on the day of the test that they were not aware of, or undetected problems in the network connection. The moral of the story is simple: MOS values are very useful, but they are not absolute.

While G.729 has a good MOS score, it is important to remember that the MOS score only ensures good voice quality. If Music on Hold (MoH) is to be used, the quality will be very poor with G.729. The compression used is aimed at voice communications and not at the variety found in musical sounds.

Another characteristic of VoIP packets that can impact performance is the packetization interval. The *packetization interval* is the duration of time between voice packet generation events. Stated differently, voice packets often contain two or more voice frames. The number of voice frames will depend on the packetization interval.

TABLE 9-3	Rating on the Five-Point Scale	Number of Participants
MOS Ratings in an MOS Test	5	4
	4	18
	3	14
	2	0
	1	1

A longer interval means that the packets will be larger, but they will carry more voice data. A shorter interval means that the packets will be shorter, but they will incur less packetization overhead. The longer interval, which results in larger packets, will reduce network management overhead. The shorter interval, which results in smaller packets, will incur more network management overhead. This difference between packetization delay and network delay is where the trade-off resides.

A final characteristic of VoIP systems that can impact performance is a feature known as *silence suppression*. Silence suppression is a feature that disables the sending of packets that are basically empty. If you'll imagine a telephone conversation with a friend or coworker, you'll see where this comes into play. Much of the time the other party is talking and you are listening. Since you are not talking, there is no need to send packets during your silence. There is also silence between when the other party finishes talking and when you begin talking. In the end, it is very common to see that the average person talks about 40 percent of the time while on the telephone call. Of course, this percentage can vary. Certainly, I speak to some people who take up 70–80 percent of the talk time in a given conversation; however, the average between the two speakers generally remains the same. Enabling silence suppression can have a processing performance impact on the media gateways, but it will usually provide a performance boost in the overall VoIP system. On another note, silence suppression may cause the beginning of words to get clipped. Clipping can impact the humans' ability to understand the message.

In summary, it is very important that you choose the right codec, packetization interval, and silence suppression settings. To make this clear, consider that the G.711 codec generate about 1280 bits of data in a 20 ms sample period. The G.729 codec only generates about 160 bits in the same exact period. This variation is very large. The G.711 codec literally generates eight times as much data to send the same audio information, and remember from Table 9-2, the MOS value for G.729 is only 3/10 lower than that of the G.711 codec.

Requirements of VoIP Packets

VoIP packets are different in their demands from standard data packets such as e-mail, FTP, or network printing. VoIP packets demand the following:

- Fast delivery
- Sequential ordering
- Accuracy of data

VoIP packets have demands that are similar to those of traditional data packets, and they have demands that are very different. The first demand is definitely not a requirement of traditional data, and that demand is fast delivery. The ultimate reasons for this demand are human involvement and the demands of the concept. By human involvement, I mean that humans are listening to the telephone conversation and they must be able to perceive what the other party is saying. If the humans using the system cannot get the results they need, they will discontinue their use of the system. When I speak of the demands of the concept, I'm simply referring to the reality that sound is a stream and if the stream is broken or delayed, its characteristics change.

For example, if you are listening to a VoIP conversation and there are gaps between the sounds (assuming the connection isn't broken completely), you will have a very difficult time listening to that conversation. You may not even be able to recognize what is being said. This reality is the reason that most VoIP systems will drop a phone call if too many packets are lost or if the delay is too great. Fast delivery is important in VoIP systems.

All data must be sequential when it is passed up to the upper layers of the OSI, and many applications accommodate this through the use of the TCP protocol. TCP will resequence the packets if they arrive out of sequence. A UDP solution does not receive the benefit of sequencing at the Transport layer. Therefore, the application must take responsibility for resequencing the data. A VoIP processor can certainly do this task, but the problem is that this resequencing adds extra processing delay and can be the factor that leads to dropped calls or poor quality. In a standard IP network, there is no way to guarantee that all packets will travel the same route and arrive in sequence at the destination; however, QoS can be used to ensure that the packets arrive quickly enough so that there is time remaining for resequencing, if it is needed.

The final requirement is that the data must be accurate. There is no time to retransmit data in a VoIP system. This requirement means that the links must be stable and error rates must be very low. QoS can only help with this insomuch that a less congested network is less likely to generate errors.

QoS Types

Now that we've established the need for QoS (the VoIP packets must arrive as quickly as possible) in a real-time system, we need to consider the different QoS types that are available. For the sake of the Convergence+ exam, you'll need to understand IP precedence, DiffServ, and 802.1p/Q.

IP Precedence

IP packets contain a header that includes various fields. One of these fields is a Type of Service (TOS) field. The first three bits of the TOS field are used to specify a priority for the packet. The levels of priority range from Normal to Network Control as represented in Table 9-4. Values other than those represented in Table 9-4 are also available (for example, values of 1–4 and 6), but the IP TOS field usage has both evolved over the years and started the process toward extinction. Today, DiffServ and other QoS tools are gaining in popularity. However, the concept of the TOS precedence value was fairly simple: give priority to packets with higher values.

DiffServ

Differentiated Services (DiffServ) is a networking architecture that enables you to manage service levels on your network based on the type of information traversing the network. DiffServ gives you the ability to prioritize voice or video data while using a best-effort model for all other data. You could say that DiffServ is the next evolutionary step in the usage of the TOS field. The DiffServ standard uses the first six bits of the TOS field instead of just the first three like IP precedence. Devices that understand DiffServ use these six bits to determine how the data should be prioritized. The data is usually placed into a queue based on its priority, and a higher-priority queue will get more "line time" than a lower-priority queue. Both DiffServ and IP precedence operate at the Network layer, as they are components of the IP protocol.

802.1p/Q

The IEEE 802.1p and IEEE 802.1Q specifications operate at the Data Link layer, or Layer 2, and provide a Layer 2 QoS solution. 802.1p is typically implemented in the NIC drivers, and 802.1Q is implemented in switches that create virtual LANs (VLANs).

TABLE 9-4	TOS Priority Value	Priority
TOS Values and Their Intention	0	Normal Data (Best Effort) – Routine
	1	Priority
	2	Immediate
	3	Flash (common default for voice signaling)
	4	Flash Override (common default for video)
	5	Critical (common default for voice bearer)
	6	Internetwork Control (Reserved)
	7	Network Management and Control (Reserved)

802.1p provides packet prioritization using queues, and 802.1Q, while primarily existing to define VLANs and bridging, includes features that allow for the encoding of priority information within data frames. In fact, 802.1p and 802.1Q are both part of 802.1D, which defines bridging.

VLANs are used to separate out voice and data traffic for logical processing. Once the data is separated into VLANs, it can be simpler to manage the QoS and bandwidth demands of the traffic.

Application of QoS Types

The question of when to use the different types of QoS is an important one. IP precedence is mostly used today to "provide QoS when it's there." By this statement, I mean that you can specify the TOS precedence values and then the transmitting devices between you and the destination can use them to prioritize the packet if those devices understand IP precedence. This method does not provide guaranteed QoS, but the result is usually better performance than not using the values. Remember that the IP precedence bits will always be there, so you are not increasing the packet size by setting them to something other than the default of 0.

When you need greater power and flexibility in your QoS management, you can move up to the DiffServ model. DiffServ provides per-hop behaviors (PHBs) that apply forwarding behaviors based on the values in the TOS field of the packet. The difference between DiffServ and IP precedence is that DiffServ uses the first six bits to define how the packet should be handled, whereas IP precedence uses only the first three. The additional three bits provide more flexibility in management and control. DiffServ specifies four standard PHBs as outlined in Table 9-5.

TABLE 9-5	Standard PHB	Definition
DiffServ PHBs	Default	No special handling; best effort only; must be supported.
	Class-Selector	First three bits set to match the packet's TOS bits; provides backward compatibility with IP precedence.
	Assured Forwarding (AF)	Traffic assigned to one of four classes; each class has 12 levels of precedence; this is an optional capability for DiffServ compliance.
	Expedited Forwarding (EF)	Assigned to a single preferred traffic flow within a DiffServ domain; the flow must receive low jitter, delay, and loss; commonly used with two classes: real-time and non-real-time.

The DiffServ standard defines a DiffServ domain as "a contiguous portion of the Internet [understood as internetwork and not the Internet] over which a consistent set of differentiated services policies are administered in a coordinated fashion."

Additional QoS types not covered by the Convergence+ exam in detail are outlined in Table 9-6 for your benefit. These QoS techniques are very common in today's enterprise networks.

TABLE 9-6	**QoS Type**	**Definition**
More QoS Technologies	IntServ	Integrated Services (IntServ) is defined in RFC 1633. IntServ is an architecture that was designed to provide predictable and guaranteed services to data streams. IntServ implements path reservation requirements.
	RSVP	Resource Reservation Protocol (RSVP), which is defined in RFC 2205, is used to provide stream reservations for the IntServ architecture networks. This stream reservation will guarantee a set amount of bandwidth through the entire network for a traffic flow.
	CBQ	Class-Based Queuing (CBQ) is a queue algorithm that divides the available network bandwidth among varied traffic classes. Classes are assigned queues, and the queue is granted a percentage of the network bandwidth.
	MPLS	Multiprotocol Label Switching (MPLS) assigns each packet a tag known as a label. The label represents the destination IP address so that routers can forward the packet based on the label rather than having to extract the IP address. This label means that packets do not have to be unencapsulated up to Layer 3 in order to forward them.
	Traffic Shaping	Traffic shaping manages inbound traffic so that it conforms to a specified flow rate. Simply stated, it is easier to manage consistency than it is to manage randomness. Traffic shaping is mostly about making the data flow consistent.

CERTIFICATION SUMMARY

In this chapter, you learned about the important of performance on a converged network. First, you learned about the fundamentals of network performance, including the common terminology used in the network performance management domain. Then you learned about the impact of the implementation of voice or video over IP on an existing network. Finally, you learned about the different QoS techniques that are available and reviewed their applications.

✓ TWO-MINUTE DRILL

Analyze Network Performance

- ❑ Bandwidth or data rate is the rate at which data can be passed across a connection.
- ❑ Throughput is the rate at which user (useful) data can be passed across a connection.
- ❑ Delay is a measurement of the time required to move data across the network; it usually includes processing time at the sender and receiver.
- ❑ Jitter is the variance that occurs in delay.
- ❑ Packet loss or loss refers to data that does not arrive at the receiver.
- ❑ Network capacity baselining is the process of calculating the actual performance of the network at a given point in time.

Demonstrate Application of Traffic Engineering Concepts

- ❑ Traffic shaping is used to provide consistency at ingress. For example, the incoming port of a router may receive more data during a brief window than you want to manage on the outgoing port. You can throttle the data and effectively shape the traffic.
- ❑ Voice traffic should be prioritized as real-time traffic.
- ❑ Two methods that can be used to improve performance of a converged network are increasing bandwidth or decreasing utilization.

Define QoS, Describe Implementation Techniques, and Show the Importance of QoS

- ❑ Quality of Service (QoS) is the generic phrase used to describe the tools and technologies used to manage network traffic in order to provide the performance demanded of the users and applications on a network.
- ❑ The Mean Opinion Score (MOS) is a rating from 1 to 5 that is given to a speech codec and reflects the quality of the resulting sound.

❑ Silence suppression is used to reduce bandwidth consumption by not transmitting packets for silent periods in a conversation.

❑ IP precedence uses the Type of Service (TOS) field in order to specify priority for IP packets.

❑ DiffServ uses six bits in the TOS field, unlike IP precedence, which uses three bits.

SELF TEST

The following questions will help you measure your understanding of the material presented in this chapter. Read all the choices carefully because there might be more than one correct answer. Choose all correct answers for each question.

Analyze Network Performance

1. Which one of the following terms refers to the useful data rate or user data rate on a network?
   A. Bandwidth
   B. Throughput
   C. Delay
   D. Jitter

2. Which one of the following terms refers to the variation in the length of time it takes for a packet to move from one point to another on a network?
   A. Bandwidth
   B. Throughput
   C. Delay
   D. Jitter

3. What is the technology that is implemented in order to alleviate the problems caused by jitter? This technology stores data temporarily in order to provide a consistent flow of data to the audio processor.
   A. Traffic modulator
   B. Jitter buffer
   C. Delay buffer
   D. Data compressor

Demonstrate Application of Traffic Engineering Concepts

4. You are implementing a router that connects to your WAN provider. You want the incoming data from the WAN to flow into the network at a more consistent rate. What method will you employ?
   A. Traffic flow analysis
   B. Virtual private networking
   C. Traffic shaping
   D. IP precedence

5. Which of the following represent two methods that can be used to improve performance on a converged network even though they are not categorized as QoS solutions? (Choose two.)

   A. DiffServ

   B. Increasing bandwidth

   C. IntServ

   D. Decreasing utilization

Define QoS, Describe Implementation Techniques, and Show the Importance of QoS

6. What voice codec rating provides a representation of perceived quality from a human perspective?

   A. MOS

   B. DiffServ

   C. IntServ

   D. DSCP

7. Which of the following reduces bandwidth consumption by taking advantage of the fact that users do not talk 100 percent of the time while on a telephone conversation?

   A. Chatter removal

   B. Noise reduction

   C. Silence suppression

   D. Silence compression

SELF TEST ANSWERS

Analyze Network Performance

1. ☑ **B is correct.** Throughput is used to reference the amount of useful data that can be transferred across a network connection.
   ☒ **A, C,** and **D** are incorrect.

2. ☑ **D is correct.** Jitter is the measurement of delay variation. Delay is the measurement of the amount of time it takes to move data from origination to destination.
   ☒ **A, B,** and **C** are incorrect.

3. ☑ **B is correct.** A jitter buffer buffers the incoming audio data so that it is sent up to the audio processor to be played on the speaker without gaps or interruptions.
   ☒ **A, C,** and **D** are incorrect.

Demonstrate Application of Traffic Engineering Concepts

4. ☑ **C is correct.** Traffic shaping is used to provide a consistent flow of data as it enters your network from a WAN or remotely connected network.
   ☒ **A, B,** and **D** are incorrect.

5. ☑ **B and D are correct.** Increasing bandwidth and decreasing utilization are valid non-QoS method used to improve performance on a network. DiffServ and IntServ are QoS methods.
   ☒ **A and C** are incorrect.

Define QoS, Describe Implementation Techniques, and Show the Importance of QoS

6. ☑ **A is correct.** Mean Opinion Score, or MOS, is the correct answer.
   ☒ **B, C,** and **D** are incorrect.

7. ☑ **C is correct.** Silence suppression prevents the transmission of data packets when there is no incoming audio.
   ☒ **A, B,** and **D** are incorrect.

10

Converged Network Management

> *Your first ten words are more important than your next 10,000. In fact, if your first ten words aren't the right words, you won't have a chance to use the next 10,000.*
>
> —*Elmer Wheeler, 1953*

Elmer Wheeler was one of the most respected and sought-after sales trainers in history. He taught sales classes to millions and sold millions of books as well. He understood the importance of first impressions, and your contact center will provide your organization with control over the first impressions your customers or constituents perceive. This chapter presents the concept of the contact center and the various components that compose such a solution. Additionally, you will learn about the many tools and tasks involved in managing a converged network. These tools include performance monitoring tools and configuration management tools. These tasks include daily monitoring tasks and periodic administrative tasks as well. Instead of spending my 10,000 words on an introduction, let's just move right along and gain an understanding of a contact center.

CERTIFICATION OBJECTIVE 10.01

Contact and Call Centers

Contact centers provide a central point for all incoming calls to be received, processed, and potentially routed. A contact center may include one or more call centers. The call centers may all share the same infrastructure, or they may only be connected through routers or special switching hardware. To fully understand the concept of a contact center, you'll need to consider the following components:

- Contact center management
- Customer interaction
- Call routing
- Call queuing
- Call recording
- Interactive voice response

Contact Center Management

Computer telephony integration (CTI) applications are used to enable contact centers. CTI applications provide the functionality a computer needs in order to manage chat, text, voice, and web communications. The software will allow you to manage call queues, record calls, and report on contact center operations and activities. Features of CTI applications include:

- *Caller ID* for caller identification and proper greeting.
- *Customer information lookup* based on caller ID. Systems supporting this feature allow the telephone representative to see more information about the caller as he is answering the telephone.
- *Direct dial* from a contact database.
- Enabling *power dialing* of contacts. This feature allows the representative to have the next number automatically dialed as soon as the current call ends.
- Full telephone control, which may be provided. This control may include answer, hold, redial, transfer, or conference calling.
- Expandability for future enhancement.

This last feature may be very important. If the CTI system you select supports expansion through scripting or APIs, you may be able to extend the life of the system far into the future. In the past, PBX systems have often endured more than a decade without major upgrades. This durability will become important in the VoIP systems of the current day as well.

Customer Interaction

One of the benefits of IP-based telecommunications is that telecommuters can answer calls for your customers or constituents as if they were onsite. Stated from the customer's perspective, the customer is not aware that the person to whom she is speaking is not at the physical location she has dialed. The fact that the Internet uses IP allows for VoIP communications to operate seamlessly for telecommuters.

on the job

As an example, at one point, JetBlue started its reservations offices by having all agents work from home or a personal location. VoIP communications were used to virtualize the centralization of the agents.

Call Routing

Call routing—the ability to direct or redirect a call to the appropriate destination based on a caller's specified request—is a typical feature of CTI systems. In fact, you could say that it is a required feature today. Call routing should be supported in two ways. The first method of call routing that should be supported is manual call routing. The second method is automated call routing.

Imagine you are working as the front desk operator for a small office. You handle the in-person visitors as well as the incoming telephone calls when a caller requests to speak to an operator. Your phone rings, and you click the answer button on your computer screen. The caller asks to be transferred to Thomas Dale. You ask the caller to hold and then click the transfer button for the call that you are currently handling. The application presents you with a list of phone extensions, and you click the option for Thomas Dale's extension. The caller is routed to that extension, and your screen notifies you of the fact that Mr. Dale has answered the phone. This entire process is manual call routing.

Now, imagine that you are the caller. You call the telephone number for the organization and are greeted by an automated response. You know the extension for Thomas Dale is 4597 and you press the appropriate buttons on your telephone to select this extension. You are immediately routed to Thomas Dale's line. This process is automated call routing.

Both manual and automated call routing are usually needed, since many callers will not know the extension of the person to whom they wish to speak. An additional feature of many CTI applications is that of extension lookup. This feature allows the caller to search for an individual's extension by his or her name. Once the extension is found, it is usually read back to the caller and then the caller is automatically routed to the appropriate line (or IP address in a VoIP implementation).

Call Queuing

Call queuing is needed in busy contact centers. Examples of such centers include customer service departments, sales departments, and even telemarketing firms. In the IT industry, the most commonly known busy contact center is the Help Desk or technical support line. When a caller dials the contact center, the caller is either routed to an available agent or placed in a queue until an agent becomes available. While a caller is being held in queue, he can be presented with Music on Hold for "comfort noise" to assure him that he has not been hung up on. The caller can also be updated on his position in the queue and estimated wait time.

The *automatic call distributor (ACD)* is responsible for locating and routing the call to an available agent. The ACD may be incorporated into the CTI system, it may be available as a plug-in for the CTI system, or it may be completely separate from the CTI system. For example, the CTI system may provide the caller with the option of being routed to different queues. Each queue may be managed by its own ACD.

Call Recording

Have you ever called a customer service department only to hear something like "Your call may be recorded for quality improvement purposes"? Many regions require that a caller be notified if the call is being recorded. For this reason, many CTI systems provide the ability to provide the caller with such a message just before the call is placed into the queue or just before the call is assigned to an agent. The recordings are indeed used for quality management, security purposes, and customer claim verifications. I remember working in a technical support call center a couple of decades ago where such recordings were made regularly. In one incident, a customer claimed that a particular support representative had spoken to him in a rude manner. The manager retrieved the tape and listened to the conversation only to learn that the customer had actually been very rude to the representative and that the agent had been as courteous as any person could be. Of course, the customer was fired (as much as a customer can be) and the agent was praised. Then again, I probably wouldn't be telling this story if I had actually "let that customer have it" as we say in West Virginia.

Some advanced third-party recording systems from organizations like Nice, Verint, and Witness have the ability to listen in on customer calls from the point of initial entry into the system. This feature can be used to evaluate customer reactions to menu prompts and then adjust them appropriately based on usability guidelines.

Interactive Voice Response

Interactive voice response (IVR) is a technology that allows a caller to interact with the database systems by responding to a menu of options. Responses are most frequently entered by pressing the buttons on the telephone keypad; however, voice recognition is more commonly integrated into the process and may provide a more useful tool. One of the most common examples of an IVR is the credit card companies' support line. You may have noticed that you can call the, usually, toll-free number and then retrieve your account balance and available credit.

How does IVR work in such cases as the credit card companies? The caller dials the number and initiates a connection to the contact center and the IVR system. The IVR system presents the customer with the option of retrieving credit card information. If the customer chooses to retrieve the information, the IVR will request the account number or some other information that will allow the system to uniquely identify the customer's account. Once the IVR system receives the response, it may look up information in the organization's database that can be used to validate the identity of the caller. An example of such information might be the caller's billing ZIP code. The IVR asks the caller to enter the validation information. Assuming it is correct, the IVR can retrieve and read the account information to the caller using text-to-speech algorithms. Newer IVR systems also provide the ability to queue calls as well as provide menu capabilities and database access for information retrieval.

CERTIFICATION OBJECTIVE 10.02

Network Performance Monitoring

An essential part of a converged network administrator's job is performance analysis and monitoring. You will not need to know how to use any specific performance monitoring tools for the Convergence+ exam, but you should be aware of the different categories of tools that are available to you and the scenarios in which you would use them. The tools you'll need to consider include:

- LAN monitoring tools
- Data and protocol analyzers
- WAN monitoring tools
- Traffic management and traffic shaping

LAN Monitoring Tools

LAN monitoring tools are used to monitor, report on, and manage devices and links within the local network. These tools include bandwidth monitoring tools and QoS monitoring tools. These tools play a critical role in converged network administration and management.

Bandwidth Monitoring Tools

General bandwidth monitoring tools are very helpful in ensuring proper network operations. Bandwidth monitoring tools may be integrated with infrastructure devices, or they may be installed as agents on the network. Integrated tools show the incoming (rx) and outgoing (tx) data rates for the device. For example, an eight-port switch may show the cumulative bandwidth from all ports as well as the individual contributions to the overall bandwidth by each port. Port-based bandwidth reporting can be very helpful in narrowing your focus to intensive nodes or networks that may be generating an inordinate amount of traffic.

Non-integrated bandwidth monitoring tools are either installed as promiscuous monitors throughout the network or used to track only specific links. Tools used to track specific links are useful when the link connects one network to another. For example, the link between two switches may be monitored in order to track the traffic traversing from one switch to the other. Alternatively, the link between two routers may be monitored in order to determine the total bandwidth being consumed by communications between the two connected networks. It is important to remember that non-integrated monitoring tools might require network downtime if they have to be inserted "mid-stream" between devices to gather their data.

An example of a bandwidth monitor is NetFlow Analyzer by ManageEngine. NetFlow Analyzer provides a web-based interface to track and report on bandwidth usage throughout an enterprise network. Cisco devices generate network bandwidth information, and this information is imported into the NetFlow engine for analysis. You can view the top consuming users, machines, applications, and networks. Additionally, NetFlow Analyzer—and other such tools—may provide threshold alerts. This alert feature allows you to configure notifications for events such as excessive web video traffic or streaming audio usage. Reducing these types of communications, when they are not necessary, is key to continually providing the needed bandwidth for your internal voice and video applications.

QoS, Voice, and Video Monitoring Tools

You will need to understand the various tools used in the process of monitoring voice and video quality on converged networks for the exam. Table 10-1 provided a list of these tools with their descriptions as well as multiple examples. Use this table as a launching pad to learn more about these tools for your practical work in converged environments. For the Convergence+ exam, you'll simply need to know the tools that are available and the basic functions they provide. The examples provided are not exhaustive, but you can search for these products and learn much about their features and benefits from the vendor's web sites.

TABLE 10-1	QoS Tools

Tool Type	Definition	Examples
Ethernet monitors	Software that is used to capture and analyze packets on the network. May be able to filter out all traffic other than voice or video traffic.	Ethereal and CommView
Integrated tools	Utilities built into the softphones or VoIP phones that can execute simple tests.	X-Lite and linphone
Monitoring tools	H.323 and SIP network monitoring and analysis software or hardware. May report on voice and video loads on the network.	WinEyeQ and NetQoS Performance Center
Network load testing or network impairment tools	Simulate network congestion, latency, packet loss, and other network problems. Excellent for stress testing a VoIP or video over IP solution.	Simena's Network emulator or NIST Net network emulator
SIP debuggers/analyzers	Software or hardware that is used to capture and analyze SIP communications on the network.	Distributed SIPFlow and Asterisk's built-in SIP DEBUG command
Traffic generators	Applications or hardware that can generate network traffic for load testing.	MyVoIPSpeed and PJSIP-perf

Data and Protocol Analyzers

A data or protocol analyzer is used to capture data packets as well as IEEE MAC layer frames and decode their contents. The information is usually presented in a very readable format that is easier to understand than simply looking at a series of ones and zeros. For example, many protocol analyzers break the frame headers down in such a way that each bit is explained in real-world terms and concepts. This process is often called frame or packet decoding.

Handheld and computer-based (usually used on laptops) protocol analyzers have existed for many years. Most of these applications are designed for wired protocol analysis. Because WLAN devices are so different (based on the chipsets used for the wireless communications), even protocol analyzers that are designed for WLANs or support WLAN analysis as well as wired are often incompatible with many WLAN network cards. For example, many early WLAN protocol analyzers supported one chipset alone. You were required to purchase that specific chipset (a network card that was based on it) if you wanted to use the protocol analyzer. Today, the situation has improved, but there are still no protocol analyzers that work with every WLAN network card. In fairness to the software vendors who create the protocol analyzers, this is—at least in part—due to the way that different operating systems interact with WLAN devices.

Whether analyzing a wired or wireless LAN, there are many protocol analyzers to choose from. Here is a partial listing:

- OmniPeek
- CommView
- Wireshark/Ethereal
- Network Instruments Observer
- Packetbone Bonelight
- nGenius Visualizer

Many of these applications come in trial versions. You can improve your understanding of protocol analyzers best by using them. If you download a few of these applications and test them on your test network (either at the office or at home), you will be even more prepared to select the right tool when the need arises.

WAN Monitoring Tools

WAN monitoring tools are similar to LAN monitoring tools; however, they are designed for monitoring the bandwidth and behavior of WAN links as opposed to internal communications. Like LAN monitoring tools, WAN monitoring tools may be implemented through software or hardware. Most tools will allow you to monitor usage trends, fire alerts if an outage occurs, report on peak loads, and more.

Traffic Management and Traffic Shaping

Finally, network performance monitoring and optimization can be assisted through the use of traffic management and traffic shaping. Traffic management and traffic shaping are not the same thing.

Traffic management is used when you want to control the way traffic traverses your network. For example, you may only want to use an expensive link when an inexpensive link is congested. This traffic management feature would allow you to reduce overall costs for communications.

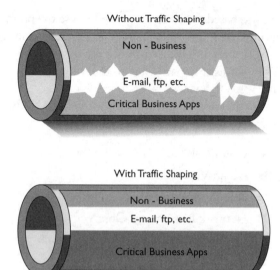

FIGURE 10-1

Traffic shaping

Traffic shaping is used when you want to eliminate or limit certain traffic types. You might think of it like traffic throttling. You could dictate that traditional web browsing can use no more than 20 percent of your Internet bandwidth and that VoIP communications can utilize the other 80 percent. This configuration would be an example of traffic shaping. Figure 10-1 shows the impact of traffic shaping visually.

CERTIFICATION OBJECTIVE 10.03

Converged Network Administration

Many of the typical network administration tasks—such as log monitoring, configuration management, and reporting—also apply to converged network administration. Special considerations must be taken for VoIP and video over IP solutions; these considerations include special log files, unique reports, admission and registration policies, and unique MAC (moves, adds, and changes) responsibilities.

Log File Monitoring

Log files are used by computer and network systems to track the activities that occur on the system. These activities may include normal operations, and they may also include configuration changes and the application of updates or patches. Any critical system should have logging enabled so that the administrator can use the log for troubleshooting and incident analysis.

I remember working with an e-mail system in the late part of 1999. Logging was enabled, but I had not looked at the logs for many months. One day I decided to open the logs and see if anything important had occurred, only to find that someone had been trying to log on as hundreds of different users with thousands of different names over the previous six months. All of the logon attempts originated from the same IP address, so I knew it was an attack. If I had not looked at the logs that day, it's very likely that the attacker would have eventually compromised the system; instead, I was able to block the IP address and I enabled alerts on the log so that I would be automatically notified of similar behavior in the future. This scenario is just one example of why log file monitoring is so important and why you should have logging enabled on all of your critical systems. As many security experts say, "If you don't log it, then it didn't happen."

Events that may be logged by various systems include:

- Successful logons
- Failed logons
- Configuration changes
- Account creation
- Account deletion
- Permission changes
- Specific activities based on the system or device performing the logging

The log entries may include the following fields, among others:

- User IDs
- Time and date stamps
- Application information (application name, version, etc.)
- Message identifiers (numeric codes associated with message types)
- Severity level codes
- Event descriptions

As an example, Avaya logs messages based on the following four error levels at this time:

- **Severe**   Indicates the loss of functionality or connectivity or a software or hardware failure
- **Warning**   Indicates a problem from which the software automatically recovered
- **Informational**   Normal activities
- **Configuration**   Configuration changes

Call Detail Records (CDR)

A specific log of importance in telephone systems is the *call detail record (CDR)*. The CDR is the log produced by the telephone exchange—either public or private—with the details about the call. A CDR includes the following fields at a minimum:

- Calling number (known as the A number)
- Called number (known as the B number)
- Data and time of the call initiation
- Duration of the call

More advanced CDR logging may include the following items, among others:

- Errors encountered
- A record ID or sequence number
- An ID for the telephone exchange that created the record
- The routes through which the call entered or exited the exchange
- Features used during the call (call waiting, caller ID, etc.)

CDRs are processed with special software known as *call accounting software*. Call accounting software is also called a billing support system. This software calculates the price for the call so that the customer can be properly billed. For internal telephone exchanges, the call accounting software may be used to report on personal calls, telephone usage times, and more. A call accounting system is the entire system, including the CDRs and the call accounting software that allows for the management of a fee-based or policy-constrained telephone network.

Reporting

Reporting features of phone systems vary greatly. Some common metrics that can be reported include:

- Long-distance calls
- Local calls
- Internal calls
- Blocked or dropped calls
- Feature utilization
- Call time per user
- Call time per group/department

The generated reports may be summary reports that are automatically created at a defined interval, or they may be on-demand reports. Many telephone systems support ad hoc reporting as well. Ad hoc reporting provides the administrators with the ability to granularly report on any metrics tracked by the system according to dynamic and changing needs. Finally, reporting systems may also provide an e-mail feature so that reports can be e-mailed to the appropriate individuals when the reports are ready.

Configuration Management

Configuration management is defined as the process of establishing and maintaining configurations that provide the needed performance and functionality from a given system or device. In a converged network, this process involves management of the PBX system, the switches and routers, and all other hardware and software that enables converged communications. Configuration management can be broken into three phases or subprocesses:

- Configuration development or identification
- Configuration change management
- Configuration auditing

The configuration development process usually involved system selection, testing, and final configuration. Once the ideal configuration has been identified, the system should be backed up for recoverability purposes. Over time, new demands will be made and the system may need to be patched or upgraded. When these changes occur, configuration change management should be used. Like change management in general applications, configuration change management is a process that involves change evaluation, change approval or denial, and change implementation. The fact that changes are not automatically applied, but are evaluated properly, leads to more stable configurations of individual systems and also the entire network.

Configuration auditing and reporting should be performed either periodically or continually in order to ensure that configurations remain as they should be. Administrators can make mistakes, and upgrades can wreak havoc. For this reason, you must have a method for ensuring that the configurations are properly set and backed up for all infrastructure hardware and software on your converged network.

Policy Management

Policies define the operational requirements of technical solutions. Policies are created for three primary reasons:

- Regulatory compliance
- Standardization enforcement
- SLA requirements

Regulatory compliance involves policies that assist you in complying with government regulations. Examples of such regulations include Enhanced 911, CALEA, and tax and regulatory compliance. Enhanced 911 allows 911 services to operate with non-traditional telephony. The Communications Assistance for Law Enforcement Act (CALEA) of 1994 grants law enforcement agencies the right to wiretap a phone line with the appropriate court orders in place. Because of the nature of VoIP, the FCC has traditionally exempted IP telephony networks from CALEA compliance requirements, but this exemption is something to watch in the future. Traditional phone networks have been subject to taxation at local, state, or federal levels. You may be required to comply with these regulations, depending on the method used to connect your VoIP network to the PSTN.

QoS Policies

Quality of Service policies document the minimum requirements for QoS support. These policies may limit the select of hardware from which you can choose. The policies may also limit the number of concurrent users that can be connected to the phone system. This may affect your buying decisions and the configuration options that you choose.

Admission Control Policies

Call admission control (CAC) is a solution that enables you to manage IP telephony call volumes if other QoS mechanisms are insufficient. Most VoIP systems support one or more of the following three CAC methods:

■ Rerouting the call over an alternate IP connection

■ Rerouting the call over the PSTN

■ Returning a busy signal

You will need to determine the scenarios where each of these should be used. For example, calls to 911 would need to be rerouted or guaranteed admission rather than returning a busy signal.

Registration Policies

Registration is the action that a user's SIP-based VoIP hardware or software performs to announce the availability of the user on the network. The phone registers the IP address, phone number, location, and availability for receiving calls. Policy issues related to registration include:

■ Allowing communications through firewalls when necessary. Port 5060 is usually used with SIP VoIP systems to register with a proxy.

■ Providing proxy servers near the users, which improves registration performance.

Moves, Adds, and Changes (MAC)

Moves, adds, and changes (MAC) must be considered as part of your administration planning. Moves include all actions that result in a user or device being moved logically or physically. Most VoIP users are only moved logically, while land line users are moved physically from one line to another. Adds include actions that result in a new user or device being added to the telephone network, and changes result in modifications to existing uses.

Most VoIP systems use a web browser–based interface for managing MAC. Features of the browser interface may include:

■ Configuring find me/follow me preferences

■ Placing a call directly

■ Viewing and listening to voicemail messages

- Working with faxes
- Call tracking such as missed calls and incoming/outgoing calls
- Setting up conference calls
- Assigning speed-dial numbers

You may also be able to create and manage hunt groups through the web-based interface. You may remember from earlier chapters that a hunt group is a configuration where a single phone number can be used to reach multiple lines. The line where the user is physically available can be answered by the user to receive calls.

CERTIFICATION SUMMARY

In this chapter, you were introduced to the general concept of a contact center and the various components that make up such a solution. You also learned about the different tools that are used to administer a converged network. You will learn more about these tools and their applications to security and troubleshooting in later chapters. Finally, you reviewed the common administrative tasks that must be performed by a converged network administrator.

TWO-MINUTE DRILL

Contact and Call Centers

❏ Computer telephony integration (CTI) applications are used to enable contact centers.

❏ VoIP provides an excellent solution for telecommuters to answer customer calls without the customer being aware that the agent is offsite.

❏ Both manual and automatic call routing are supported by most CTI applications.

❏ Manual call routing involves human interaction.

❏ Automatic call routing is performed using caller input and database lookups.

❏ In a multiple-agent contact center, the automatic call distributor (ACD) is responsible for routing calls to an available agent or providing a queue to hold the call until an agent is available.

❏ Interactive voice response (IVR) allows the caller to retrieve information from a database such as credit card information, travel schedules, or other basic information sets.

❏ IVRs can retrieve data for a caller either through touch tones pressed on their phone or through Automated Speech Recognition (ASR), which allows for voice responses to system queries.

Network Performance Monitoring

❏ Many infrastructure devices have built-in bandwidth monitoring tools.

❏ SIP debuggers are used to capture and analyze SIP packets on a VoIP network.

❏ A protocol analyzer can capture information related to all or filtered protocols.

❏ Traffic management is used when you want to control the way or the route through which traffic traverses your network.

❏ Traffic shaping is used when you want to control or eliminate certain traffic types from your network.

Converged Network Administration

❑ The log produced by the telephone exchange that contains details about calls is known as the call detail record (CDR).

❑ Call accounting software is used to process the CDR log.

❑ Configuration management includes configuration development, configuration change management, and configuration auditing.

❑ Policies are written documents that define operational procedures for network equipment and software.

SELF TEST

The following questions will help you measure your understanding of the material presented in this chapter. Read all the choices carefully because there might be more than one correct answer. Choose all correct answers for each question.

Contact and Call Centers

1. The receptionist at the front desk receives a call. The caller asks to be transferred to you and the receptionist performs this act in the CTI application. What kind of call routing is described?

   A. Automatic

   B. Manual

   C. IVR

   D. CDR

2. A caller dials a group telephone number and the PBX system automatically routes the caller to an available agent in the group. What component of a contact center has performed the routing operation?

   A. ACD

   B. CDR

   C. DTS

   D. Receptionist

3. What kind of application is used to enable a contact center?

   A. CDR

   B. IVR

   C. Protocol Analyzer

   D. CTI

Network Performance Monitoring

4. What specific application can be used to capture and analyze SIP packets only on a converged network?

   A. Protocol analyzer

   B. SIP debugger

   C. Ethernet monitor

   D. Spectrum analyzer

5. You want to capture packets related to SIP and HTTP using the same application; however, you want to capture only these packets. What monitoring tool can you use?

   A. Protocol analyzer

   B. SIP debugger

   C. Spectrum analyzer

   D. Traffic generator

6. You need to ensure that the web browsing performed by the valid users on your network does not interfere with the needed bandwidth for VoIP communications across the Internet. There can be up to 20 telecommuters using VoIP at any given time. What technology will allow you to ensure that the users can browse the Internet while not allowing these users to hinder the operations of the telecommuters?

   A. Traffic management

   B. Firewall filters

   C. Traffic shaping

   D. Port blocking

Converged Network Administration

7. What is the name given to the software that is used to analyze CDR logs?

   A. Automatic call distributor

   B. Interactive voice response

   C. Call administration and policy limiter

   D. Call accounting software

8. You've been asked to develop a configuration management plan for your network in order to ensure that the stability of the network is never sacrificed. Which of the following are commonly implemented phases or subprocesses in a configuration management plan? (Choose all that apply.)

   A. Configuration auditing

   B. Configuration intervention

   C. Configuration development

   D. Configuration change management

SELF TEST ANSWERS

Contact and Call Centers

1. ☑ **B** is correct. This is manual call routing, since the receptionist is actually involved in the routing process.
   ☒ **A, C,** and **D** are incorrect.

2. ☑ **A** is correct. The ACD, or automatic call distributor, is responsible for this routing.
   ☒ **B, C,** and **D** are incorrect.

3. ☑ **D** is correct. The CTI (computer telephony integration) application will enable a contact center.
   ☒ **A, B,** and **C** are incorrect.

Network Performance Monitoring

4. ☑ **B** is correct. A SIP debugger is used to monitor only SIP packets. The protocol analyzer and Ethernet monitor options are incorrect because they do not monitor SIP packets only.
   ☒ **A, C,** and **D** are incorrect.

5. ☑ **A** is correct. The protocol analyzer can filter down to the SIP and HTTP packets only. A SIP debugger would not work, in this case, because you want to capture HTTP packets as well as SIP packets.
   ☒ **B, C,** and **D** are incorrect.

6. ☑ **C** is correct. Traffic shaping will provide you with the ability to control the amount of bandwidth consumed by various technologies based on ports, source addresses, destination addresses, and other parameters. Traffic management will allow you to reroute traffic across different connections; however, it will not allow you to throttle the traffic.
   ☒ **A, B,** and **D** are incorrect.

Converged Network Administration

7. ☑ **D** is correct. The call accounting software is used to analyzer the CDR logs and report on call durations, costs, and more.
   ☒ **A, B,** and **C** are incorrect.

8. ☑ **A, C,** and **D** are correct. The configuration development process usually involves system selection, testing, and final configuration. Over time, new demands will be made and the system may need to be patched or upgraded. When these changes occur, configuration change management should be used. Configuration auditing and reporting should be performed either periodically or continually in order to ensure that configurations remain as they should be.
   ☒ **B** is incorrect.

11

Troubleshooting Converged Networks

> *Each problem that I solved became a rule which served afterwards to solve other problems.*
>
> —Rene Descartes

Troubleshooting is a science. Troubleshooting is an art. The logic of science and the patience of art must be applied to resolve the most difficult of technical problems. This chapter begins by presenting the science of troubleshooting. First, you will learn about the different problem solving and analysis processes and thinking tools that will make you a better troubleshooter. Next, you will read about the various symptoms and problems that occur on a converged network. Finally, you will investigate some common troubleshooting tools.

CERTIFICATION OBJECTIVE 11.01

Problem Solving and Analysis

Humans have been solving problems as long as we have existed. Troubleshooting is problem solving; however, systematic troubleshooting is a very organized problem solving approach. As a converged network administrator, you should understand different troubleshooting methodologies and the benefits that come from using them.

Troubleshooting Methodologies

A methodology can be defined as a standard way of doing something; therefore, a troubleshooting methodology is a standard way to troubleshoot. The benefits of methodologies are multifaceted. First, they help to ensure that you do all the things you need to do to complete a task or set of tasks. Second, they provide you with collective knowledge when developed over time. Collective knowledge is that knowledge that you do not possess yourself and yet benefit from. For example, most of us have never researched the statistics to know if wearing a seat belt is safer than not wearing a seat belt when we drive; however, we trust the research done by others and, therefore, we wear our seat belts (it also helps to motivate us when the law requires it). A good troubleshooting methodology will cause you to take steps for which you may not fully understand the purpose, but you will still get the benefit.

You may not know all the benefits of documentation, but you will still reap those benefits if you perform it.

I will share five different methodologies with you in this section as well as one concept that will help you in the troubleshooting process. The five methodologies are:

- REACT
- The OSI model
- The Hardware/software model
- Symptom, diagnosis, and solution
- The Convergence+ process

REACT

Early in my Information Technology career, I worked as a help desk analyst and a telephone troubleshooter. I found that I would frequently forget to do an important thing in the troubleshooting process that would cost me minutes or even hours of time—not to mention the added stress. For this reason, I developed an acronym to remind me of the stages I should go through when troubleshooting a problem. This way, I can work through the acronym until I reach a solution. I always reach a solution by the end of the acronym. The reality is that sometimes the solution is a complete reload of firmware and settings for some devices and sometimes it is a complete reload of the operating system on a client computer; however, more often than not, a simpler solution is found.

The REACT acronym stands for five stages or phases of troubleshooting:

- Research
- Engage
- Adjust
- Configure
- Take note

I'll cover each one briefly in the following sections so that you can understand how they fit together and why I go through these stages.

Research I remember, it was probably 1997, that I was trying to resolve a problem with a Microsoft Access 95 database. Every time the user tried to open the database, she received an error that read, "A device attached to the system is not functioning." Now, I don't know about you, but when I see an error about a device, the first thing

I think of is hardware. I spent more than two hours trying to verify that all the hardware was functioning properly, and of course, it was. By this time, the end of the day had arrived and I went home tormented by my failure to resolve the problem (hopefully I'm the only one who suffers like this when I can't fix a computer problem).

The next day, I decided to do some research, so I opened the MSDN CD (that's right, it used to fit on one CD and have plenty of space left over). I searched for the error message and found that the error could be generated if VBRUN300.DLL was corrupt. If you haven't been around long, VBRUN300.DLL was used by all Visual Basic 3.0 applications. The only problem was that Microsoft Access and this database did not rely on the Visual Basic 3.0 runtime for anything; however, my mind was racing. The jungle of my mind was suddenly clear and I realized the implications: If the corruption of VBRUN300.DLL could cause the error, maybe any corrupt DLL could cause the error. I reinstalled Microsoft Access and the error went away.

You are probably wondering what the moral is of this intriguing story. The moral is that I could have saved the first two hours of work with a few minutes of research. My new standard became: Research at least fifteen minutes before moving to the adjust stage with any new problem that requires troubleshooting. Not all problems require troubleshooting, and the confusion that usually comes into play is the result of a misunderstanding of what troubleshooting really is. Here is my favorite definition:

Troubleshooting is the process of discovering the unknown cause and solution for a known problem.

You see, if I know the solution to a problem, I am repairing and not troubleshooting. I will start with research only when it is a real troubleshooting scenario according to this definition. In the end, by researching for just fifteen minutes, I find that the cause and solution are often learned without spending any time adjusting various settings and parameters to resolve the issue. For example, if my VoIP client cannot connect to the wireless LAN and is receiving a specific error message when the client software first loads, I will search for this error message at Google.com (or another search engine) to see what I can discover. I may also search the web site of the client vendor and the wireless LAN vendor. If I don't find the cause or solution, I will usually get direction so that I can focus on the right area as I move to the Engage or Adjust stage.

on the job *I have found that some of my best search results lead me to discussion forums where others have experienced similar difficulties. The forum thread may not provide a direct solution, but it can often help me focus my thinking in the right direction.*

At the end of the Research phase, I may have a solution to my problem. If that is the case, I will resolve the problem and move to the final phase so that I can document the problem and solution for future reference. If my research has only given me direction or has yielded no results after a few minutes, I will move on to the next phase: Engage.

Engage   While you may be eager to move from the Research phase to the Adjust phase, you should engage the user if he or she is involved in the problem. Avoid the temptation of jumping right in and making changes.

When you engage the user, ask a question like, "Do you know if anything has changed about your system in the past few days?" Notice I didn't say, "Did you change anything?" The latter question will usually cause people to become defensive and fail to get you any valuable information. The users do not usually have any knowledge of what caused the problem, but when they do, it can save you hours or even days of trouble. Always engage the user. Other questions that might be beneficial include:

- Have you seen any strange activity in the area lately? (rogue APs, hackers roaming around)
- Has the problem only recently begun, or has it been happening for some length of time?
- Are you aware of any others experiencing similar difficulties?
- When was the last time it worked?
- Is it turned on? (Seriously.)
- Is the network cable plugged in? (Seriously.)

"Is it turned on?" is probably the most commonly asked question in the history of technical support, and it is still a very important question. With modern complex systems, it is very possible that the user has powered on one component of the system and not another component. Sometimes you can save the user from embarrassment by asking him or her to power cycle everything. This way, if the user did not have the hardware powered on, he or she will not have to admit it. The big key is to always handle users cautiously because you will definitely find it easier to work with users that enjoy being around you.

Now that the Research and Engage phases are complete, it's time to move on to the Adjust phase if the problem is still unresolved. If the problem is solved, you can skip ahead to the Take Note phase and document the problem and solution or simply close the trouble ticket if the problem is a common recurring issue.

Adjust Interestingly, we've just now arrived where I see many techs, convergence administrators, or network administrators in general, beginning. Don't feel bad—I've done it many times myself. Through years of experience, I've found that I can save hours of difficulty by moving through the Research and Engage phases before the Adjust phase. Keep in mind that I'm going to commit ten to twenty minutes at most to the first two phases. In most cases, I gain the information I need in only five to ten minutes. At one point, I tracked the results of using this methodology with all of my troubleshooting cases for six months. During this time, I had to troubleshoot 206 unique problems (there were hundreds of common problems that I dealt with during the same time period). With 117 of the problems, I found the solution in the first two phases. This result meant that I solved more than half of the problems in less than ten to twenty minutes.

What did I do with the remaining 89 problems? I moved on to the Adjust phase. This phase is where you begin trying different things to see if you can track down the source of your problem. You might try updating the firmware on an AP or installing new device drivers for a client adapter in order to make them more compatible with the VoIP software you are using. You could also change settings or disable features to see if the problem goes away. This point is where you begin the "technical" side of troubleshooting. While many techs begin here, you will have an advantage because you are starting into this phase with a more accurate view of the problem.

Once you've completed these first three stages, you've always come to a solution. Again, this solution is sometimes reinstalling the application or operating system, but things are working again. You may have your solution after the Research phase, or you may have it after the Engage phase. Whether you make it all the way to the Adjust phase or you solve the problem earlier in the process, you are now ready to move to what I call the ongoing stability phases: Configure and Take Note.

Configure This phase is the first of the two ongoing stability phases. In the Configure phase, you ensure that the systems and devices are configured and are operating according to your standards before leaving the physical area (or remote management tool). This phase allows you to maintain a standardized environment, and a standardized environment is usually more stable. Of course, with a reinstallation you will need to reinstall according to the original specifications for the installation and then apply any configuration changes that have been approved or processed since that time. The primary goal of this phase is to provide consistency in your environment and improve troubleshooting processes moving forward.

In a consistent environment, you can assume certain things. For example, you can assume that a device has the newest firmware or you can assume that an operating system has a particular service pack loaded. The real world, however, introduces great difficulty in the area of consistency. For an environment to be and remain consistent, the following things must all be true:

- The users must be unable to make configuration changes that affect the consistency of the configurations.
- All administrators, who have the power to make configuration changes, must be consistent in the changes they make.
- Some automated system must be in place that adjusts the configurations of all machines simultaneously.

As you can see from these three requirements, it is very difficult to have a truly consistent environment because of the human factor. However, even an environment that is mostly consistent makes troubleshooting much easier because an administrator is less likely to implement a bad configuration than a user or customer. This ratio is not based on some superior intellect of the administrator; it is simply based on the training and expertise of the administrator. In the end, configuring your machines and devices to a standard specification generally reduces support costs by reducing support incidents.

Take Note   This final phase completes the process and ensures that you get the greatest benefit out of this methodology going forward—it is the second ongoing stability phase. By documenting your findings, what I call *Take Note*, you provide a searchable resource for future troubleshooting. For example, the situation I shared earlier where the device error was generated should be documented. I suggest documenting the following at a minimum:

- The problem, with any error messages, if they existed
- The cause, concisely explained
- The solution, with any necessary step-by-steps
- Any learned principles, such as a DLL being referenced as a device by the operating system

If your organization does not provide a centralized trouble ticket tracking system or help desk solution, I encourage you to consider creating your own database. You can use any desktop database application like Microsoft Access or FileMaker. Just be sure you can document the needed information and query it easily. I tracked the

206 unique problems in an Access 2.0 database back in the late nineties. It worked fine for me and a similar solution can still work for a small organization. However, web-based trouble ticket or call tracking systems seem to be the direction in which help desk applications are going, and I certainly find them very useful as well. One benefit of a web-based system is that you may be able to access it from home or any location where you have web access. This way, you can check in to see if there are new problems at any time. If you have remote access to systems, you may even be able to resolve the problem without ever going onsite.

This first methodology, REACT, is my own methodology. It is the one that I use for every new and unique problem I face. Through the years, I've incorporated some of the concepts of the OSI model methodology and systems thinking into my processes as well, but REACT still provides the basic framework in which I operate. Next, you will learn about the OSI model methodology.

OSI Model

The OSI model is an excellent thinking tool for designing and understanding networks; however, it can also be used for troubleshooting purposes. When using the OSI model as a troubleshooting methodology, you will walk up or down the OSI model, analyzing the system at each layer. This process allows you to break the problem into logical sections and then verify that the system is operating properly in each of these sections (OSI layers). You may choose only to analyze Layers 1 through 3, or you may choose to evaluate all seven layers (or even eight if you're considering the User layer, which is sometimes colloquially called Layer 8).

Layer 1 Layer 1 is the Physical layer, and this layer would mean evaluation of the LAN client devices or infrastructure devices to ensure that they are working properly at the hardware and possibly driver levels. For example, is the radio still functioning appropriately in a wireless NIC or not? With client devices, this can be tested quickly by removing the wireless LAN client device (assuming it's a USB, CardBus, or other kind of removable device) and installing it in another computer that is using the exact same model device and is working. If the new computer stops working too, the wireless LAN device is most likely at fault at the hardware level or the internal software level (firmware). If it is an AP and you have a spare, you could load the identical configuration on the spare and place it in the same exact location to see if the problem persists. If it does not persist with the new AP, again, the radio—or some other hardware/firmware configuration—may be failing.

An additional key is to evaluate the hardware used by Layer 1 (sometimes called Layer 0). Once you get to the wired network, you may also evaluate patch panels, connectors, and cabling at this point. Another area where failure may occur in wireless VoIP connections in relation to Layer 1 is the Ethernet connection between the AP and the wired LAN. If your clients are authenticating and associating with the AP, but they cannot obtain an IP address from your DHCP server (assuming you're using one), the Ethernet connection may not be working. Go to a wired client and try to connect to the AP via the Ethernet connection. Can you? If not, verify that the Ethernet cable is good, that the port on the switch is working, and that the Ethernet port in the AP is still functioning (by connecting directly to the AP and trying to connect).

These same guidelines apply whether you are troubleshooting a wired or wireless connection. Consider replacing cables as well as NICs to test the Physical layer. Also, keep in mind that many Physical layer problems will exist in hardware other than the end nodes, such as switches and routers. You will also want to consider testing at higher layers first. If you can PING a device on the network but your VoIP softphone is not working, the Physical layer is not the likely problem.

Layer 2   The second layer you will usually evaluate is Layer 2. This layer is where the bridges and switches perform their magic. Make sure the switch ports are still working properly (though this is really a Layer 1 issue), that VLANs are configured appropriately, and that port security settings are set accurately on your switches. Check the configuration in your bridge or bridges to ensure that they are configured correctly. Be sure the wireless link on all wireless bridges is still working and that the signal strength is still acceptable for operations. Since bridges evaluate incoming frames and forward them according to information in the frame, be sure that your bridging rules or filters are set up appropriately. Of course, with both bridges and switches, the problem can be at the cable and connector level, which you will check at Layer 1.

The same rule applies to Layer 2 when it comes to troubleshooting this layer. If you can PING a device on the network but your VoIP softphone does not work, the Data Link layer (Layer 2) is most likely functioning just fine. When working with hardware VoIP phones, you may not have the ability to perform PING tests. In these scenarios, you can simply attempt to connect with a different phone to see if the problem is resolved. If it is resolved, it is likely a hardware failure or a configuration problem somewhere in the settings of the phone. While the OSI model of troubleshooting usually involves a logical flow up or down through the layers, it is also important to think of the seven layers as a systematic whole. All of the layers work together to achieve actual communications.

Layer 3   If you've evaluated the radios, NICs, cabling, and connectors at Layer 1 and you've checked the bridges and switches at Layer 2 and you've still not found a solution, you might have to move on to Layer 3. Here you'll need to check the routing tables to ensure proper configuration. Make sure any filters applied are accurate. Using common desktop operating system tools like *ipconfig, ping,* and *arp,* you can ensure that you can route data from one location on the network to another.

Layer 3 is all about the network topology at a logical level. This logic means that you will be considering the network from a segmentation and routing perspective. You will need to ensure that the VoIP device is properly configured with default router settings, proper DNS server settings, and a valid IP address/subnet mask pair.

Upper Layers   Finally, if you've tested the first three layers and can't find a problem there, the network infrastructure is probably working fine. It's time to move to the upper layers. Look at the configuration settings in your applications and client software for VoIP utilization. Be sure that the authentication mechanisms are installed and configured correctly. Try using different tools and software that provide the same basic functionality. Do they work? If so, there may be a compatibility problem with the specific application you're using and the hardware on which it is operating.

One important lesson to take away from this section is that you cannot become narrow in your focus. If you insist on investigating only one layer of the OSI model, you will likely fail to find a solution or even the cause of your problem. This expanded focus is why many techs will begin by attempting basic PING commands and traceroute commands in order to verify the functionality at Layers 1 through 3 quickly. If these commands work, the problem preventing appropriate communications is likely an upper-layer issue.

As you can see, the OSI model of troubleshooting can help you both focus and move through a sequence of testing procedures until you find the true source of the problem. I've only touched on the concept here, but you can take it farther by learning more about what happens at each layer and the tools that can be used to test at that layer. For example, you can use a spectrum analyzer to test and troubleshoot the Physical layer and a protocol analyzer to inspect the Data Link layer of LANs.

Hardware/Software Model

The hardware/software model is a troubleshooting methodology that is used in an attempt to narrow the problem to either hardware or software. Certain problems are commonly hardware problems and others tend to be software problems. Many administrators will attempt to troubleshoot software first and then hardware. Others will do the opposite. In most cases, the situation will help you to determine which should be your first point of attack. If everything is working in a system except one application, that is often a good sign that the software is the problem. If multiple applications that use the same hardware are experiencing the same problem, that is often a good sign that the hardware is the problem. These are not absolute rules, but they are good general guidelines.

You will also want to consider the speed with which you can "swap out" the hardware. For example, a laptop PC card can be exchanged very quickly if you have a handy spare. It may be faster to swap the card and see if the problem goes away than to begin troubleshooting software settings. If the problem is resolved by swapping the hardware, you have resolved the issue. If it has not been resolved, then you can begin the more time-consuming process of software testing. Of course, some hardware cannot be swapped so easily, and in those cases you will likely perform some hardware tests through vendor utilities or you will inspect the software configuration.

Hardware Problem   In LANs, there are hardware problems that present specific symptoms. While I cannot provide you with an exact list of symptoms mapped to problems, the list in Table 11-1 is a good place to start.

TABLE 11-1	Problem	Symptoms
Hardware Problems and Symptoms	Client adapter failed	Device driver will not load, client cannot connect, OS reports errors loading the device
	Firmware outdated	No support for newer security features, poor performance, reduced stability
	Improper antenna installed or antenna disconnected with a wireless device	RF coverage in the wrong place, signal too weak at a distance
	Bad cables	Could cause improper power for PoE devices or low SNR for antennas; could cause intermittent connection problems for wired links

TABLE 11-2	Problem	Symptoms
Software Problems and Symptoms	Client software misconfigured	Client cannot connect, cannot receive an IP address, unable to browse the network
	Improper passphrase entered	Client associated with the AP but cannot log on to the wireless network; client connected to the wired LAN, but cannot place phone calls
	DHCP server down	Client connects physically, but cannot acquire an IP address
	RADIUS server down	Client associates with the AP but cannot log on to the network when using a Robust Security Network for a wireless LAN

Software Problems   Table 11-2 lists common software problems and their symptoms.

Symptom, Diagnosis, and Solution

Because certain symptoms usually surface with specific problems, many issues can be resolved in a similar way to human health issues. Look at the symptom, identify the most likely cause (diagnosis), and then treat it (solution). Repeat this process until the problem is resolved.

Symptom   Defining the symptoms means gathering information about the problem. What is happening? Where is it happening? What technology is involved? Which users, if any, are involved? Has it always been this way? Answering questions like these will help you determine the various details about the problem. Good questions are at the core of effective problem definition.

Diagnosis   Based on the information gathered from your symptoms analysis, what is the most likely cause or what are the most likely causes? You can treat one or all, but you will most likely learn more by treating one cause at a time. Try one solution based on your diagnosis first and evaluate the results. This ability gives you expert knowledge over time or what some call "intuition."

Solution   The solution is the potential fix for the problem. You may try replacing a CardBus PC wireless card because you determine that, based on the symptoms, the most likely cause is a failed card. After replacing the card you note that you are experiencing the exact same problem. Next, you may decide to try both cards in another machine that is currently using the same card model and is working. When you do this, both cards work in the other computer. Next, you may attempt to reload the drivers in the malfunctioning computer, but this step doesn't help either. In the end, you discover that the CardBus port is experiencing intermittent failures in the malfunctioning laptop. You send it to the vendor for repairs.

This illustration demonstrates the diagnosis and solution method—what I call the Adjust phase in my REACT methodology. You make changes and try different tactics until something solves the problem. You document the solution for future reference, but you also mentally document it. This memory is called experience, and as you get more and more of it, you eventually approach the level of expertise that helps you solve problems more quickly. For this reason, I look at problems as stepping stones to a better future, because solving this network or computer problem today will only make me more able to solve similar and different problems tomorrow.

Systems Thinking

Systems thinking is the process of analyzing all interdependent components that compose a system. In other words, it is the opposite of being narrow-minded in the troubleshooting process. I've seen administrators blame everything from network connectivity to application errors on an operating system or a particular brand of PC instead of looking for the actual problem. While some operating systems and some PC brands may seem more prone to problems than others, the reality is that there are probably thousands of individuals out there who have had the opposite experience as you. In other words, if you like the computers from company A because they are very stable and you don't like the computers from company B because of your experience with them, there is likely someone (or thousands) out there who feels exactly the opposite because of his or her experience.

I remember when Packard Bell computers were very popular. IT professionals tended to dislike the machines because they were integrated. The sound card was built-in. The video was built-in. The drive controller was built-in. Do you see a pattern that is similar to almost every computer on the market today? There are very few systems that do not have integrated sound, video, and drive controllers today; yet, this configuration was the thing that gave Packard Bell a bad name—at least in part. If memory serves me correctly, I think it really had more to do with support than anything, but you can see how blinders can keep us from troubleshooting the real problem. Obviously, the integrated components were not the real culprit, since most machines are designed that way today. Of course, the Packard Bell PCs also had a serious problem with out-of-the-box DOA machines, but if a machine works from the start, it shouldn't be classified as a bad brand; instead, it's worth troubleshooting.

The point is simple: rather than focusing on a vendor that I do not like, I must focus on the actual problem and seek a solution. When I focus on the problem and not the vendor, I'm less likely to just reinstall every time a problem comes up. I want to ask questions like:

- What are the systems or devices between this device and the network or device with which it is attempting to communicate?
- What other devices are attempting to communicate with the same system at this time?
- What has changed in the environment within which the system operates?
- Has the system been physically moved recently?

Asking these kinds of questions causes you to evaluate factors that are more related to the actual system you have in place and less related to the vendors that have provided the components. Indeed, if a vendor has provided you with bad components over a period of time, you will likely discontinue partnering with that vendor. However, blaming the problem on a single vendor every time does not help me solve the problems I am facing right now. For that, I need systems thinking and a good methodology.

Whether you adopt one or more of these methodologies, pursue another methodology, or create one of your own, you should consider how you troubleshoot problems and then be sure it is an efficient and effective process.

Now that you've been exposed to several troubleshooting methodologies and thinking processes, I will cover the troubleshooting flow that is presented in the Convergence+ objectives and relate this flow to what you've learned so far.

The Convergence+ Process

The Convergence+ objectives outline a troubleshooting process that encompasses the concepts that were introduced in this chapter. The process includes the following stages of problem resolution:

- Log the problem
- Confirm the problem
- Troubleshoot the problem
- Escalate if required
- Close the log

The following sections will outline the efforts required at each stage.

Log the Problem

A good call tracking or problem tracking system will help to provide excellent service to the user community you support. The information logged should include the following items at a minimum:

- Name of the user
- Any hardware involved
- Current date and time
- A problem statement
- When the problem first started happening
- Name of the call recipient
- Comments

Additional items that you might consider logging include:

- Operating system in use
- Configuration ID (if you track configurations as unique entities)
- Network configuration of the devices involved
- Priority of the incident

Confirm the Problem

When a user calls or visits your office, it is important that you confirm the actual problem. Due to the user's lack of training, he may not express the problem well and you could find yourself troubleshooting a problem that does not exist. This confusion is not only a waste of your time, but it is also frustrating to the user. In order to confirm the problem, consider testing the device directly if you can. If you must work over the telephone, walk the user through some basic steps that will allow you to verify that the problem described is the problem that exists.

Sometimes the problems are simple. For example, a user may indicate that she cannot access the Internet. You can ask her to try again and read any errors to you that show up on the screen. Re-creating the problem is the first step to problem resolution. Many times, when you confirm the problem, you will have an "oh, I know what's wrong" moment. These moments are always best because they prevent you from spending unnecessary time troubleshooting the wrong thing. Always try to confirm the problem.

This stage of the Convergence+ process is similar to the Engage phase in my REACT methodology, but it is not exactly the same. Confirming the problem should really come before you even start into the REACT methodology. In the Engage phase, you are discovering what the user may know about changes to the system that could have caused the confirmed problem. As you will see, the entire REACT methodology is confined in the "Troubleshoot the problem" stage of the Convergence+ process.

Troubleshoot the Problem

Once you've logged the problem in a tracking system, which is used to verify that the problem gets resolved, and you've confirmed the problem, you are ready to begin the actual troubleshooting process. Remember to take advantage of the wealth of knowledge at vendor web sites and support forums through research. Engage the user in order to learn of any relevant information that may show how the problem began. At this point you can begin the troubleshooting process of adjusting the system, which will lead to an isolation of the cause in most cases. Finally, ensure that you leave the system configured according to the configuration standards of the organization and that you document what you've learned while troubleshooting the problem.

Escalate If Required

In some cases, the difficulty may be outside your realm of authority or expertise. You may need to escalate the problem to another support individual or group. The process usually involves updating the problem log and passing the incident on to the other party. This ability may be a feature of the problem tracking application, or you may simply need to copy the information you've gathered into another application that is used by the other party. Either way, it is important that you pass on the information you've already gathered so that the user is not required to communicate it all over again.

on the Job

I want to emphasize the importance of background communications and the impact it has on the users' perspectives. When you communicate effectively with your peers, it helps them communicate more effectively with the users. One of the biggest challenges IT groups face is that of internal peer communications. Transferring problem information is one action that can go a long way in resolving this issue and presenting a strong image to the user community.

Close the Log

The final step is a simple one and needs little elaboration: close the log. Once the problem is solved and the user's system is functioning properly again and you've documented your findings, there is little to do but close out the problem log and move on to the next issue. Great job! You've solved the problem and learned valuable information that will help you in the future.

CERTIFICATION OBJECTIVE 11.02

Common Symptoms and Problems

It is important, both for the exam and your efforts in troubleshooting networks, that you understand many of the common symptoms and problems that occur in converged networks. The Convergence+ objectives list eleven symptoms specifically and fifteen problems. All of these objectives are addressed in this section.

Common Symptoms

The first piece of information you have when troubleshooting is a collection of one or more symptoms. Understanding these symptoms and the common problems that can cause them is very important. You will learn about the symptoms first, and then we will move on to the common problems.

Poor Voice Quality

Voice or call quality is a subjective term that describes how accurately the human voice is digitized and transmitted as a voice signal. It is a subjective term because voice quality is based on the perception of the hearer and is impacted by that hearer's experience. For example, if the hearer has been exposed to very high-quality audio communications, he or she will be more likely to notice poor voice quality on a VoIP conversation. By the same token, one who has not been exposed to such quality is less likely to perceive the exact same voice signal as being poor. Even with this subjectivity, an industry standard for judging voice quality has been developed. This standard is known as the Mean Opinion Score (MOS). The MOS is based on responses from a group of participants and attempts to remove some of the subjectivity through the process of averaging the participants' responses.

Factors that impact voice quality include clarity, delay, and echo. Clarity is a perceptive measurement of an original sound compared to a digitized signal. Delay causes the perception of a continuous voice signal (like the continuous sound waves in a face-to-face communication) to be lost. Echo causes the voice signal to be reflected back toward the speaker. This echo can result in overlapping sounds and can reduce quality.

Clipping

Clipping occurs when parts of the sound waves are lost during the digital communications. This loss can occur as a result of many different problems:

- Processing overload on the encoder due to insufficient hardware
- Dropped packets between the sender and receiver due to media errors or other problems
- Overflows or underflows on the jitter buffer
- Network latency causing too much delay and forcing dropped packets
- Using a half-duplex speakerphone

When full-duplex speakerphones are used, they usually sample the environment to build a digital image of the noise floor within the area. This image is used to distinguish between a human speaker and the background noise. Half-duplex speakerphones do not do this imaging and may interpret the background noise as a human speaking, which may in turn result in clipping.

Whatever the cause, remember that clipping is the loss of portions of sound within the digital stream of communications.

Echo

Echo occurs when the person speaking hears what he or she said a few milliseconds after having said it. The effect is very similar to what you experience when you speak loudly into a cave. When echo is involved in a telephone conversation, it can become very frustrating to the participants; however, echo is a normal occurrence on the PSTN and even private voice networks. The goal is to remove or minimize as much echo as possible.

Echo can be resolved through impedance matching in the cabling systems, reducing delay as much as possible, and using echo suppression and/or cancellation.

Echo cancellers are installed at a local site to prevent an echo from returning to the remote site. If you have a VoIP network that interconnects multiple locations, you may consider installing an echo canceller at each site. Echo cancellers are usually implemented on the VoIP gateway's PSTN interface.

The good news is that the problems with echo, while not completely solved, have become less of an issue with the later versions of IP telephony equipment. The codecs and firmware continue to be refined and improved. The issue of "hybrid echo," where there is a two-wire–to–four-wire conversion happening in an analog network, is still an issue; however, this last issue is a Telco or PSTN problem at the CO and not within VoIP systems.

Delay

Delay is the time it takes for a packet to travel across a network. As a packet traverses the network, it must pass through switches, routers, and other devices. Each of these devices can introduce delay. Delay can cause loss of call clarity, clipping, and other problems on VoIP networks, so you should attempt to minimize delay as much as possible. Implement paths between calling partners that introduce as few handling delays as possible. Use the fastest technology your budget will allow (for example, use Gigabit Ethernet instead of 100 Mbit whenever possible).

Audio codecs will introduce different transcoding delays as well. Table 11-3 provides a listing of common delays introduced by the different codecs.

The reality is that the codec-induced delay is directly related to the amount of compression. G.729 compresses the audio very tightly, whereas G.711 doesn't compress the audio at all. It is important, when looking at these delays introduced by transcoding, that you do not forget the reduction in delay provided by the smaller data size.

TABLE 11-3	Transcoding Delay (ms)	Codec
Codec-Introduced Delay	0.125	G.722
	0.75	G.711 PCM audio
	1	G.726 and G.727
	3–5	G.728
	10	G.729 and G.729A
	30	G.723 and G.723.1

No Dial Tone

If a user takes the phone off-hook and does not receive a dial tone, the problem could be one of many possibilities. The possibilities include:

- Configuration issues
- A digital signal processing (DSP) problem
- A software bug or malfunction
- Hardware failure

Cross Talk

When an electrical signal in one cable causes a voltage fluctuation in another cable, cross talk occurs. With telephone signals, cross talk may be detected as a faint background conversation. When the signal from another line bleeds into your line, the conversation on the other line may come through.

Longer cables are more susceptible to cross talk, and poorly terminated cables may also increase its occurrence. Cross talk is rated as near-end cross talk (NEXT). NEXT is measured by intentionally introducing a signal on one line and measuring the strength of that signal on an adjacent line. The different between the proper signal on the main line and the cross talk signal on the adjacent line is known as signal-to-noise ratio (SNR). A higher SNR is desirable.

A recent example involved a group of FXO circuits in a router that sporadically introduced a humming sound to calls. The customer suggested that they had been experiencing the issue for some time now. During inspection, the technician noted a large transformer next to the router. The transformer main cable was running in parallel to the FXO circuits, and this caused the interference. In this scenario, the transformer could be moved, or the FXO could be changed to a loop start instead of the ground start for which it was configured.

Dropped Calls

A call can be dropped for a number of reasons. One of the users may have inadvertently pressed the End or Hang-up button. A switch may fail to forward packets quickly enough, and the call may be dropped due to inactivity. A call could be dropped because routers are reconfiguring routing paths or routing tables and this process can detract from the ability to forward the voice packets. In addition, all of the typical reasons network connections fail will apply to VoIP connections.

Blocked Calls

When a call is blocked, the dialer receives a busy signal. Grade of service (GoS) expresses the probability that a single call will be blocked. A system has a high GoS if a call goes through on the first try. When some callers receive a busy signal, the GoS is lower; however, it is usually acceptable to have some calls receive a busy signal. The question is, how many calls per hundred or thousand can you tolerate being blocked?

GoS, then, is the probability that a given call will be blocked on the first try during the busy hour of phone operations. The formula for calculating GoS is

$$GoS = bC/cC$$

where bC is equal to the number of blocked first-attempt calls in the busy hour and cC is the number of completed first-attempt calls in the same time window. If 312 calls are completed on the first try and 17 calls receive a busy signal, the calculation is as follows:

$$GoS = 17/312 \text{ or } GoS = 0.054$$

A GoS of 0.054 means that 54 calls out of every 1000 calls, or 5.4 percent of calls, will usually be blocked on the first try during the busy hour. GoS and busy hours are a very important part of sizing a call center.

Loss of Features

Cisco IP phones stay in touch with the Cisco CallManager software by transmitting KeepAlive messages every five seconds or so. Three sequential failed KeepAlive messages, within an active call, will cause the message "CM Down, Features Disabled" to appear on the phone. This state will prevent features such as conference calling, call parking, and call holding from functioning. The cause may be a network problem between the phone and the CallManager, or it may be a problem within the server or phone software. While Cisco IP phones can register with up to three CallManager servers, they will use only one at any given time for call processing. If their primary CallManager goes down (or there is a network issue in-line between the client and the server), the phone will lose any of the previously mentioned features for the duration of the call. When that call is ended, the phone will then begin using its secondary CallManager (or its tertiary one, if the secondary isn't available) for call processing and all the ancillary services.

Poor Video Quality and Video Frame Loss

Like voice packets, video packets arriving out of sequence or with high levels of delay will cause poor quality in the rendered video. When using the H.323 protocol, the required bandwidth for video is about 384 Kbps plus 25–30 percent more for signaling overhead. This number is usually rounded to approximately 480 Kbps of required bandwidth. To resolve video quality and frame loss problems, you might consider using multiple parallel flow paths for the data transfer. You can also implement load balancing or load splitting for the video processing.

H.323 has some very tight constraints for video transfer. The packet loss for H.323 video must be less than 1 percent. Jitter has to be less than 30 milliseconds, and latency must be between 150 and 200 milliseconds in one direction.

To assist in the reduction of negative outcomes on packet-based networks, consider classifying video with QoS tools. You can use IP precedence solutions like DiffServ and queue the video packets with Low Latency Queueing (LLQ) or Weighted Fair Queueing (WFQ) so that the packets move out of the network nodes as quickly as possible. You could also put the video traffic in the next queue down to make sure that it doesn't interfere with voice traffic.

Common Problems

Now that you have learned the basics of the different trouble symptoms that you are likely to encounter on a converged network, you can explore the problems that may cause these symptoms to arise. While it is important to have and master excellent troubleshooting tools, it is equally important to understand common points of failure so that you can focus your efforts. Table 11-4 provides a listing of the common problems you are likely to encounter.

TABLE 11-4   Problems and Definitions

Problem	Definition
Media Errors	Bad optic terminations or RF connectors. Poor shielding or damaged cables.
Packet or Data Loss	A measurement of the number of packets lost compared to the total sent.
Protocol Mismatch	Using an inefficient protocol or improperly labeling packets for Label Switch Routers (LSRs).
Framing and linecode mismatch	Using the wrong framing or line code for a T1 circuit can cause packet loss, slips, and errors that can cause a voice connection to have poor quality.
Jitter	Variableness in the rate of delay.
Port Settings	Port settings include perimeter firewall ports and client firewall ports. H.323 requires that UDP ports 1718 and 1719 be open. It also requires that TCP ports 1720 and 1503 be open. RTP and RTCP video and audio streams use dynamic ports from 1024 to 65,535 on TCP or UDP.
Configuration Settings	Improper configuration settings can wreak havoc on a VoIP network.
Packet Re-Ordering	When routers are congested or traffic is improperly prioritized, VoIP packets may arrive out of sequence more often than they normally would. This disorder results in packet resequencing, which can impact voice quality. Asynchronous routing could also cause packet re-ordering if differing paths to the same endpoint are available and can be used.
MTU Issues	When a small maximum transfer unit (MTU) size is configured, it can cause fragmentation. This fragmentation can result in extra processing delay.
Bandwidth Restrictions	VoIP requires consistency in bandwidth availability in order to operate smoothly. Bursty network communications can impact this availability and lower voice quality.
Router Misconfigurations	It is not uncommon to see routers running multiple routing protocols (like RIP, OSPF, or IGRP) when only one or two of the protocols are actually needed. These extra routing protocols can degrade network performance.
QoS Tags Being Dropped	Not all devices support QoS tags, and as a router extracts the IP packet out of an Ethernet frame, the QoS information can be lost during re-encapsulation. Be sure to select infrastructure devices that support the QoS technology you plan to implement. In addition to verification of QoS support, you must verify that QoS is turned on. The Cisco Catalyst 6509s needed to have trust of QoS enabled; otherwise, they would just ignore the QoS tags.
IP Packet Loss	Due to UDP's connectionless nature and the fact that many VoIP protocols rely on it, IP packets can be lost. This loss could be due to slow routers and other slow devices on the infrastructure.
Backup Over the Network	Many organizations are now performing data backups across the network. This activity can generate a large amount of network traffic. The backup packets should be tagged as a lower-priority transfer than voice and video. You can configure the interleaf ratio so that five or more voice packets are sent for every one of the backup packets.
Hardware Failure	In order for any networking technologies to function, all intermediate hardware devices must continue to function. A common problem causing VoIP and other networking communication symptoms is simple hardware failure.

Troubleshooting Tools

It's one thing to have a good methodology and understand the common symptoms and problems that occur on a converged network. It's another thing to know how to use the right tools to analyze and discover what's really happening on your network. This section will introduce a number of IP troubleshooting tools that you may find useful when you're analyzing your next big problem. The tools covered include:

- PING
- TraceRT
- PathPING
- ARP
- NSLookup
- Hostname
- NetStat

PING

PING (the term stands for Packet InterNet Groper) is a command used to test IP connectivity. It is very useful when you are unsure if your network connection is even operating at the Network layer. The PING command varies slightly from one operating system to another, but most operating systems that support TCP/IP also support the PING command. Even Windows for Workgroups 3.11 (many years ago) included a PING command when you installed the TCP/IP protocol. The following exercise will walk you through using PING on a Windows machine.

EXERCISE 11-1

Using the PING Command

In this exercise, you will use the PING command to verify that a remote system is functioning on the network. You will also use the command to perform a continuous PING in order to test for intermittent problems. Finally, you'll see how you can increase and decrease the size of the PING request.

Verify a Remote System Is Functioning

1. Open a Command Prompt window by clicking Start | All Programs | Accessories and selecting Command Prompt.

2. Type **PING remote_system_IP**.

   For example, if you are testing the remote system at 10.10.10.100, you would type **PING 10.10.10.100**.

Perform a Continuous PING

1. Open a Command Prompt window.

2. Type **PING –t localhost** and press ENTER.

You will notice that the PING command requests a response from the local machine periodically and without ceasing. If you want to test for intermittent network problems, use the PING -t command against a remote machine.

Increase or Decrease the Size of the PING Request

1. Open a Command Prompt window.

2. Type **PING –l 10240 localhost** and press ENTER.

The default PING request size is 32 bytes. The –l switch allows you to increase or decrease this size. Using the –l switch with the –t switch can be very useful when troubleshooting latency or dropped packet problems.

TraceRt

The TraceRt (trace route) command provides you with a simple method for testing the intermediate devices between two endpoints. For example, if you are working at a computer with the IP address of 10.10.10.200 and you want to test the route to 10.10.50.200, you would type the following command:

```
TRACERT 10.10.50.200
```

The TraceRt command also provides several switches for manipulating the command. Table 11-5 outlines these switches.

TABLE 11-5	Switch	Description
TraceRt Switches	-d	Do not resolve addresses to hostnames. This switch can speed up the TraceRt processing.
	-h maximum_hops	Specify the maximum number of hops to trace.
	-j host	Force TraceRt to route through a given host.
	-w timeout	Specify the timeout value, in milliseconds, for each reply.
	-R	Trace the round-trip path for IPv6 connections.
	-S source_address	Specify the source address to use for IPv6.
	-4	Force the use of IPv4.
	-6	Force the use of IPv6.

Remember that operating systems implement similar tools in different ways. For example, the Mac OS has a TraceRoute command as opposed to the Windows TraceRt command.

PathPING

PathPING is a newer command-line tool for Windows systems that allows you to PING the devices along the path to a destination. Unlike TraceRt, PathPING also reveals the number of sent packets compared to the number of dropped or lost packets. For this reason, PathPING is an excellent tool for troubleshooting dropped packets in voice networks. The command uses the switches listed in Table 11-6.

TABLE 11-6	Switch	Description
PathPING Switches	-g host	Same functionality as the -j switch in TraceRt.
	-h maximum_hops	Specifies the maximum number of hops across which the connection should be traced.
	-i address	Tells PathPING to use the specified source IPv6 address.
	-n	Same functionality as -d in TraceRt.
	-q number_of_ queries_per_hop	Specifies the number of PINGs to execute against each device in the path.
	-w timeout	Specifies how long to wait, in milliseconds, for each query reply.
	-4	Forces the use of IPv4.
	-6	Forces the use of IPv6.

ARP

The Address Resolution Protocol (ARP) is used to resolve IP addresses to MAC addresses for LAN packet delivery. Directly linked Ethernet devices communicate with each other based on their MAC addresses; however, routers receive packets destined for IP addresses. Some mechanism must be used to resolve these IP addresses to MAC addresses, and this is where the ARP cache is utilized.

When a machine needs to resolve an IP address to a MAC address, it will first look in the ARP cache. The ARP cache contains a listing of previously resolved IP-to-MAC mappings. To see the ARP cache on Windows:

1. Open a Command Prompt window.

2. Type **ARP –a** and press ENTER.

Figure 11-1 shows sample output from the ARP command on a Windows Vista client. You can see the MAC addresses of the local network interfaces as well as the devices to which the machine has connected. For more detailed information, type **ARP –a –v**, which will provide you with a verbose listing of the ARP cache. To clear the ARP cache, simply type **ARP –s *** and all entries will be deleted.

NSLookup

If you are finding it impossible to connect to devices using their DNS or hostnames, you can use the NSLookup tool to query the DNS server. This tool can perform

FIGURE 11-1

ARP output

```
Administrator: C:\Windows\system32\cmd.exe

C:\Users\Tom>arp -a

Interface: 10.0.50.105 --- 0x9
  Internet Address      Physical Address      Type
  10.0.50.1             00-50-e8-01-2a-b6     dynamic
  224.0.0.22            01-00-5e-00-00-16     static
  224.0.0.252           01-00-5e-00-00-fc     static

Interface: 192.168.58.1 --- 0xe
  Internet Address      Physical Address      Type
  192.168.58.255        ff-ff-ff-ff-ff-ff     static
  224.0.0.22            01-00-5e-00-00-16     static
  224.0.0.251           01-00-5e-00-00-fb     static
  224.0.0.252           01-00-5e-00-00-fc     static

Interface: 192.168.115.1 --- 0x10
  Internet Address      Physical Address      Type
  192.168.115.255       ff-ff-ff-ff-ff-ff     static
  224.0.0.22            01-00-5e-00-00-16     static
  224.0.0.251           01-00-5e-00-00-fb     static
  224.0.0.252           01-00-5e-00-00-fc     static

C:\Users\Tom>
```

FIGURE 11-2

NSLookup in
interactive mode

```
Administrator: C:\Windows\system32\cmd.exe - nslookup

C:\Users\Tom>nslookup
Default Server:  cns.cmc.co.denver.comcast.net
Address:  68.87.85.98

> www.sysedco.com
Server:  cns.cmc.co.denver.comcast.net
Address:  68.87.85.98

Non-authoritative answer:
Name:     www.sysedco.com.nomadix.com
Address:  0.0.0.1

> mail.sysedco.com
Server:  cns.cmc.co.denver.comcast.net
Address:  68.87.85.98

Non-authoritative answer:
Name:     mail.sysedco.com.nomadix.com
Address:  0.0.0.1

>
```

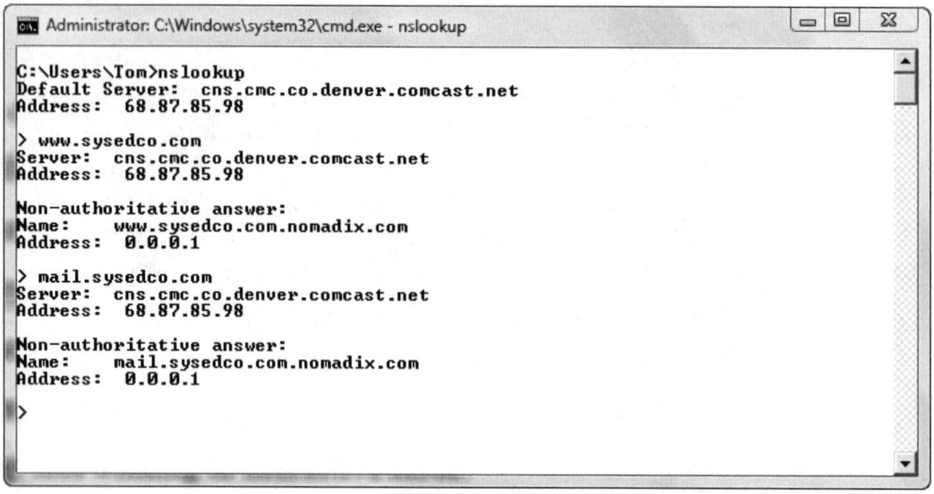

both forward and reverse lookups, as well as special operations for service records. For example, to resolve the IP address of a web site, type **NSLOOKUP** *website_domain_name*. The result will be the IP address of the web site and the name of the DNS server that resolved the hostname to the IP address. You can also use NSLookup in interactive mode by typing **NSLookup** and pressing ENTER. Figure 11-2 shows the NSLookup tool working in interactive mode.

Hostname

Hostname is a simple command-line tool that reveals the local machine name. Typing **hostname** at the Command Prompt will return the hostname of the computer. You can get the same information by typing **ECHO %computername%**. While the latter method has been available for many years in Windows systems, the former method is more recent.

NetStat

NetStat is a command-line tool that reveals network interface statistics. You can view both statistics related to communications and the currently open connections with this tool. Figure 11-3 shows the command-line options available for NetStat.

One of the useful features of the NetStat command is the ability to view statistics for a specific protocol. For example, you can view only TCP statistics, only IP statistics,

FIGURE 11-3

NetStat options

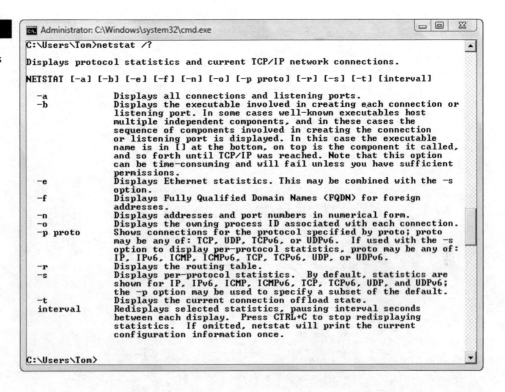

```
Administrator: C:\Windows\system32\cmd.exe

C:\Users\Tom>netstat /?

Displays protocol statistics and current TCP/IP network connections.

NETSTAT [-a] [-b] [-e] [-f] [-n] [-o] [-p proto] [-r] [-s] [-t] [interval]

  -a            Displays all connections and listening ports.
  -b            Displays the executable involved in creating each connection or
                listening port. In some cases well-known executables host
                multiple independent components, and in these cases the
                sequence of components involved in creating the connection
                or listening port is displayed. In this case the executable
                name is in [] at the bottom, on top is the component it called,
                and so forth until TCP/IP was reached. Note that this option
                can be time-consuming and will fail unless you have sufficient
                permissions.
  -e            Displays Ethernet statistics. This may be combined with the -s
                option.
  -f            Displays Fully Qualified Domain Names (FQDN) for foreign
                addresses.
  -n            Displays addresses and port numbers in numerical form.
  -o            Displays the owning process ID associated with each connection.
  -p proto      Shows connections for the protocol specified by proto; proto
                may be any of: TCP, UDP, TCPv6, or UDPv6. If used with the -s
                option to display per-protocol statistics, proto may be any of:
                IP, IPv6, ICMP, ICMPv6, TCP, TCPv6, UDP, or UDPv6.
  -r            Displays the routing table.
  -s            Displays per-protocol statistics. By default, statistics are
                shown for IP, IPv6, ICMP, ICMPv6, TCP, TCPv6, UDP, and UDPv6;
                the -p option may be used to specify a subset of the default.
  -t            Displays the current connection offload state.
  interval      Redisplays selected statistics, pausing interval seconds
                between each display. Press CTRL+C to stop redisplaying
                statistics. If omitted, netstat will print the current
                configuration information once.

C:\Users\Tom>
```

or only UDP statistics. The NetStat -s -p TCP command, for example, will only display statistics for TCP communications. Figure 11-4 shows the output for viewing TCP, IP, and UDP statistics respectively.

CERTIFICATION SUMMARY

In this chapter, you reviews common troubleshooting methodologies and studied the specific methodology recommended in the Convergence+ objectives. You learned about the importance of research and documentation. Then you reviewed the common symptoms and problems that arise in converged networks. Finally, you explored many of the command-line tools available on Windows clients for troubleshooting IP networks, including VoIP networks.

FIGURE 11-4

Viewing protocol-specific statistics

```
Administrator: C:\Windows\system32\cmd.exe                              □ □ ☒

C:\Users\Tom>netstat -s -p tcp

TCP Statistics for IPv4

    Active Opens                    = 494
    Passive Opens                   = 119
    Failed Connection Attempts      = 20
    Reset Connections               = 176
    Current Connections             = 8
    Segments Received               = 15392
    Segments Sent                   = 14919
    Segments Retransmitted          = 109

Active Connections

    Proto  Local Address          Foreign Address        State
    TCP    10.0.50.105:49679      cg-in-f104:http        CLOSE_WAIT
    TCP    10.0.50.105:49680      cg-in-f104:http        CLOSE_WAIT
    TCP    10.0.50.105:49685      cg-in-f147:http        CLOSE_WAIT
    TCP    10.0.50.105:49686      cg-in-f147:http        CLOSE_WAIT
    TCP    127.0.0.1:49227        Tom-PC:49228           ESTABLISHED
    TCP    127.0.0.1:49228        Tom-PC:49227           ESTABLISHED
    TCP    127.0.0.1:49244        Tom-PC:49245           ESTABLISHED
    TCP    127.0.0.1:49245        Tom-PC:49244           ESTABLISHED

C:\Users\Tom>netstat -s -p ip

IPv4 Statistics

    Packets Received                    = 13938
    Received Header Errors              = 0
    Received Address Errors             = 827
    Datagrams Forwarded                 = 0
    Unknown Protocols Received          = 0
    Received Packets Discarded          = 7782
    Received Packets Delivered          = 28708
    Output Requests                     = 22060
    Routing Discards                    = 0
    Discarded Output Packets            = 193
    Output Packet No Route              = 9
    Reassembly Required                 = 0
    Reassembly Successful               = 0
    Reassembly Failures                 = 0
    Datagrams Successfully Fragmented   = 0
    Datagrams Failing Fragmentation     = 0
    Fragments Created                   = 0

C:\Users\Tom>netstat -s -p udp

UDP Statistics for IPv4

    Datagrams Received    = 5341
    No Ports              = 1347
    Receive Errors        = 6442
    Datagrams Sent        = 6292

Active Connections

    Proto  Local Address          Foreign Address        State

C:\Users\Tom>
```

TWO-MINUTE DRILL

Problem Solving and Analysis

❏ When a problem is reported, you should log the problem before doing anything else.

❏ Take some time to confirm the problem by recreating it if possible.

❏ If you can recreate the problem, begin troubleshooting at that point.

❏ When you cannot resolve the issue yourself, escalate it to the appropriate individual or group.

❏ When the problem is resolved, close the log to indicate that the issue no longer exists.

Common Symptoms and Problems

❏ Poor voice quality can be caused by dropped packets, high latency, and poor codecs.

❏ Clipping may occur due to insufficient hardware, dropped packets, and poor jitter buffer operations.

❏ Echo occurs when electrical signals from one line are injected into an adjacent line.

❏ Echo problems can sometimes be resolved through careful impedance matching.

❏ Codecs that introduce more delay in the encoding process may also significantly reduce the size of the voice packets, resulting in less overall delay.

❏ It is possible for two VoIP phones to maintain a connection with each other even though neither phone can contact the call manager.

❏ Grade of Service (GoS) is a statement of the probability that a given call will be blocked during busy hours.

❏ GoS and busy hours are vital components in call center sizing.

❏ Media errors are often caused by poor connectors or faulty cables.

❏ H.323 uses TCP ports 1720 and 1503 as well as UDP ports 1718 and 1719.

❏ Small MTU sizes can result in higher fragmentation levels, which may increase packet delay rates.

❏ The PING command is used to verify that a remote machine is operating on the network and that the local machine can successfully communicate with that remote machine.

SELF TEST

The following questions will help you measure your understanding of the material presented in this chapter. Read all the choices carefully because there might be more than one correct answer. Choose all correct answers for each question.

Problem Solving and Analysis

1. You've received a call from a user on a land line indicating that the user cannot place calls with her VoIP softphone. She says that she can check her e-mail. What is the first thing you should do?

    A. Attempt to PING her machine.

    B. Instruct here to PING the call manager.

    C. Log the problem.

    D. Execute the ARP -s -v command and verify that the ARP cache is accurate.

2. You have been working on a particular video conferencing problem for two days. You've logged the problem and successfully recreated it. You have attempted everything you know to do in order to resolve the problem. What is your next step?

    A. Close the log.

    B. Escalate the problem.

    C. Use Google.com to look for help.

    D. Tell the user that the problem cannot be solved.

3. A user sends you an e-mail indicating that his VoIP phone is not working. He says that he can turn it on, but it will not communicate with the network. What should you do next?

    A. PING the phone's IP address.

    B. Verify that the call manager is up and running.

    C. Gather more information.

    D. Close the log.

Common Symptoms and Problems

4. Which of the following may result in poor voice quality? Choose all that apply.

    A. Dropped packets

    B. Using codecs with low MOS ratings

    C. That the router between the two phones is down

    D. High latency on the network

5. You are selecting cabling and connectors to use on your converged network. What common problem can be partially minimized by ensuring that the cables and connectors do not have an impedance mismatch?

   A. Echo

   B. Dropped calls

   C. Loss of features

   D. No dial tone

6. You are implementing the G.729 codec for VoIP. This codec has an encoding delay of approximately 10 ms, while the G.711 codec, which doesn't use compression, only incurs a 0.75 ms delay. Will the G.729 implementation definitely have more delay between the initial beginning of transmission and the actual recipient of audio at the remote speaker than the G.711 codec would?

   A. No

   B. Yes

7. You need to verify that a client computer, which acts as a VoIP softphone, can resolve several different domain names to IP addresses by performing direct DNS lookups. Which of the following command-line tools will allow you to verify this? Choose one.

   A. NetStat

   B. ARP

   C. NSLookup

   D. PING

8. You are implementing a VoIP network using H.323 as the primary voice management protocol. In the process, you need to ensure that a firewall will allow the H.323 communications through. Which of the following ports must be opened in the firewall? (Choose all that apply.)

   A. 1718 TCP

   B. 1720 TCP

   C. 1719 UDP

   D. 1503 UDP

SELF TEST ANSWERS

Problem Solving and Analysis

1. ☑ **C** is correct. Before you begin troubleshooting the problem, you should log the problem. This log can be used to trace the problem from first contact to resolution.
   ☒ **A, B,** and **D** are incorrect.

2. ☑ **B** is correct. According to the Convergence+ troubleshooting process, you should escalate the problem. You may be able to hand the problem over to another individual or group, or you may be able to gain assistance from the software or hardware vendors.
   ☒ **A, C,** and **D** are incorrect.

3. ☑ **C** is correct. You should gather more information. There is not sufficient information in the e-mail to troubleshoot the problem. If you attempt to troubleshoot the problem without gathering more information, you will most likely waste a large amount of time.
   ☒ **A, B,** and **D** are incorrect.

Common Symptoms and Problems

4. ☑ **A, B,** and **D** are correct. Dropped packets, poor codecs, and high latency can all result in poor voice quality.
   ☒ **C** is incorrect.

5. ☑ **A** is correct. Ensuring that the impedances of cables and connectors match is the starting point for reducing the occurrence of echo on your VoIP network. While impedance mismatches could potentially cause dropped calls, this would be a very rare occurrence.
   ☒ **B, C,** and **D** are incorrect.

6. ☑ **A** is correct. The correct answer is no. While the G.729 codec may incur more overhead from encoding, the packets are carrying much more audio data with much smaller packets. For this reason, G.729 will likely have less delay than G.711.
   ☒ **B** is incorrect.

7. ☑ **C** is correct. NSLookup can be used to query DNS servers. PING will attempt to resolve domain names to IP addresses or IP addresses to domain names when you PING a device, but this is not a direct name resolution tool.
   ☒ **A, B,** and **D** are incorrect.

8. ☑ **B** and **C** are correct. H.323 uses TCP ports 1720 and 1503 as well as UDP ports 1718 and 1719. The answers 1718 TCP and 1503 UDP are incorrect because they specify the right port, but the wrong protocol.
   ☒ **A** and **D** are incorrect.

12

Security Vulnerabilities

> *Become a student of change. It is the only thing that remains constant.*
>
> *–Anthony J. D'Angelo*

Some people call them hackers. I prefer to call them crackers or attackers; however, regardless of the name they are assigned, they are an evolving and morphing collective. This group of technically savvy and intensely creative individuals continues to surprise us each month as they develop new techniques for penetrating our networks and systems. While we are spending our days implementing, maintaining, and troubleshooting these networks and systems, the attackers have no such constraints in many cases. The attackers are spending the time we are spending to enable our users to get business results for very different ends. Since such a time investment disparity exists, it is essential that we also collaborate to gather our collective protection knowledge as a unit.

The preceding paragraph represents, to the best of my recollection, the opening remarks of a speaker at a security conference I attended. Even if I don't agree with every statement, I must consent to the fact that attackers are developing new methods nearly every day. As network administrators and security practitioners, we must evolve with them. We may not have the time to perform the research that leads to discovering vulnerabilities and protecting against those vulnerabilities, but we must make the time to learn of these vulnerabilities and solutions through books, web sites, magazines, and conferences.

It is for this reason that I have broken the security section of this book into two chapters. In this chapter, you will learn about the threats, vulnerabilities, and exploits to which all networks—including converged networks—are vulnerable. You will also learn about some specific vulnerabilities that apply only to VoIP and video over IP networks. In the next chapter, you'll learn how to protect against these vulnerabilities.

CERTIFICATION OBJECTIVE 12.01

Explain Concepts and Components of Security Design

Information and network security technologies exist to protect our data and systems from attackers or crackers. You can only truly understand why the security technologies are needed if you understand the attack methods and tools.

This section will not be an exhaustive treatment of the cracking world, but it should give you enough information to help you understand why the technologies covered in the next chapter are so important. Here, I'll cover the following major topics:

- Importance of security
- Threats, vulnerabilities, and exploits defined
- Attack points
- Hacking examples
- Zero-day hacks

Importance of Security

The importance of security varies by organization. The variation exists due to the differing values placed on information and networks within organizations. For example, organizations involved in banking and health care will likely place a greater priority on information security than organizations involved in selling greeting cards. However, in every organization there exists a need to classify data so that it can be protected appropriately. The greeting card company will likely place a greater value on its customer database than it will on the log files for the Internet firewall. Each of these data files has value, but one is more valuable than the other and should be classified accordingly so that it can be protected properly. This process is at the core of information security and it can be itemized as follows:

1. Determine the value of the information in question.
2. Apply an appropriate classification based on that value.
3. Implement the proper security solutions for that classification of information.

As an example, your organization may choose to classify information in three categories: internal, public, and internal sensitive. Information classified as internal information may require only appropriate authentication and authorization. Information classified as public information may require neither authentication nor authorization. The internal sensitive information may require authentication, authorization, and storage-based encryption.

You can see why different organizations have different security priorities and needs from this very brief overview of information classification and security measures. It is also true, however, that every organization is at risk to certain threats. Threats like Denial of Service (DoS), worms, and others are often promiscuous in nature. The attacker does not care what networks or systems are damaged or made less effective in a promiscuous attack. The intention of such an attack is often only to express the attacker's ability or to serve some other motivation for the attacker, such as curiosity or need for recognition. Since many attacks are promiscuous in nature, it is very important the every organization place some level of priority on security regardless of the intrinsic value of the information or networks it employs.

Statistics

Various organizations perform surveys and gather statistics that are useful in gaining a perspective on the need for security. One such organization is *InformationWeek* magazine. Its 2008 security survey ("2008 Security Survey," June 2008) showed that complexity is the greatest difficulty in securing systems. In fact, 62 percent of respondents cited complexity as the biggest security challenge. Administrators deal with varied data types, and that data is often unclassified. Without classification, it's difficult to determine how to protect the data. There is good news, however, in *InformationWeek*'s survey: solutions exist that can help reduce the likelihood of a security incident. According to the survey, the following solutions were selected by the indicated percentage of respondents:

- Firewalls (63%)
- Antivirus (59%)
- Encryption (46%)
- VPNs (45%)
- Strong passwords (40%)
- Spam filtering (35%)
- E-mail security (34%)

Another organization that reports on the state of information and systems security is the Computer Security Institute (CSI). CSI has performed its annual security survey for more than ten years, and the statistics show that security should be a very important part of any organization's budget and plans. For the five years preceding the 2007 survey, the results showed a drop in the average organization's losses due to cybercrime; however, the 2007 survey reported a significant rise in estimated losses. The good news is that the spike in the 2007 survey results in losses that are still lower than those reported in 2002. This continued lower loss rating may indicate that we are doing a better job of securing our data and assets, or it may only indicate that we are spending less on hardware and software and, therefore, losing less when these assets are stolen or compromised.

The following statistics represent just a few of the important reports from the 2007 CSI Computer Crime and Security Survey:

- Twenty-five percent of responding organizations spend between 6 and 10 percent of the annual IT budget on security.
- Sixty-one percent of responding organizations still outsource no security functions.
- Forty-six percent indicated that they had experienced a security incident in the previous 12 months, and 10 percent indicated that they were unsure.
- Twenty-six percent of the total responding pool indicated that there had been more than 10 incidents in the previous 12 months.
- Only 36 percent indicated that they accrued no losses due to insider threats, which means that 64 percent experienced an insider attack that led to losses.
- The most common type of attack was the simple abuse of Internet access by valid users.
- Viruses were also a common attack problem, with 52 percent reporting such attacks.
- Only 5 percent reported telecom fraud and 13 percent reported system penetration; however, it is important to know that some experts estimate as many as 85 percent of all attacks go undetected.

It is very clear from these statistics that threats are real and security is important. The statistics show us what is happening, but the theory can help us gain an understanding of why these attacks occur.

Theory

Why does a seemingly unprovoked attacker attack? This question is an important one, but it is very difficult to answer with certainty. We are, after all, dealing with human nature in these circumstances. It is very easy to understand why an employee who is terminated decides to attack: that employee is not thinking rationally. It is even easy to understand why a competitor might attack: to gain the upper hand on your organization. But why does a script kiddy (one who lacks deep technical understanding, but has the ability to run scripts or follow instructions) choose to attack your organization?

One theory says that they don't choose your organization. Instead, the suggestion is that the attacker is promiscuous. The attacker does not care who the target is, but will attack any target that is vulnerable to a particular exploit. Attacks from script kiddies often fall into this category. These attackers will scan hundreds or even thousands of networks looking for any network that is vulnerable. When a vulnerable network is found, the attacker will launch other scripts or utilities against the network to penetrate it and gain access to data and resources. This method may also be used by skilled crackers who wish only to gain control of the network and resources so that an attack may be launched against a primary target using these easily penetrated resources. A distributed DoS (DDoS) attack would be an example of just such an attack. This attack is more than a theory, however; it is a reality. Script kiddies exist in the many thousands and are a prime threat for any organization. This theory indicates that one form of attacker is the promiscuous attack, but another theory explores the motivations of the attackers more deeply and suggests that there may be underlying motives that move the attacker against your organization.

Whether the thinking is correct or incorrect doesn't matter. All that matters is that the attacker perceives your organization to be a threat to something he or she values. These values may include environmental concerns, freedom of speech concerns, freedom from government, or any other value that the attacker holds in high esteem. If the attacker sees your organization as a threat to the realization of these values, this perception may be the motivation for the attack. Depending on the attacker's value system, he or she may attempt only to deface your web site or else be bent on complete destruction of your data and systems. Either way, we must protect against these individuals as well.

What is the difference between these two attacker types, and why does it matter? The big difference is the answer to the question, "Why?" Why does the attacker want to attack your network or systems? If the attack is promiscuous in nature, traditional protection mechanisms will likely suffice. If it is targeted, the attacker will most likely be willing to spend much more time attempting to penetrate your network, and stronger security mechanisms will be needed. You will need to evaluate your organization's risk of being an intentional target based on strong motivations or a promiscuous target based on weak motivations. Additionally, you must remember that even an attack that is promiscuous in nature may be intended to harm another organization through the utilization of your resources.

Reality

All of these statistics and theories, which I've only covered here in part, lead us to an important reality. Every organization must deal with information, network, and systems security. Protection methods must be considered for the information. These methods will include authentication, authorization, accounting, and encryption.

For the network systems, you will need to implement authentication and authorization to ensure that only the assigned personnel may administer the devices. For the systems security, you should ensure secure management of your application code base and secure programming practices. This picture is the big picture. Throughout the rest of this chapter, you'll learn the finer details of how attackers gain access to your networks, your systems, and ultimately your information. In the next chapter, you'll learn how to implement the technologies to protect against these attacks.

Threats, Vulnerabilities, and Exploits Defined

A *threat* is defined as an individual, group, circumstance, or event with the potential to cause harm to a system. The only requirement for a person or event to be considered a threat is the potential for harm. Certainty is not required. Threats fall into two general categories: intentional and unintentional. Intentional threats include all threats that have intellect behind them. Stated differently, intentional threats are those threats that are planned and executed by an individual or a group of people. Unintentional threats include those events or circumstances that are often called acts of God. Lightning strikes, hurricanes, accidents of any kind, and other similar events are unintentional threats; however, these unintentional threats must be accounted for as well.

A *vulnerability* is defined as a weakness in a system or object. The object may be part of a system, or it may be an independent entity. For example, an RFID chip may be considered as an independent entity or as part of a larger networked system. As an independent entity, the chip may have vulnerabilities; however, if you plan to implement the chip as part of a larger system, that system must be checked for vulnerabilities as a whole. New vulnerabilities that were non-existent in individual objects are often discovered when those objects are used together as a system. For example, there may be no vulnerabilities in a given software module, but when that module communicates with another module, the communication channel may introduce a new vulnerability.

The discovery of vulnerabilities is known as *vulnerability analysis*. Vulnerability analysis may be performed by a software or hardware vendor in order to test its solutions. It may also be performed by organizations implementing the solution in order to ensure the privacy and protection of their data. In most cases, it will be performed by both the vendor and the implementing organizations. This dual testing is needed because the implementing organizations will be deploying the solution in an environment that is foreign to the vendor and may, therefore, introduce new vulnerabilities.

An *exploit* is a specific method used to expose and take advantage of a vulnerability. Exploits introduce threats because of vulnerabilities. An exploit may be a procedure that an attacker must perform, or it may come in the form of source code that must be executed.

When an attacker wishes to gain access to a network, he will go through the following basic steps:

1. Scan for devices on the network.
2. Scan for services on those devices.
3. Discover the versions of the running services.
4. Research vulnerabilities.
5. Launch an exploit based on one or more vulnerabilities.

This step-by-step process shows that attacking a network is a simple process. You simply have to have the right tools. For instance, on a Windows systems, you could use nMap or Angry IP Scanner to find the devices, services, and versions. These Windows tools are free to download. Next you can search the Internet for known vulnerabilities, and then you can take advantage of those vulnerabilities through exploits. In many cases, you can download free applications that are specially designed to launch the exploit. As an illustration, AirCrack is a program designed specifically for cracking WEP keys on wireless networks.

When the tools are easy to get and the instructions are easy to follow, the threat increases. This increase is due to the fact that script kiddies can easily launch the exploit. For this reason, WEP cracking must be considered a valid threat to all organizations, since promiscuous attackers can use the exploit against them. WEP cracking is used as an example here, but any other exploit that is similar in nature (it can be acquired and executed without in-depth technical knowledge) should be considered a threat, and protection against it should be part of all security policies and procedures.

Attack Points

When you are considering the threats to a converged network, a good starting place is to document the various attack points. An *attack point* is an entry point to the system or a location within the system than an attacker may attempt to penetrate. At a minimum, the following attack points should be considered:

- Networks
- Servers

- Storage
- Authentication systems
- Encryption systems

Networks

The earliest converged networks were wired only; however, with the standardization of wireless technologies in the late 1990s, wireless VoIP has become very popular. We must consider both wired and wireless networks and the potential vulnerabilities they introduce.

Wired Wired networks may be exploited by gaining access to an insecure port or by penetrating the network through a secured port. If the network is connected to the Internet, the Internet connection may also be exploited. The last method of exploit is through dial-up connections. Dial-up connections are becoming increasingly rare, but they do still exist.

An insecure wired port is an Ethernet (or some other wired network standard) port that is enabled and not protected with authentication. IEEE 802.1X is a standard that defines mechanisms for securing such a port. Some organizations choose to implement 802.1X, while others choose to resolve the issue by disabling any unused ports until they are needed. The latter method leaves the network vulnerable to human error or forgetfulness. The 802.1X solution is preferred as long as a secure Extensible Authentication Protocol (EAP) implementation is used.

If a port is insecure, an attacker may connect to the port and begin scanning and ultimately attacking the network. Prime targets include ports in conference rooms, unused offices, and remote areas of warehouses or manufacturing plants. These ports should certainly be secured or disabled any time they are not in use.

One of the most commonly used attack points is an organization's Internet connection. Many administrators have noted more than one thousand attack attempts in a single day. If you have an Internet connection (hopefully a good firewall) and the connection attempts can be logged, you should enable this logging. After a few days you can look at the log to see how many connection attempts are being made against ports that are commonly attacked. You may be surprised by the number of attempts.

Wireless Wireless networks are vulnerable to penetration through Internet-facing connections just like wired networks; however, wireless networks also introduce entirely new vulnerabilities. Instead of focusing on ports, you must focus on connections. Wireless networks allow client devices to connect to the network without the use of pre-assigned ports. For this reason, disabling ports is not an option. MAC filtering has been used in the past in an attempt to accomplish security at the same level as port management; however, MAC filtering is very weak, since an attacker may monitor the network and discover valid MAC addresses. Once the valid addresses are known, the attacker may reconfigure her device to use an allowed MAC address. For this reason, you should consider MAC filtering as a security myth and not as a security solution.

In fact, there are many myths associated with wireless security. I'll cover just a few of them here, including:

- MAC filtering
- SSID hiding
- All modern equipment uses "Better WEP"
- Wireless networks can't be secured

MAC Filtering

Vendors of wireless devices and books on wireless networking often provide a list of the "Top 5" or "Top 10" things you should do to secure your WLAN (wireless LAN). This list usually includes MAC filtering and SSID hiding or cloaking. The reality is that neither of these provides a high level of security. MAC addresses can easily be spoofed, and valid MAC addresses can be identified in just a few moments. For example, an attacker can eliminate the AP in an infrastructure basic service set (BSS) by looking for the MAC address that sends out Beacon frames. This MAC address will always be the AP in the BSS. With this address filtered out of the attacker's protocol analyzer, he has only to find other MAC addresses that are transmitting with a destination MAC address equal to that of the AP. Assuming the captured frames are data frames, the attacker now knows a valid MAC address.

There is no question that MAC filtering will make it more difficult for an attacker to access your network. The attacker will have to go through the process I've just outlined (or a similar process) in order to obtain a valid MAC address to spoof. However, you are adding to your workload by implementing such MAC filtering, and you have to ask, "Am I getting a good return on investment for my time?" The answer is usually no. Assuming you are using TKIP or CCMP with a strong EAP type for authentication (or even preshared keys), these methods will be so much more secure than MAC filtering could ever hope to be that it makes the extra effort of MAC filtering of minimal value. I recommend that you do not concern yourself with MAC filtering in an enterprise or SMB implementation. It may be useful in a SOHO implementation, but I question its value even then.

SSID Hiding

Hiding or cloaking the SSID of your WLAN falls into a similar category as MAC filtering. Both methods provide very little in the way of security enhancement. Changing the name of your SSID from the vendor defaults can be very helpful, as it will make dictionary attacks against preshared key implementations more difficult. This attack is possible because the SSID is used in the process of creating the pairwise master key. Hiding the SSID only makes it difficult for casual eavesdroppers to find your network.

Hiding the SSID also forces your valid clients to send out probe requests in order to connect to your WLAN, whether using the Windows Wireless Zero Configuration (WZC) utility or your vendor's client software. This activity means that, when the user turns on his or her laptop in a public place, the laptop is broadcasting your SSID out to the world. This could be considered a potential security threat, since a rogue AP of any type can be configured to the SSID that is being sent out in the probe requests. Software-based APs can respond to random SSIDs generated by WZC, but hiding your SSID effectively makes every WLAN client in existence vulnerable to such attacks, since they will all have to send probe requests with the SSID.

I always recommend changing the SSID from the default, but I never recommend hiding the SSID for security purposes. Some people will hide the SSID for usability purposes. Turning off the SSID broadcast in all APs' Beacon frames will prevent client computers from "seeing" the other networks to which they are not supposed to connect. This configuration may reduce confusion, but SSID hiding should not be considered a security solution.

All Modern Equipment Uses "Better WEP"

When the initial scare hit, many vendors looked for solutions to the weak initialization vectors (IVs) used in the WEP implementations that existed at the time. Eventually, many vendors began implementing newer WEP solutions that attempted to avoid the weak IVs. As early as 2003, I noticed people posting on the Internet and saying that the newer hardware didn't have this problem. In fact, I have a network-attached storage device that was purchased in 2005 that includes a built-in AP. This device is running the most recent firmware from the vendor (D-Link, in this case), and I can connect a brand new Intel Centrino chipset laptop to the device using WEP. While monitoring from another computer, I am able to capture weak IVs and crack the WEP key in a matter of minutes. You simply cannot trust that a vendor has actually implemented algorithms that protect you against WEP weaknesses just because the hardware is newer. Instead, you would need to monitor the communications with the device in order to determine if weak IVs are being used. It's easier to implement WPA or WPA2, so I recommend that.

Wireless Networks Can't Be Secured

Don't allow these last few false security methods to keep you from implementing a WLAN. WLANs can be implemented in a secure fashion using IEEE 802.11i (Clause 9 of IEEE 802.11-2007) and strong EAP types. In fact, WLANs can be made far more secure that most wired LANs, since most wired LANs do not implement any real authentication mechanisms at the node level. If you buy into the concept that WLANs cannot be secured and you decide not to implement a WLAN for this reason, you will likely open your network up to more frequent rogue AP installations from users that desire to have wireless access to the network. The simplest way to avoid or at least diminish the occurrence of user-installed rogue APs is to implement a secure WLAN for the users. In the end, WLANs can be secured, but you must be aware of the security myths surrounding them.

Servers

The second attack point that I will cover is the server or servers on your network. Servers are used to store data, provide services to users, or provide services to other systems. Many servers are running Linux or Windows operating systems, and these systems are heavily targeted by attackers because of their heavy use. Attack methods include:

- Exploiting known vulnerabilities
- Exploiting configuration errors
- Exploiting running services

Attackers can locate known vulnerabilities using search engines, discussion forums, and other web sites. Common web sites used to discover vulnerabilities include:

- Microsoft.com/security
- MilW0rm.com
- Zone-H.org

■ HackerWatch.org

■ Secunia.com/products

As a network administrator, you should visit these web sites regularly to keep your knowledge up-to-date on the hardware, operating systems ,and applications you are utilizing.

Configuration error exploits can often be avoided by implementing a strong security management process. The process would usually include threat and vulnerability analysis, security policy development, and policy implementation. By implementing configurations based on solid security policies, you reduce the likelihood of configuration errors. However, it does require a team effort because each technician must abide by the policies when configuring a device. An attacker requires only one misconfigured device to gain entry to the network. Auditing and security assessments may also be used to ensure proper configuration.

Many services are insecure regardless of the implementation method used. For example, Telnet sends authentication credentials as clear text when implemented according to the standards. FTP also sends the username and password in the clear. Passwords sent as clear text can be easily retrieved using protocol analyzers. This statement is particularly true of wireless networks that do not implement encryption, such as wireless hotspots and older networks.

Storage

Many storage attacks are really authentication attacks. The attacker performs password guessing, password sniffing, or offline password cracking in order to gain access to the storage location.

In addition to authentication attacks, an attacker may take advantage of vulnerabilities inherent in the embedded operating system of the device. For example, many network-attached storage (NAS) devices use embedded Linux. Since the operating system is implemented through firmware, the administrator may fail to update the operating system as often as on normal computers running the same operating system. This delay can result in vulnerabilities being exposed for longer periods of time. The morale of the story is simple: update the firmware on your devices any time a security vulnerability is patched and the firmware does not introduce problems into the system. If you cannot update the firmware because the vendor is no longer updating it, consider placing the device behind a router or firewall that can be used to block all traffic that may result in exploitation of the vulnerability.

Authentication Systems

Authentication systems are used to validate user identities and allow for authorization of the users for access to resources. Authentication systems are based on credentials. Credentials can include any of the following three types:

- Something you know
- Something you have
- Something you are

Something you know includes passwords and personal identification numbers (PINs). Something you have includes keys, smart cards, and RFID chips. Something you are includes biometrics such as fingerprint scanners, retina scanners, and even weight measurements.

Authentication systems can be attacked by exploiting weak protocols, weak credential stores, or weak credentials. Weak protocols are protocols that are implemented poorly. The passwords may be sent in clear text across the network, or another vulnerability may be inherent in the system. Weak credential stores are exploited by cracking the encryption used on the store or simply accessing the credential store when no encryption is implemented. Weak credentials are usually weak passwords. Today, weak passwords are passwords that are less than eight characters and those passwords that do not include multiple character types. However, a password such as "thehorsejumpedoverthemoononabroom" is very secure even though it does not contain multiple character types. The ultimate determiner of the strength of a password, with few exceptions, is the size of the password pool. The unusually long password referenced here is more than 30 characters long. A 30-character password with only lowercase letters is part of a password pool that includes 254,186,582,832,900,000,000,000,000,000,000,000 possible passwords. This number represents more than 254 undecillion passwords. To put it into perspective, you could guest 100 trillion passwords each second and it would take more than 40 quadrillion years to guess the password on average. These numbers assume a blind brute-force attack, but clearly even with advanced methods including rainbow tables and intelligent algorithms, it would take far too long to make it worth the attempt.

My point with this example is simple: if you have strong enough passwords, you will be safe enough for most data. If you feel passwords cannot be made strong enough, due to the human element (think, writing passwords on sticky notes), you should consider other authentication methods such as smart cards or biometrics.

Encryption Systems

Sensitive data should be encrypted in two places: during transit and during storage. Encryption for transmitted data is processor intensive and may introduce additional processing delay in VoIP systems; however, the trade-off may be worth it if security is of the utmost importance to your organization. In-transit encryption solutions are vulnerable to various sniffing attacks. For example, WEP encrypts traffic for WLANs, but the algorithm and keys were improperly implemented, resulting in the ability to easily crack the WEP key and then gain access to the transmitted data.

Storage encryption is most frequently attacked by attacking the key store. You may have noticed that I'm not mentioning brute-force methods with encryption. Brute force is seldom used to crack any encryption scheme due to the time required. Even DES (Digital Encryption Standard) keys with 40-bit key lengths take too long for most attacks. For this reason, attackers will usually look for vulnerabilities in the key store or the method used to access the key store. I give an example of this in the next section of this chapter, "Hacking Examples," under the heading "Encryption Hacks."

Hacking Examples

With an awareness of the common attack points, you're ready to investigate a few hacking examples. You will improve your understanding of security by learning about specific hacking methods. In the next few pages, I'll present various hacks that can be used against a selection of the attack points previously discussed.

Network Hacks

Cracking WEP is a perfect example of a network hack. The Wired Equivalent Privacy (WEP) protocol is used to encrypt data on WLANs and authenticate users to the WLAN based on the fact that the user knows the WEP key. There are numerous problems with the WEP protocol that result in the ability to crack it easily.

An understanding of the basic WEP process will help you to understand the weaknesses that are covered next. The WEP process starts with the inputs to the process. These inputs include the data that should be encrypted (usually called plaintext), the secret key (40 bits or 104 bits), and the IV (24 bits). These inputs are passed through the WEP algorithms to generate the output (the ciphertext or encrypted data).

Since WEP is a Layer 2 security implementation, it doesn't matter what type of data is being transmitted as long as it originates above Layer 2 in the OSI model. In order to encrypt the data, the RC4 algorithm is used to create a pseudorandom string of bits called a keystream. The WEP static key and the IV are used to seed the pseudorandom number generator used by the RC4 algorithm. The resulting keystream is XORed against the plaintext to generate the ciphertext. The ciphertext alone is transferred without the keystream; however, the IV is sent to the receiver. The receiver uses the IV that was transmitted and the stored static WEP key to feed the same pseudorandom number generator to regenerate the same keystream. The XOR is reversed at the receiver to recover the original plaintext from the ciphertext.

WEP was never intended to provide impenetrable security, but was only intended to protect against casual eavesdropping. With the rapid increase in processor speeds, cracking WEP has become a very short task and it can no longer be considered for protection against any organized attack. The weaknesses in WEP include the following:

- Brute-force attacks
- Dictionary attacks
- Weak IV attacks
- Re-injection attacks
- Storage attacks

In late 2000 and early 2001, the security weaknesses of WEP became clear. Since then many attack methods have been developed and tools have been created that make these attack methods simple to implement for entry-level technical individuals.

The *brute-force* attack method is a key guessing method that attempts every possible key in order to crack the encryption. With 104-bit WEP, brute force is really not a feasible attack method; however, 40-bit WEP can usually be cracked in one or two days with brute-force attacks using more than 20 distributed computers. The short time frame is accomplished using a distributed cracking tool like jc-wepcrack. jc-wepcrack is actually two tools: the client and the server. You would first start the tool on the server and configure it for the WEP key size you think the target WLAN uses and provide it with a pcap file (a capture of encrypted frames) from that network. Next, you launch the client program and configure it to connect to the server. The client program will request a portion of the keys to be guessed and will attempt to access the encrypted frames with those keys. With the modern addition of Field Programmable Gate Arrays (FPGAs), which are add-on boards for hardware acceleration, the time to crack can be reduced by more than 30 times. In fairness, the 20 computers would have to be P4 3.6 GHz machines or better. If you chose to go the FPGA route, you would be spending a lot of money to crack that WEP key. Since smart enterprises will no longer be using WEP, you will not likely gain access to any information that is as valuable as your hacking network.

The *dictionary attack* method relies on the fact that humans often use words as passwords. The key is to use a dictionary cracking tool that understands the conversion algorithm used by a hardware vendor to convert the typed password into the WEP key. This algorithm is not part of IEEE 802.11 and is implemented differently by the different vendors. Many vendors allow the user to type a passphrase that is then converted to the WEP key using the Neesus Datacom or MD5 WEP key generation algorithms. The Neesus Datacom algorithm is notoriously insecure and has resulted in what is sometimes called the Newsham-21-bit attack because it reduces the usable WEP key pool to 21 bits instead of 40 when using a 40-bit WEP key. This smaller pool can be exhausted in about six–seven seconds on a P4 3.6 GHz single machine using modern cracking tools against a pcap file. Even MD5-based conversion algorithms are far too weak and should not be considered secure because they are still used to implement WEP, which is insecure due to weak IVs as well.

The *weak IV attacks* are based on the faulty implementation of RC4 in the WEP protocols. The IV is prepended to the static WEP key to form the full WEP encryption key used by the RC4 algorithm. This method means that an attacker already knows the first 24 bits of the encryption key, since the IV is sent in clear text as part of the frame header. Additionally, Fluhrer, Mantin, and Shamir identified "weak" IVs in a paper released in 2001. These weak IVs result in certain values becoming more statistically probable than others and make it easier to crack the static WEP key. The 802.11 frames that use these weak IVs have come to be known as *interesting frames*. With enough interesting frames collected, you can crack the WEP key in a matter of seconds. This vulnerability reduces the total attack time down to less than five–six minutes on a busy WLAN.

on the
!
◑ o b

The weak IVs discovered by Fluhrer, Mantin, and Shamir are now among a larger pool of known weak IVs. Since 2001, another 16 classes of weak IVs have been discovered by David Hulton (h1kari) and KoreK.

What if the WEP-enabled network being attacked is not busy and you cannot capture enough interesting frames in a short window of time? The answer is a *re-injection attack*. This attack usually re-injects ARP packets onto the WLAN.

The program aireplay can detect ARP packets based on their unique size and does not need to decrypt the packets. By re-injecting the ARP packets back onto the WLAN, it will force the other clients to reply and cause the creation of large amounts of WLAN traffic very quickly. For 40-bit WEP cracking, you usually want around 300,000 total frames to get enough interesting frames, and for 104-bit WEP cracking, you may want about 1,000,000 frames.

Storage attacks are those methods used to recover WEP or WPA keys from their storage locations. On Windows computers, for example, WEP keys have often been stored in the Registry in an encrypted form. An older version of this attack method was the Lucent Registry Crack; however, it appears that the problem has not been fully removed from our modern networks. An application named *wzcook* can retrieve the stored WEP keys used by Windows' Wireless Zero Configuration. This application recovers WEP or WPA-PSK keys (since they are effectively the same, WPA just improves the way the key is managed and implemented) and comes with the Aircrack-ng tools used for cracking these keys. The application only works if you have administrator access to the local machine, but in an environment with poor physical security and poor user training it is not difficult to find a machine that is logged on and using the WLAN for this attack.

WEP makes up the core of pre-RSNA security in IEEE 802.11 networks. I hope the reality that WEP can be cracked in less than five minutes is enough to make you realize that you shouldn't be using it on your networks. The only exception would be an installation where you are required to install a WLAN using older hardware and you have no other option. I've encountered this scenario in a few churches where I've assisted in their network implementation. The problem was not with the infrastructure equipment in any of the scenarios. The problem was with the client devices that the church members wanted to use to connect to the WLAN. These devices did not support WPA or WPA2, and we were forced to use either WEP or no security at all. While WEP can certainly be cracked quickly, at least it has to be cracked. Open System authentication with no WEP, WPA, or WPA2 security is just that: open.

In the end, businesses and organizations that have sensitive data to protect must take a stand for security and against older technologies. This use of new technology means that you should not be implementing WEP anywhere in your organization. When you have the authority of a corporation, the government, or even a non-profit oversight board, you can usually sell them on the need for better security with a short (five minutes or less) demonstration of just how weak WEP is. If you're implementing Voice over WLAN, these insights will be tremendously valuable.

Password Hacks

Most computer access controls are based on passwords. Weak passwords are one of the most serious security threats in networking, for obvious reasons. Intruders easily guess commonly used and known passwords, such as "password," "admin," etc. Short words or strings of characters are often at risk from a brute-force password attack program, and passwords made from words found in the dictionary can be guessed using dictionary attacks.

All of this information is common knowledge to security administrators, but what is not commonly considered is that passwords flow from client to server across unsecured networks all the time. In the past, there was a common misconception that wired networks were secure, but wireless LANs have opened the eyes of many administrators and attackers that networking systems using passwords passed in clear text across any medium are vulnerable to interception. For this reason, password encryption has become very popular along with security mechanisms, such as Kerberos, that implement such encryption. Two auditing tools often used by administrators and hackers alike to view clear-text passwords are WinSniffer and ettercap.

WinSniffer WinSniffer is a password capture utility capable of capturing FTP, HTTP, ICQ, Telnet, SMTP, POP3, NNTP, and IMAP usernames and passwords in a shared-medium networking environment such as wireless APs or wired hubs. WinSniffer is installed on a Windows-based computer, usually a laptop being used to audit wireless networks. In a switched network, WinSniffer can only capture passwords that originate from either the client that sent the password or the server that sent the client the information directly. WinSniffer can be used to capture your own passwords (when saved in applications) when you forget them. Sample output from WinSniffer is shown in Figure 12-1.

FIGURE 12-1

Sample password output from WinSniffer

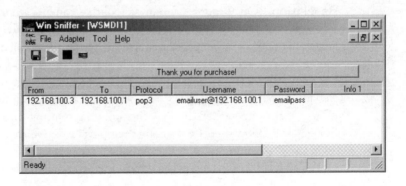

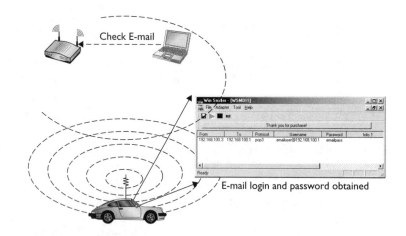

FIGURE 12-2

Obtaining passwords from unsuspecting users

Consider Figure 12-2, in which the user is checking e-mail over an unencrypted wireless LAN segment. An attacker is scanning the wireless segment using a password sniffer and picks up the user's e-mail login information and the domain from which the user is checking the e-mail. The attacker now has access to the user's e-mail account and can read all of the user's e-mail.

Public access wireless networks (hotspots) such as those found in airports or in metropolitan areas are some of the most vulnerable areas for user attacks. Users that are not familiar with how easy it is to obtain their login information through a peer-to-peer attack unknowingly check their e-mail or access their corporate network—even VoIP systems—and end up giving access to their accounts to a hacker. Once the hacker obtains a valid login to a corporate account, that hacker is now well equipped to try to obtain further access into the network to locate more sensitive information.

Revelation On Windows systems, a tool that can be used to discover passwords is Revelation. This program will allow you to drag a cursor over a password field in any login dialog or a web page and have the password revealed. Of course, to use this tool, the user would have to have left his or her computer logged on and you will have to have the ability to run the tool. However, with users saving their passwords in web forms so frequently today, this tool can reveal passwords for many situations. To protect against it, you can disallow the tool from running through Windows Group Policies or disallow users from saving their passwords. Neither method will provide complete protection, but they can provide extra protection and make it more difficult for the attacker. For example, the attacker would have to use a hex editor to modify the binary file (revelation.exe) in order to get around the hash-based Group Policies in Windows Server 2003 and supported by Windows XP clients. Revelation can be used on wired and wireless systems in order to discover passwords.

ettercap   Written by Alberto Ornaghi and Marco Valleri, ettercap is one of the most powerful password capture and auditing tools available today. ettercap supports almost every operating system platform and can be found at http://ettercap .sourceforge.net. ettercap is capable of gathering data even in a switched environment, which far exceeds the abilities of most other audit tools. ettercap uses as a menu-style user interface, making it user friendly. Some of the features available in ettercap are:

- **Character injection into an established connection**   A user can inject characters to a server (emulating commands) or to a client (emulating replies) while maintaining a live connection.
- **SSH1 support**   A user can analyze usernames and passwords, and even the data of the SSH1 connection. ettercap is the first software capable of analyzing an SSH connection in full-duplex mode.
- **HTTPS support**   A user can sniff HTTP-SSL data even if the connection is made through a proxy.
- **Remote traffic through a GRE tunnel**   A user can analyze remote traffic through a GRE tunnel from a remote router.
- **PPTP broker**   A user can perform man-in-the-middle attacks against PPTP tunnels.
- **Plug-ins support**   A user can create his own plug-in using the ettercap's API. Many plug-ins are included in the base package.
- **Password collector for**   TELNET, FTP, POP, RLOGIN, SSH1, ICQ, SMB, MySQL, HTTP, NNTP, X11, NAPSTER, IRC, RIP, BGP, SOCKS-5, IMAP4, VNC, LDAP, NFS, SNMP, HALF LIFE, QUAKE 3, MSN, & YMSG.
- **Packet filtering/dropping**   A user can configure a filter that searches for a particular string (even hex) in the TCP or UDP payload and replace it with a new string or drop the entire packet.
- **OS fingerprinting**   A user can fingerprint the operating system of the victim host and its network adapter.
- **Killing a connection**   From the connections list, a user can kill all the connections he or she chooses.

- **Passive scanning of the LAN**   A user can retrieve information about any of the following: hosts in the LAN, open ports, services versions, host type (gateway, router, or simple host), and estimated distance (in hops).

- **Checking for other poisoners**   ettercap has the ability to actively or passively find other poisoners on the LAN. These would be devices that have hacked the ARP cache to point to improper devices in what is known as ARP poisoning.

- **Binding sniffed data to a local port**   A user can connect to a port on a client and decode protocols or inject data.

In addition to these features, the newer versions of ettercap support internal WEP decryption for wireless packets. When you provide the WEP key, which you must know or have previously cracked, the packets can be decrypted on-the-fly for storage and later viewing.

L0phtCrack   In many cases, operating systems implement password authentication and encryption at the application layer. Such is the case with Microsoft Windows file sharing and NetLogon processes. The challenge/response mechanism used by Microsoft over the years (and over several operating system and service pack upgrades) has changed from LM (weak), to NTLM (medium), to NTLMv2 (strong). Before NTLMv2, tools such as L0phtCrack could easily crack these hashes. It is important to properly configure your Windows operating system to use NTLMv2 and not to use the weaker versions. This process must be accomplished manually, and instructions can be found at www.technet.com.

L0phtCrack, also known by the newer name LC5 (short for L0phtCrack version 5), is a password auditing and recovery tool created by L0pht Heavy Industries, now owned by @stake. L0phtCrack is used to audit passwords on Windows operating systems. There are many different ways that L0phtCrack can capture password hashes, but two in particular that auditors frequently attempt are file share authentication and network logons. L0phtCrack can capture these challenge/response conversations and derive the password. The stronger the challenge/response mechanism used, the more difficult it is for L0phtCrack to crack it. The output of a password recovery session in L0phtCrack version 4 (LC4) is shown in Figure 12-3.

Once the intruder has captured the targeted password hashes (as many as are deemed appropriate in a given audit), the hashes are imported into LC4's engine, and a dictionary attack automatically ensues. If the dictionary attack is unsuccessful, a brute-force attack automatically begins thereafter. The processor power of the computer doing the audit will determine how fast the hash can be broken. L0phtCrack has many

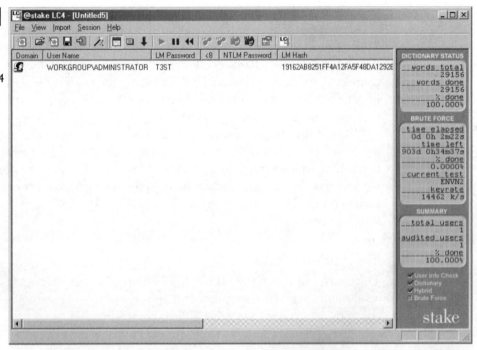

FIGURE 12-3

Sample password auditing output from L0phtCrack 4

modes for capturing password hashes and dumping password repositories. One mode allows for "sniffing" in a shared medium (such as wireless), while another goes directly after the Windows Security Access Manager (SAM).

Windows 2000 Service Pack 3 introduced a new feature called "SysKey," which is short for System Key. This feature, implemented by running the syskey.exe executable file, encrypts the SAM such that L0phtCrack cannot extract passwords from it as was possible before it was encrypted. L0phtCrack has the capability of letting the auditor know that he or she is auditing a SAM that has been encrypted so the auditor will not waste much time attempting to extract that password.

on the job *While L0phtCrack will still work for password auditing, Symantec acquired @stake in 2004 and discontinued support for L0phtCrack version 5 (LC5) and all other versions. They site U.S. Government regulations on export as the reason for the dismissed support.*

LRC Proxim Orinoco PC Cards store an encrypted hash of the WEP key in the Windows Registry. The Lucent Registry Crack (LRC) is a simple command-line utility written to decrypt these encrypted values. The problem is getting these values from another computer—one that has the WEP key that the attacker wants to obtain installed. This task is accomplished through a remote Registry connection. The attacker will make a remote Registry connection to the target computer using the tools in Window's Registry Editor on his own computer. Once the hacker is connected, the hacker must simply know where the key is located in the Registry in order to copy and paste it into a text document on his or her computer. These encrypted strings are stored per profile, as shown in Figure 12-4.

LRC can then be run against this encrypted string to produce the WEP key. The decryption process takes about one to two seconds. Once the attacker has the WEP key, it is a simple matter of plugging it into his or her computer to gain access to the network. For this reason, wireless end users should implement peer attack safeguards such as personal firewall software or IPsec policies. The LRC operation process is shown in Figure 12-5.

Encryption Hacks

The Encrypting File System (EFS) in Windows 2000 and later operating systems is an example of storage encryption. It is also an example of potential weaknesses in encryption systems. EFS is vulnerable to key store attacks.

Orinoco WEP key in encrypted form stored in the Windows Registry

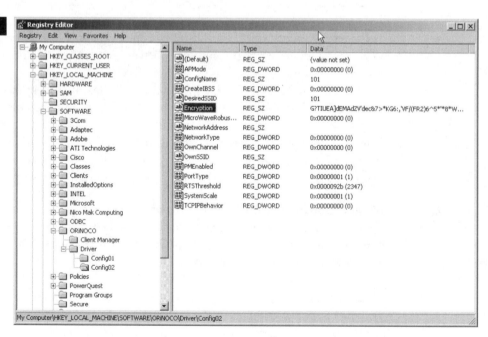

FIGURE 12-5

LRC cracking the
encryption used
for safeguarding
WEP keys

```
A:\>
A:\>
A:\>lrc -d "G?TIUEA]d5MAdZU'deb_49:[26:,'UF/(FR2)6^5*'*8*W6;+GB>,7NA-'ZD-X&G.H2J
/8>M0(JP0XUS1HbU29.Y3>:\3YF_4IRb56"
Lucent Orinoco Registry Encryption/Decryption
Version 0.3
Anders Ingeborn, iXsecurity 2001 <ingeborn@ixsecurity.com>
Hex mode by Don French, 2002 <french_don@yahoo.com>
Input value = G?TIUEA]d5MAdZU'deb_49:[26:,'UF/(FR2)6^5*'*8*W6;+GB>,7NA-'ZD-X&G.H
2J/8>M0(JP0XUS1HbU29.Y3>:\3YF_4IRb56
Decrypted key is: 98765 (or 3938373635 in hex)

A:\>
```

In any encryption system, the most difficult thing to do is protect the key store. The problem is found in the method used to access the keys. If a user needs to decrypt data she previously encrypted, she must be able to retrieve the encryption key. The EFS encrypts the file encryption key (FEK) with the user's public key. This process means that the user's private key will be needed in order to decrypt the FEK, which will be used to decrypt the file. The question is this: how does the user access her private key? The answer is simple: automatically.

By default, when the user opens a file that is encrypted by EFS, the user's private key is automatically retrieved and the FEK is then decrypted followed by the decryption of the data file. As long as the user is logged on, it all happens automatically. This process reveals the potential weakness: the user's authentication credentials.

The EFS uses a solid encryption algorithm with a sufficient key length; however, the user's password may be very weak and this reality introduces an important vulnerability into the system. If the attacker can guess the user's password, all of the data encrypted by that user will now be accessible to the attacker in many, if not most, scenarios. However, this problem is not unique to EFS. If a user implements the very popular Pretty Good Privacy (PGP) Desktop encryption system and uses weak passphrases, the data may be equally vulnerable.

VoIP Eavesdropping

Eavesdropping in VoIP is different from the traditional eavesdropping in data networks, but the basic concept remains the same. In a data network, you capture data packets and reassemble data files. In a VoIP network, you must intercept the signaling and associated media streams of a call and then reassemble this data into an audio stream or file for listening. The signaling messages use separate network protocols (i.e., UDP or TCP) and ports from the media itself. Media streams usually are carried over UDP/RTP. The process basically involves three steps:

1. Capture and decode the RTP packets.
2. Rebuild a session.
3. Save the rebuilt session as an audio file.

All three steps can be performed in Ethereal, which is a free protocol analyzer. To protect against this, you must use encryption or a network topology and security mechanisms that prohibit network sniffing (such as a switched infrastructure that is properly secured); however, encryption is the best choice. Encryption is best because eavesdropping is passive and very difficult to detect.

Social Engineering

Social engineering is defined as persuading someone to give you or tell you something that they should not give you or tell you through the manipulation of human or social interactions. Successful social engineering attacks occur because the target might be ignorant of the organization's information security policies or intimidated by an intruder's knowledge, expertise, or attitude. Social engineering is one of the most dangerous and successful methods of hacking into any IT infrastructure. If defeating WEP has stumped the hacker, the hacker might try to trick an employee who is authorized to have the WEP key into giving up this information. Once the hacker has the WEP key, the hacker will enter it into his own computer and use the available monitoring tools or protocol analyzers to capture sensitive data in real time, just as though there were no security. For this reason, social engineering has the potential of rendering even the most sophisticated security solution useless.

Hackers are not always like they are portrayed in movies: the cigarette-smoking, caffeine-loaded teenager in a dark room in a basement with multiple high-speed connections to the Internet, loud music, and plenty of spare time. Many times the most successful and damaging network intrusion is accomplished in broad daylight through clever efforts of someone who walks into a business like he owns it. In the very same manner, a hired professional security auditor should openly attempt intrusion as one tactic of testing security policy adherence.

There are some well-known targets for this type of attack:

- The help desk
- On-site contractors
- Employees (end users)

The Help Desk   The help desk is in place to assist those individuals who need help with some aspect of a computer or network. It becomes quite awkward in many situations for the help desk not to provide answers to questions when the person on the other end of the line seems to know what they need. It is not an easy task to train help desk personnel not to be helpful in certain situations; nevertheless, this type of education is crucial to corporate network security. The help desk should be trained to know exactly which pieces of information related to the wireless network should not be given out without the proper authorization or without following specific processes put in place by security policy. Items that might be marked for exclusion are:

■ SSID of access points

■ WEP key(s)

■ Physical locations of infrastructure devices

■ Usernames and passwords for network access and services (i.e., e-mail)

■ Passwords and SNMP strings for infrastructure devices

■ VoIP solutions in use

The auditor should (and the hacker *will*) use two particular tactics when dealing with help desk personnel:

■ Forceful, yet professional language

■ Playing dumb

Both of these approaches have the same effect: getting the requested information. Help desk personnel understand that their job is to help people with their problems. They also understand that their manager will not be happy with them if their customers are not happy with the service they are receiving. By threatening to speak with, or write a letter to, the manager, the social engineer can get the help desk person to give over the requested information just to appease and settle down the social engineer. Some people are just naturally inept at handling personal conflict, and some people are easily intimidated by anyone with an authoritative voice. Both of these situations can be used to the advantage of the social engineer. The human factor has to be overcome with training, discipline, and repetitively following documented procedures.

Playing dumb is a favorite of many social engineers. The help desk person is usually disarmed and stops paying attention when they figure out that the person to whom they are speaking knows very little. This situation is exacerbated when the "dumb" customer is overly polite and thankful for the help. It's important that a help desk person be alert to this tactic at all times. A social engineer is likely to call over and over, hoping to speak with different representatives, and taking different approaches with each.

Contractors   IT contractors are commonplace at many businesses today, and very few, if any, are put through organizational security training. Few are given a copy of the company security policy or required to sign privacy agreements. For this reason, and because IT contractors, like the help desk, are there to help, IT contractors can be especially good targets for social engineers. Contractors are aware of the specific details about network resources because they are often on-site to design or repair the network. In wanting to be helpful to their customer, contractors often give out too much information to people who are not authorized to have such information. For this reason, strong security solutions that rely on multifactor authentication are recommended.

Employees   Since people spend many hours each day with each other at their work location, they often share private information—such as network login information—with one another. It is also common to see that same login information on sticky notes under keyboards and on monitors. Another problem is that most computer users are not computer network or security savvy. For this reason, they might not recognize spyware, hacking attempts, or social engineering.

VoIP technology is still very new to many organizations. Employees who are not educated about network security may not realize the dangers that unauthorized access via the network can pose to the organization and to them personally. Specifically, non-technical employees who use the network should be aware of the fact that their computers can be attacked in a peer-to-peer fashion at work, at home, or on any public wireless network, if the device uses wireless networking. Social engineers take advantage of all of these facts and even engineers elaborate stories that would fool almost anyone not specifically trained to recognize social engineering attacks.

Similar to social engineering is *shoulder surfing*. Shoulder surfing is a non-technical way of capturing information. As its name implies, you will simply watch over the user's shoulder to see what information you can gather. Frequently, users enter their passwords slowly enough that you can see what they are typing. At other times, you can see configuration screens with information about SSIDs for WLANs, pre-shared keys when WPA-Personal is being utilized, VoIP configuration settings, encryption passwords, and more.

Zero-Day Hacks

In the end, you must stay up-to-date on the various vulnerabilities that may post a threat to your network. You may have noticed that most of this chapter was not specific to VoIP networks. VoIP networks are vulnerable to the same exploits as traditional wired and wireless networks. This similarity between networks does make your efforts somewhat easier if you're already familiar with network security; however, you must remember that the impact of security technologies can be detrimental on VoIP networks. As an example, the implementation of encryption for VoIP calls could be just enough to take your network latency to an unacceptable level. Balancing between security and performance is an important issue in VoIP networks.

The phrase zero-day hacks is a reference to the newest attack methods in use. You will need to frequent web sites mentioned in this chapter in order to keep your knowledge fresh.

CERTIFICATION SUMMARY

In this chapter, you focused on the dark side of security. You learned about the different types of attackers and the methods they use. This chapter provides the foundation for the next. You'll better understand the need for the security solutions covered in Chapter 13 after reviewing the hacking methods here.

Remember that VoIP networks, also known as unified communications, are vulnerable to the same attacks as traditional IP networks. Additional concerns include compliance issues. Organizations must comply with various federal and local regulations that may impose constraints on the implementation methods they chose for their converged networks.

✓ TWO-MINUTE DRILL

Explain Concepts and Components of Security Design

❑ Attacks may be passive or active. Eavesdropping is an example of a passive attack. Password cracking is an example of an active attack.

❑ Attackers motive may be to target your organization specifically, or they may attack promiscuously.

❑ Password cracking and password attacks are used to gain access to valid credentials for network penetration.

❑ A brute-force password attack is an attack where every possible password is attempted using a random or logical algorithm.

❑ VoIP networks may be vulnerable to eavesdropping attacks when encryption is not used.

❑ Social engineering is used to gain access to data and information without using technical methods.

❑ A zero-day hack is a fresh or newly discovered hack.

❑ Wireless networks may introduce vulnerabilities, such as eavesdropping, that are more easily exploited than wired networks.

❑ Many wireless security recommendations are actually myths, such as the supposed benefits of SSID hiding, MAC filtering, and WEP encryption.

SELF TEST

The following questions will help you measure your understanding of the material presented in this chapter. Read all the choices carefully because there might be more than one correct answer. Choose all correct answers for each question.

Explain Concepts and Components of Security Design

1. You are selecting the policies for passwords in your converged network. Which of the following passwords would be considered secure for most applications? Choose all that apply.
   A. Password1
   B. fortycowshidoutsidethebarnforhay
   C. B7ybtro3
   D. Nimda911

2. Which of the following are valid security solutions without significant weaknesses for wireless networks? Choose all that apply.
   A. SSID hiding
   B. MAC filtering
   C. WEP
   D. None of the above

3. In hacker terms, what is a newly discovered hack called?
   A. Sweet
   B. Zero-day
   C. Dangerous
   D. A wild one

4. You want to scan your network for vulnerabilities. What is this process called?
   A. Vulnerability training
   B. Threat analysis
   C. Vulnerability analysis
   D. Risk analysis

5. Eavesdropping is an example of what kind of attack?
   A. Passive attack
   B. Active attack
   C. Promiscuous attack
   D. Unintentional attack

SELF TEST ANSWERS

Explain Concepts and Components of Security Design

1. ☑ **B** and **C** are correct. The passwords represented in B and C are either long enough or complex enough to be considered secure for most systems by today's standards.
   ☒ A and D are incorrect.

2. ☑ **D** is correct. All three of the listed security solutions are either not security solutions or contain significant weaknesses.
   ☒ A, B, and C are incorrect.

3. ☑ **B** is correct. Zero-day hacks are those that are most likely not prevented in target systems. They are so new that most networks or system will not have implemented protection against them.
   ☒ A, C, and D are incorrect.

4. ☑ **C** is correct. Vulnerability analysis is the process used to discover vulnerabilities. Threat analysis is the process used to analyze potential threats to your organization. Risk analysis may ultimately include vulnerability or threat analysis, but it is a higher-level task. Vulnerability training would be used to build awareness among the IT professionals in your organization of security issues, but it is not usually considered a formal process in most organizations.
   ☒ A, B, and D are incorrect.

5. ☑ **A** is correct. Eavesdropping is a passive attack. The attacker is not required to actively penetrate your network, but he or she can listen in on the traffic.
   ☒ B, C, and D are incorrect.

13
Security Solutions

> *Every time someone makes a decision about security . . . he makes a trade-off.*
>
> —*Bruce Schneier*

I f you look at the Computer Security Institute's annual security surveys, you will notice one startling thing: every year a large percentage of respondents indicate that no security incidents occurred within their organizations. I suppose you could indicate such, if an incident is defined as a security breach that causes known losses; however, if you remove the word known, so that it is defined as a security breach that causes losses, it seems to me that the only responses to the survey question could be yes or unsure. If I am sure that a security incident has not occurred, I probably don't understand security.

The real issue here seems to be related to the definition you choose to use for the word incident. The term *incident* commonly means a single or distinct event, and there is no indication that the event was actually observed, analyzed, or detected by anyone. Many security resources define an incident as a detected security event that adversely affected systems or data. This definition is not effective, since an incident may occur without detection. For this reason, I prefer to define incident using the Sans.org security terminology glossary's definition: an adverse network event in an information system or network or the threat of the occurrence of such an event.

This thinking comes from another quote by Bruce Schneier, one of the world's foremost experts on security. He says, "There's no such thing as absolute security." You can rephrase this quote with a different focus to say, "If your system is completely secure, your users can't use it." If you decide to implement a firewall filtering rule that disallows traffic destined for TCP port 21, you've just made it impossible for users to connect to FTP servers through that firewall. The decision has forced a trade-off: you aren't susceptible to direct FTP server attacks, but you can't use the FTP server through the firewall.

With this understanding, you will explore the different security solutions available to you. You will learn how they help to protect your converged network, but unlike many books, this one will describe the reality of how they can impact the network negatively. Stated differently, you'll learn about the trade-offs. These trade-offs become very important in VoIP networks, as performance is paramount to success. By keeping the following concepts in mind, you will make better decisions related to network security in general and VoIP security in particular:

- A system cannot be made absolutely secure and still provide a useful function to the users.
- Incidents occur whether you detect them or not.
- Higher security requirements usually result in lowered usability or lowered performance or both.

CERTIFICATION OBJECTIVE 13.01

Explain Concepts and Components of Security Design

I've chosen to cover the Convergence+ objectives that are related to security by investigating four categories:

- Network design and security
- Perimeter security solutions
- Connectivity solutions
- Security monitoring

In addition to these categories, this chapter will also review specific VoIP tools and technologies that provide solutions to security issues in converged networks.

Network Design and Security

Security should be designed into a system or network. It should not be an afterthought. You will have a more secure network when security is designed into the implementation from the start. Instead of thinking of specific attacks and specific countermeasures alone, you may want to consider security as a system. A system is defined as a group of independent but interrelated elements that form a whole.

A good example of the fact that security is a system is a bank vault. A bank vault is usually thought of as a single entity that helps to protect valuables, but it is actually a group of independent and interrelated elements. The vault combination lock is combined with procedures and policies as well as alarms and response mechanisms to form the whole of the bank vault. Additionally, many vaults are layered: one door opens the vault and smaller doors may open compartments within the vault.

Security design is about building systems and implementing layers that help to protect valuable assets. When you design a security system, you are designing a unique system that is aimed at keeping certain actions—attacks—from working. You are designing a system to protect against intelligent, intentional, and malicious attacks. This process is very different than safety management, where you are protecting against unintentional problems that occur randomly. Security attacks may be intentional and occur at specially selected times that provide the attacker with the greatest opportunity.

Two key principles assist in security design: layered security and isolation. *Layered security* implies that more than one protection mechanism is used between an attack point and a valued resource. Layered security is sometimes called defense in depth. *Isolation* provides virtual or literal separation of one set of users or services from another set of users or services.

Demilitarized Zones and Perimeter Networks

A *demilitarized zone (DMZ)* is a concept borrowed from military operations. It defines a portion of the network that is not as secure as the rest of the network. The DMZ is usually located between the private network and the Internet or another external network. DMZs are also known as perimeter networks, since they exist at the edge of the private network. The DMZ acts as a location for Internet service servers and as a point of inspection and authentication for access into the internal or private network.

Most organizations will choose to place a firewall between the Internet and the DMZ. An additional firewall will usually be placed between the DMZ and the private network. This dual-firewall implementation allows for reduced restrictions at the ingress to the DMZ from the Internet and increased restrictions at the ingress from the DMZ to the private network. Figure 13-1 provides an example of a DMZ or perimeter network.

VLANs

Virtual LANs or VLANs are used to segment a physical network into multiple logical networks. VLANs operate within the switches and routers on your network, and client computers are usually unaware of their participation in a VLAN. To the client computers, the VLANs look and operate just like a physically segmented LAN. For this reason, VLANs can be used to provide increased security on converged networks.

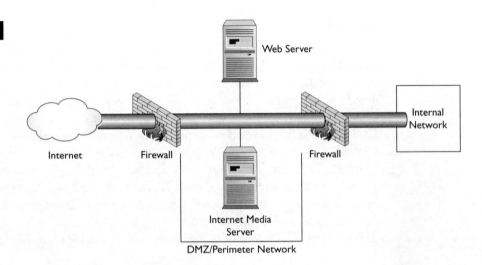

FIGURE 13-1

Demilitarized zone

Web Server

Internet

Firewall

Internal Network

Firewall

Internet Media Server

DMZ/Perimeter Network

The most common way to use VLANs with VoIP is to configure at least two VLANs. The first VLAN will be used for data traffic and the second for voice traffic. Since VLANs are a logical grouping of devices that may or may not be physically near each other, they can be used to group both directly connected and indirectly connected devices into logical arrangements.

If you've worked with VLANs, you know that devices in one VLAN cannot communicate with devices in another VLAN without the configuration of some sort of trunking protocol or routing solution. However, you should not assume that the segregation provided is a solid security solution by itself. VLAN protocols were not designed with security as the primary intent and can be compromised with the right knowledge.

Since VLAN protocols do not provide sufficient security by themselves, you should consider requiring VLAN communication to route through a firewall or a policy-based router. These devices can filter traffic based on TCP or UDP port destinations and help provide greater security for VLAN traffic.

Perimeter Security Solutions

Should you decide to implement a perimeter network or DMZ, you will need to understand the technologies used to implement it. These technologies include firewalls, proxies, and network address translation devices. It is important that you understand the functions provided by each of these devices and the different implementation methods that they offer.

Firewalls

A *firewall* may be defined as a logical or physical break in the network links where network packets may be accepted, rejected, or stored for further evaluation. Firewalls come in different types, and each type provides different filtering processes. Some firewall devices can implement several types of filtering. Firewalls may be deployed as hardware appliances or as services running on operating systems such as Linux or Windows. Table 13-1 provides a listing of the common firewall types.

Without a firewall, any network connected to the Internet is more vulnerable to attacks. The firewall will prevent entry of specific types of communications or all communications from or to specific IP addresses. Additionally, many vendors provide rule sets that will automatically configure your firewall to protect against many common attacks. However, it is up to the administrator to continue managing the firewall in order to protect against any new threats that may be introduced.

on the job

A newer approach to security is the unified threat management (UTM) method. UTM devices and applications provide traditional firewall functions while also implementing spam filtering, virus detection, and intrusion detection. Since most UTM solutions are single-box solutions, the complexity of implementing multiple security devices is removed.

Proxies

Proxy servers act as intermediaries between two communicating devices or networks. Proxy servers allow caching of web pages for improved efficiency in browsing and filtering of web content. Proxy servers are often used by organizations to prohibit

TABLE 13-1 Firewall Types	Firewall Type	Description
	Packet Filtering	A firewall that evaluates the received packets and determines the appropriate action based on rules.
	Stateful Inspection	A firewall that can eliminate packets originating outside the network that are not part of a current session or connection.
	Application Layer Firewall	A firewall that can filter communications based on application types and that is aware of applications regardless of the port (TCP or UDP) used. Also known as a proxy firewall.
	Desktop	A desktop firewall that runs on the client computers connected to your network. Each computer runs a local copy of the firewall and may have different rules.

access to particular web sites, and many organizations disallow direct access to the Internet. In a converged network, proxy servers may introduce problems with VoIP communications if the servers do not support VoIP protocols. For this reason, many organizations allow VoIP communications that need to traverse the Internet to bypass the proxy and use the proxy server only for web browsing.

Network Address Translation and Port Address Translation

Instead of providing public IP addresses to each VoIP phone or desktop computer on your network, you may decide to implement private IPv4 addressing. The following addresses are reserved for internal private use:

- 10.*x.x.x*
- 172.16.*x.x*–172.31.*x.x*
- 192.168.*x.x*

For example, you could use the IP addresses ranging from 10.10.10.1 to 10.10.10.254 to implement a 254-node network. The devices using these IP addresses will not be able to communicate on the Internet. Instead, they will need to communicate through a device that can translate private addresses to Internet addresses. *Network Address Translation (NAT)* servers perform this operation. A NAT server sits between your network and the Internet or between your network and your DMZ in some cases. The internal users communicate with the NAT server, and the NAT server maintains sessions for those users with the Internet locations the users are requesting. NAT is one of the major reasons that IPv4 has continued as the most popular communications protocol. Without NAT, or a similar technology, we would have been forced to upgrade to IPv6 with the explosion of nodes connected to the Internet.

Do not confuse NAT with PAT. *Port Address Translation (PAT)* is very different from NAT. While NAT is about managing IP address translation, PAT can perform IP address translation as well as Layer 4 port translation. PAT servers usually have only one Internet-facing IP address, and all communications coming out of the network appear to originate from that one address. NAT servers may have multiple Internet-facing IP addresses. With NAT, the Internet-facing addresses may be assigned to an internal IP address for the duration of a session or they may be used in a round-robin fashion. With PAT, the single IP address is used for outbound communications and the inbound communications are automatically redirected to the appropriate internal IP address based on the PAT routing table.

Because VoIP protocols use two communication channels for the calls, NAT and PAT servers may introduce problems. The NAT server can usually initiate the call, but the audio may only work in one direction or the audio may not work at all. This failure is caused by communicating through the NAT device. The NAT device does not know that the two streams (signaling and audio) are for the same communication. You can resolve this issue in several ways:

- Don't use NAT or PAT.
- Implement IP tunneling through the NAT device to the remote network and run the VoIP connection through the tunnel. This will introduce extra latency.
- Use VoIP-aware NAT or PAT devices.

Connectivity Solutions

One of the key security solutions that should be implemented on a converged network is secure connectivity. Secure connectivity includes authentication to the network, confidentiality of network communications, and accounting.

Authentication

One of the important components of a security strategy is an identity management system. An identity management system provides a storage location for identity objects, usually called user accounts, and one or more methods for connecting to the storage location and proving identity ownership—a process known as authentication. User accounts are objects that identify and are owned by users. These objects provide properties for use by authentication systems and network operating systems. In addition to user accounts, certificates, biometrics, tokens, and other credentials may also be used for authentication and identity management.

Without a clear understanding of authentication and identity management, you will have difficulty installing a secure converged network. There are both basic and advanced authentication systems, and many systems include the ability to support both. For example, Windows server systems allow for advanced authentication mechanisms through the Internet Authentication Service (Microsoft's RADIUS implementation) and basic authentication using simple passwords against the Active Directory database. Both methods serve a valid purpose and are best for certain scenarios. Determining which method is right for your scenario is the first step to secure authentication.

In addition to selecting advanced or basic authentication methods, you must determine whom to authenticate. Do you wish to authenticate the clients only? Do you need to validate the authentication server as well? When both the client and the authentication server or authentication device are authenticated, this is known as mutual authentication. Mutual authentication helps prevent the introduction of rogue authentication devices to your network. Client-only authentication allows the network to "feel" secure, whereas mutual authentication allows both the clients and the network to have confidence and trust in the connections.

Authentication should not be confused with authorization. *Authentication* can be defined as proving a person or object is who or what he or it claims to be. *Authorization* is defined as granting access to a resource by a person or object. Authorization assumes the identity has been authenticated. If authentication can be spoofed or impersonated, authorization schemes fail. From this, you can see why authentication is such an integral and important part of network and information security.

Advanced authentication systems generally utilize stronger credentials and better protection of those credentials than basic authentication systems. The strength and protection of the credential is determined by the effort it takes to exploit it. A password-protected credential is usually considered weak when compared with biometric-protected credentials. This belief, in some cases, is a misconception, because strength of authentication really depends on how the authentication information (the credential and proof of ownership) is sent across the network. If you were to implement a biometric system, such as a thumb scanner, and the client sent the credentials and proof of ownership (a unique number built from the identity points on the user's thumb) to the server in clear text, that would be no more secure than a standard password-based system; however, I am not aware of any biometric authentication system that sends the authentication data as clear text.

The key element, which will provide a truly strong authentication pathway, is the encryption or hashing of the user credentials, or at least the proof of identity information. This encryption can be accomplished with virtual private networking (VPN) technology or with well-designed authentication systems. One example of a well-designed authentication system is 802.1X with a strong EAP type. 802.1X and EAP types are used to secure both wired and wireless connections.

You use authentication every day of your life. For example, when you are at a seminar and the speaker says he is an expert on the topic of his speech, you use authentication mechanisms to verify this information. In other words, you listen to the information he delivers and use it to determine if he is truly an expert. In addition, suppose someone walks up to you and says, "Hi, my name is Bill and I am tall." You would look at him and compare his height with a height you consider to be tall and authenticate whether he is truly tall or not. If he is not tall, by your standards, he will lose *credibility* with you.

Remember the word *credentials*? Consider other important "cred" words: credit, credibility, and credentials. Do you see how they are all related? They all have to do with having proof of something. When you have good credit, you have proof of your trustworthiness to pay debts. When you have credibility, you have proof that you are authentic, persuasive, and dynamic. When you have credentials, you have an object or the experience that proves your skill or identity. Authentication results in the verification of credentials.

Advanced authentication is more secure that basic authentication because advanced mechanisms are used to protect the user's credentials. This task usually means protecting a user name and password pair, but it can also include protecting a user/certificate combination, a user/machine combination, or any other user/object combination used to identify a specific user. In addition to the extra protection offered by advanced authentication systems, when 802.1X-based systems are used, you have the benefit of standards-based technology. This standard means that hardware from many different vendors is likely to support the authentication process. Sometimes driver or firmware upgrades are required, but there is often a path that can be taken to implement the authentication mechanism.

In order to increase security, you should avoid basic authentication mechanisms such as the Password Authentication Protocol (PAP), which sends the password as clear text across the network. Another example of a basic authentication protocol would be Basic Authentication to web servers. This authentication mechanism is part of the HTTP standard, but the passwords are sent across the network using a reversible encryption algorithm. Use an authentication system that provides for better protection of credentials and you will be a long way toward a more secure network.

Credentials   There are many different credential solutions available for securing your networks. It's important to select the right system for your needs. In this process you will consider the primary features of a credential solution and whether you need a multifactor authentication system. In addition, you should be aware of the various credential types available to you.

A credential solution should provide a means of user or computer identification that is proportional to your security needs. You do not want to select a credential solution that places unnecessary burdens on the users and results in greater costs (of both time and money) than the value of the information assets you are protecting. You should evaluate whether the selected authentication solution provides for redundancy and integration with other systems such as Active Directory or Lotus Notes. The system should also support the needed credential types such as smart cards and/or biometrics. In addition, consider the following factors when selecting a credential solution:

- The method used to protect the credentials
- The storage location of the credentials
- The access method of the credential store

If an authentication system sends the credentials as clear text, a protection method is effectively non-existent. Advanced authentication systems will protect the user credentials by encrypting them or avoiding the transmission of the actual credentials in the first place. Instead of transmitting the actual credentials, many systems use a hashing process to encode at least the password. Hashing the passwords means that the password is passed through a one-way algorithm, resulting in a fixed-length number. This number is known as the hash of the password or the message digest. The hash is stored in the authentication database and can be used as an encryption key for challenge text in a challenge/response authentication system.

The credentials, both user name and password (or hash) or certificates, must be stored in some location. This storage location should be both secure and responsive. It must be secure to protect against brute-force attacks, and it must be responsive to service authentication requests in a timely fashion. Certificates are usually stored in a centralized certificate store (known as a certificate server or certificate authority) as well as on the client using the certificate for authentication. Both locations must be secure, or the benefit of using certificates is diminished. In addition to the standard certificate store, users may choose to back up their certificates to disk. These backups are usually password protected, but brute-force attacks against the media store may reveal the certificate, given enough time. For this reason, users should be well educated in this area and understand the vulnerability presented by the existence of such backups.

Access methods vary by authentication system and storage method, but there are standards that define credential access methods. One example is LDAP (Lightweight Directory Access Protocol). LDAP is a standard method for accessing directory service information. This information can include many objects, but is usually inclusive of authentication credentials. LDAP is, or can be, used by Lotus Notes, Novell's Directory Services, and Microsoft's Active Directory, among others.

Sometimes, one type of authentication alone is not sufficient. In these cases, *multifactor authentication* can be used. Multifactor authentication is a form of authentication that uses more than one set of credentials. An example of a multifactor authentication process would be the use of both passwords and thumb scanners.

Usually, the user would place her thumb on the thumb scanner and then be prompted for a password or PIN (personal identification number) code. The password may be used for network authentication, or it may only be used for localized authentication before the thumb data is used for network authentication. However, in most cases the password and thumb data are used to authenticate to the local machine and then the network or just to the network alone. A common example of multifactor authentication would be your ATM card. You have the card and you know the PIN (something you have and something you know).

There are many common credential types. They include:

- Username and password
- Certificates
- PACs (Privilege Attribute Certificates)
- Biometrics
- Tokens

Username and password pairs are the most popular type of credential. This combination is used by most network operating systems, including Novell Netware, Linux, Unix, and Windows. Due to the human factor involved in the selection of the password, they often introduce a false feeling of security. This caveat is because the chosen password is usually too weak to withstand dictionary attacks and, depending on the length of the password, certainly brute-force attacks. In addition, the passwords are often written down or stored in plain text files on the system and then changed infrequently, resulting in a longer attack opportunity window.

An alternative to username and password pairs is certificates. In order to use certificates throughout an organization, a certificate authority must exist. This certificate authority can be operated by the organization or an independent third party. In either case the costs are often prohibitive to widespread use due to the need for an extra server or even a hierarchy of servers. Small and medium-sized organizations usually opt for server-only certificates or no certificates at all because of the cost of implementation. A full PKI (Public Key Infrastructure) would usually consist of more than one certificate authority. Each certificate authority would be a single server or cluster of servers. The PKI is the mechanism used for generation, renewal, distribution, verification, and destruction of user and machine certificates.

The Privilege Attribute Certificate (PAC) is used by the Kerberos authentication protocol in Windows 2000 and higher (Windows XP and Windows Server 2003 as well as newer versions). The PAC contains the authorization data for the user and includes group memberships and user rights. This feature means that the user's group assignments and right assignments are transferred as a portion of the ticket-granting ticket (TGT)—a feature of the Kerberos authentication protocol.

Yet another authentication credential is you. Biometrics-based authentication takes advantage of the uniqueness of every human and uses these differences for authentication purposes. For example, your thumb can be used as a unique identifier, as can your retina. The balancing of cost and security is important with biometric credentials. While hair analysis could potentially be used to authenticate a user, the cost and time involved are still too high for practical use. Today, both thumb scanners and retina scanners are becoming more popular.

There are two common types of authentication tokens: software-based and hardware-based. Software-based authentication token systems often run on PDAs or cell phones. The use of software allows the user to launch the application on her mobile device, and retrieve an authentication code. This code is usually used in conjunction with a password or PIN, essentially creating a two-factor authentication system. The hardware-based token systems usually provide a keychain-sized device that, with the press of a button, will show the current authentication code. This code, again, is used with a password or PIN. Most token systems work off of a time-synchronized (or some other synchronization point) algorithm that generates proper codes in the software system or hardware device. This method of generating unique codes means that a valid code today will not work tomorrow and provides greater security.

Virtual Private Networks

A *virtual private network (VPN)* is implemented by creating an encrypted tunnel between two endpoints over a public network. The same technologies originally introduced for public network use are also used on private networks today. Since VPNs are used on both public and private networks, the qualification of running on a public network is no longer necessary. Today a connection is considered a VPN connection if it uses VPN technologies such as authentication and encryption, which are the two requirements of a secure VPN solution. Figure 13-2 shows an example implementation of a VPN connection across the Internet. In this case, it is a site-to-site VPN. All traffic between the sites is routed through the VPN connection.

FIGURE 13-2
Site-to-site VPN

VPN Router   Firewall   Internet   Firewall   VPN Router

In addition to the site-to-site VPN connection, you may choose to implement any of the following VPN solutions:

- **Device-to-device** These VPN connections are made directly between two devices on a network in order to provide additional security.
- **Remote client connections** Remote users can use a VPN client program to connect to the organization's network across the Internet.
- **Wireless security VPNs** VPNs are used for wireless security when the wireless client connects to an insecure public hotspot.

The last bullet point warrants some further explanation. The user may connect his laptop to a coffee shop hotspot that uses no encryption. Next, the user can launch a VPN client connection across the Internet to his organization's headquarters. Now all Internet traffic will be routed through the VPN connection to the home office and then back out from the home office to the Internet. This solution prevents wireless eavesdropping when no actual wireless security is available. Once the VPN is established, the user may also launch a software VoIP application and initiate calls without fear of someone listening in on the calls.

Many VPN technologies are available, but the two most common solutions are the Point-to-Point Tunneling Protocol (PPTP) and the Layer 2 Tunneling Protocol (L2TP) with IPsec. The latter is considered most secure. The key decision factors include the authentication method and encryption used by the VPN solution. Weak authentication equals a weak VPN, regardless of the encryption strength. Weak encryption equals a weak VPN, regardless of the authentication strength. Both factors must be strong in order for the VPN solution to be considered secure. For this reason, early PPTP implementations are not recommended, and even the recent implementations are not considered as secure as a solid implementation of L2TP with IPsec. The variance is partly because L2TP and IPsec provide you with many options for authentication and encryption. You can choose the level of security you need with more granularity than with PPTP.

Encryption

The foundation of a VPN is encryption. Once the user is authenticated, if no encryption is used, you still do not have a VPN. For example, you could create an L2TP tunnel between two endpoints without using encryption. This configuration would result in tunneled communications, but it would not be a VPN. In most cases, IPsec is used to establish the encryption for the L2TP tunnel in an L2TP/IPsec VPN solution.

Encryption is the process of encoding information so that it can be unencoded given the appropriate algorithm and inputs. Most encryption systems use an algorithm with a key and some data as the inputs. When encrypting, the system will use the key and the data to generate the cypherdata. When decrypting, the system will use the cypherdata and the key to regenerate the original data. Consider the following algorithm:

$$3 * k + D * k = cypherdata$$

where k is the encryption key and D is the original data. This algorithm could be used to encrypt any base 10 number. For example, assume you want to encrypt the number 12 with a key of 7. The algorithm would look like this:

$$3 * 7 + 12 * 7 = 105$$

The resulting cypherdata, 105, can be decrypted to retrieve the original data as long as the proper algorithm is applied and the correct key is known. Here's the decryption algorithm:

$$(cypherdata - 3 * k) / k = original data$$

The resulting algorithm would look like this for our example:

$$(105 - 3 * 7) / 7 = 12$$

While this example is not an implemented encryption algorithm (the actual algorithms are binary and much more complex), it shows how you can use the same algorithm with different encryption keys to encrypt data. Having the algorithm does not automatically allow you to decrypt the data with certainty. You must know the key that was used for encryption.

Many encryption standards exist. Table 13-2 lists the most common encryption algorithms.

TABLE 13-2   Encryption Standards

Encryption Algorithm	Type	Applications
Digital Encryption Standard (DES)	Block	Data encryption; communications encryption.
Advanced Encryption Standard (AES)	Block	Data encryption; communications encryption.
Triple DES (3DES)	Block	Data encryption; communications encryption. Used to increase the strength of DES on systems that cannot implement AES.
RC4	Stream	Data encryption; communications encryption.
Message Digest 5 (MD5)	Hashing	Digital signatures; password storage; data integrity.
Secure Hashing Algorithm 1 (SHA-1)	Hashing	Digital signatures; password storage; data integrity.

VoIP Encryption Solutions

The encryption solutions will vary, depending on the VoIP technology you implement. SIP uses standards from the IETF, and H.323 uses standards from the ITU. Table 13-3 lists the ITU specifications known as the H.235 hierarchy. These standards are used in H.323 VoIP networks when security is required. One of the major enhancement to H.323 version 6 and the H.235 standards at that level is the inclusion of SRTP. SRTP, the Secure Real-Time Transport Protocol, provides the same functionality as RTP while adding secure communications.

SIP uses the standards created by the IETF. These standards implement existing security standards for VoIP. The standards include Transport Layer Security (TLS), Secure/Multipurpose Internet Mail Extensions (S/MIME), and the SRTP. SIP essentially relies on an existing security layer beneath the VoIP communications in the OSI model rather than implementing security within the VoIP protocols themselves. TLS is used to protect the signaling and SRTP is used to protect the voice payload. Additionally, S/MIME may be used to secure both the signaling and the payload.

Accounting

The term accounting is used to reference any method or system that provides record keeping and activity tracking. In network security, an accounting system at its most basic level is a logging system that logs the activity occurring on the network.

TABLE 13-3   H.323 Security Solutions Using H.235

Standard	Definition
235.1	Basic or baseline security profile. Includes only authentication and integrity of signaling streams.
235.2	Signature security profile. Specifies methods for distribution of digital signatures used to secure the signaling stream.
235.3	Hybrid profile. Used to combine 235.1 and 235.2.
235.4	Direct and selective routed call security. Implements symmetric key management.
235.5	Secure authentication in RAS using weak shared secrets.
235.6	Voice encryption profile. This profile must be supported if the VoIP system requires encryption of the actual voice data.
235.7	Usage of the MIKEY key management protocol for SRTP.
235.8	Key exchange for SRTP using secure signaling channels. May include end-to-end security or only security in sensitive areas of the network.
235.9	Support for security gateways.

Authentication gets you onto the network. Authorization gets you into things on the network. Accounting keeps a record. Accounting is important for incident response and is also important as a tool for baseline analysis and anomaly detection. Advanced accounting systems are called intrusion detection systems or intrusion prevention systems.

Security Monitoring

The security technologies presented in this chapter can help to protect your network from an attack; however, new attack methods are continually being developed, and you must have a solution that allows you to monitor for both the older and newer attacks. The technologies that assist you with this effort include:

- Intrusion detection and intrusion prevention systems
- Antivirus and antispyware solutions

Intrusion Detection and Intrusion Prevention Systems

An *intrusion detection system (IDS)* detects many security-related incidents and logs the information. An IDS may notify an administrator of suspect activity. Incidents that may be detected by an IDS include unwanted connections, high bandwidth consumption, attacks based on signatures, and anomalies in network activity. Signature-based detection relies on patterns that exist within attack scenarios. Anomaly-based detection relies on comparisons with the baseline (normal operations) of network activity.

An *intrusion prevention system (IPS)* goes one step further than the IDS solution. Intrusion prevention systems may prevent an attack by disallowing connections from suspect devices or even shutting down services that are under attack.

Antivirus and Antispyware Solutions

A computer program with the ability to regenerate itself is called a *virus*. A virus may or may not harm the infected computer. Viruses may lay dormant for some period of time before they attack the infected host machine. A *worm* is a self-replicating application that requires no user action for reproduction. Viruses usually require human interaction in some way, whereas worms do not.

Another type of malware is the Trojan Horse. Named after the fabled gift of ancient fame, the Trojan Horse enters the computer under the guise of a useful program or utility. Once in the machine, it may infect the machine with a virus or worm, or it may download other Trojans.

Similar to the Trojan Horse is the spyware or adware villain. Spyware is installed on your computer and reports back to the source. Adware is installed on your computer and causes unwanted ads to display on your screen. Additionally, spyware and adware combinations are common.

To protect your network from these malware applications, you will need to run antivirus and antispyware applications. There are two basic types of antimalware applications: ingress and host-based. Ingress applications reside at the entry point of the data, and host-based antimalware applications run on the host devices. An example of an ingress antimalware application would be an e-mail server scanner. This software would scan e-mail messages as they enter (and possibly exit) the e-mail server. If malware is detected, the message can be rejected, flagged as malware infected, or passed on without attachments.

Antivirus software must be maintained. You will need to download and apply new definition files frequently. Many antivirus applications include automatic update features so that the definitions can be maintained without the need for user interaction. The definition file includes the signatures that are used to identify known malware.

CERTIFICATION SUMMARY

In this chapter you learned about the basic security technologies that can be used to protect your network. You learned about firewalls, proxies, and NAT/PAT devices. Authentication systems were explored and encryption technologies were introduced. You also learned how VLANs can be used to help protect your networks and the importance of perimeter networks such as DMZs. Finally, you explored the use of antimalware applications that help to protect your systems from viruses, worms, Trojan Horses, and other malicious applications.

TWO-MINUTE DRILL

Explain Concepts and Components of Security Design

- ❑ Firewalls are used to control or limit the traffic at the ingress/egress to your network.
- ❑ Authentication systems prove the identity of users and devices.
- ❑ Proxies are used as intermediaries between clients and servers or between internal clients and Internet locations.
- ❑ A virtual private network (VPN) is used to secure traffic passing across an insecure network by encrypting the data.
- ❑ Encryption is used to encode data so that it is not easily readable.
- ❑ VLANs can be used to enhance security on VoIP networks by separating the voice traffic from the data traffic.
- ❑ Intrusion detection systems detect intrusions, but do not disable services or connections.
- ❑ Intrusion prevention systems can disallow connections or disable services in order to prevent an attack.
- ❑ Antivirus software must be maintained through definition updates.

SELF TEST

The following questions will help you measure your understanding of the material presented in this chapter. Read all the choices carefully because there might be more than one correct answer. Choose all correct answers for each question.

Explain Concepts and Components of Security Design

1. Which of the following are the three categories of authentication credentials? (Choose three.)
   A. Something you are
   B. Something you heard
   C. Something you know
   D. Something you have

2. You want to implement a solution that will automatically disallow connections if an attack is suspected. Which technology should you choose?
   A. IDS
   B. IPS
   C. IRS
   D. SRTP

3. You are implementing a VPN using L2TP. You have chosen to use L2TP alone without any additional encryption protocols. What result will you achieve?
   A. Tunneled network traffic without security
   B. Tunneled network traffic with security
   C. Secure network traffic without a tunnel
   D. Insecure network traffic without a tunnel

4. Which of the following are types of firewalls?
   A. Application layer firewalls
   B. Packet filtering firewalls
   C. Stateful inspection firewalls
   D. Proxy firewalls

5. What is the number code associated with security standards for H.323 networks?
   A. 235
   B. 208
   C. 335
   D. 308

SELF TEST ANSWERS

Explain Concepts and Components of Security Design

1. ☑ **A, C,** and **D** are correct. Credential types include something you know, something you have, and something you are. An example of something you know is a password. An example of something you have is a smart card. An example of something you are is a thumb scanner.
   ☒ **B** is incorrect.

2. ☑ **B** is correct. An intrusion prevention system (IPS) can take automatic actions in order to prevent a network attack.
   ☒ **A, C,** and **D** are incorrect.

3. ☑ **A** is correct. The Layer 2 Tunneling Protocol (L2TP) does not provide encryption when used alone. The result of this scenario will be tunneled traffic with no security.
   ☒ **B, C,** and **D** are incorrect.

4. ☑ **A, B, C,** and **D** are correct. All four of these are valid firewall types.
   ☒ **None** are incorrect.

5. ☑ **A** is correct. H.235.0 through H.235.9 currently define security standards for use on H.323-based VoIP networks.
   ☒ **B, C,** and **D** are incorrect.

A

About the CD

T he CD-ROM included with this book comes complete with MasterExam and the electronic version of the book. The software is easy to install on any Windows 2000/ XP/Vista computer and must be installed to access the MasterExam feature. You may, however, browse the electronic book directly from the CD without installation. To register for a second bonus MasterExam, simply click the Online Training link on the Main Page and follow the directions to the free online registration.

System Requirements

Software requires Windows 2000 or higher and Internet Explorer 6.0 or above and 20MB of hard disk space for full installation. The electronic book requires Adobe Acrobat Reader.

Installing and Running MasterExam

If your computer CD-ROM drive is configured to auto run, the CD-ROM will automatically start up upon inserting the disk. From the opening screen you may install MasterExam by pressing the MasterExam button. This will begin the installation process and create a program group named LearnKey. To run MasterExam use Start | All Programs | LearnKey | MasterExam. If the auto run feature did not launch your CD, browse to the CD and click on the LaunchTraining.exe icon.

MasterExam

MasterExam provides you with a simulation of the actual exam. The number of questions, the type of questions, and the time allowed are intended to be an accurate representation of the exam environment. You have the option to take an open book exam, including hints, references, and answers; a closed book exam; or the timed MasterExam simulation.

When you launch MasterExam, a digital clock display will appear in the bottom right-hand corner of your screen. The clock will continue to count down to zero unless you choose to end the exam before the time expires.

Electronic Book

The entire contents of the Study Guide are provided in PDF. Adobe's Acrobat Reader has been included on the CD.

Help

A help file is provided through the help button on the main page in the lower left hand corner. An individual help feature is also available through MasterExam.

Removing Installation(s)

MasterExam is installed onto your hard drive. For best results removing programs, use the Start | All Programs | LearnKey | Uninstall option to remove MasterExam.

Technical Support

For questions regarding the technical content of the electronic book or MasterExam, please visit www.mhprofessional.com or e-mail customer.service@mcgraw-hill.com. For customers outside the 50 United States, e-mail international_cs@mcgraw-hill.com.

LearnKey Technical Support

For technical problems with the software (installation, operation, removing installations), please visit www.learnkey.com, e-mail techsupport@learnkey.com, or call toll free at 1-800-482-8244.

Glossary

access control   The prevention of access to unauthorized resources by users and systems.

access device   A common term for a networking device through which client devices connect to a network, for example, a typical access point device.

ad hoc network   An alternate term often used to reference an independent basic service set (IBSS).

Advanced Encryption Standard (AES)   The encryption standard that replaced the Digital Encryption Standard (DES) in order to improve encryption strength. It is based on the Rijndael algorithm.

Application layer   The layer of the OSI model (*see* OSI model) that provides access to the lower OSI layers for applications and provides information to applications from the lower OSI model layers. Also known as Layer 7.

area code   A three-digit prefix identifying the geographic region to which calls should be routed.

authentication   A process that results in the validation or invalidation of user or system credentials.

authentication server   A device that provides 802.1X authentication services to an authenticator.

authenticator   A device at one end of a point-to-point LAN segment that facilitates the 802.1X authentication of the device at the other end.

bandwidth   Either the difference between the upper and lower frequencies used by a wireless channel, or the sheer number of bits per second that may be communicated through the channel, including all frame protocol overhead. Sometimes the number of payload bits that can be communicated through the channel in a second—more properly called throughput.

bit   An individual information element that can be equal to a 1 or a 0. A single bit can represent any two values to an application.

byte   A collection of bits. Usually eight bits in computer systems.

call admission control (CAC)   A system that allows you to manage VoIP call volumes when Quality of Service controls are insufficient.

call detail record (CDR)   A record of the details of a call, including time, duration, and source.

capacity   The amount that can be contained or managed. System capacity is usually a measurement of storage space or data space. In communications systems, capacity is a reference to the amount of data that can be transferred through the system in a given window of time.

central office (CO)   A location including one or more PSTN switches. The CO may be connected with other telephone systems.

channel   An instance of communications medium (radio frequency bandwidth) used to pass information between two communicating stations (STAs).

circuit switching   A switching method used to reserve a route or path between the two endpoints that need to communicate.

coaxial cable   Networking cable that is implemented with a center conductor made of copper surrounded by a shielding made of some type of plastic. An additional mesh coating surrounds the plastic shielding and is used for grounding. Also known as *coax*.

codec   An abbreviation for coder/decoder. A method used to convert analog signals to digital signals and vice versa.

contact center   A location or call center that processes PSTN, VoIP, e-mail, instant messaging, and other forms of customer contact.

convergence   The process of bringing voice, multimedia, and data communications together on a shared network.

core device   A common term for a networking device that neither connects to client devices nor provides services to the network other than packet forwarding.

coverage A term used to refer to the physical space covered by an access point. An administrator might say that she needs to provide coverage in the Accounting department and mean that she needs to be sure the RF signal from the access point is of acceptable quality in that physical area that contains the Accounting department.

Data Link layer The layer of the OSI model (*see* OSI model) that is responsible for physical network management such as the detection of errors in the physical media and for locating and transferring data on the physical medium such as Ethernet. This is where MAC frames and data reside. Also known as Layer 2.

data rate The instantaneous rate at which bits are communicated in a WLAN during a single-frame transmission. In a wired LAN, it is the rate at which bits are communicated. This rate may vary in a WLAN.

demilitarized zone (DMZ) A network, usually at the perimeter, that resides between a private network and a public network such as the Internet.

Denial of Service (DoS) An attack that is used to prevent valid users from accessing a network or system.

distribution device A common term for a networking device that does not connect to client devices, but provides services to the network such as packet filtering and forwarding; sometimes used to describe access point–to–access point bridging.

encapsulation The process of enveloping information within headers so that the information can be passed across varied networks.

encryption The process of obfuscating data so that it cannot be viewed by non-intended systems or individuals.

ENUM The E.164 Number Mapping protocol defined in RFC 3761. Translates E.164 numbers (standard telephone numbers) into SIP URIs.

Extensible Authentication Protocol (EAP) A standards-based model used to implement various authentication types such as certificates and preshared keys or passphrases.

FCC   The Federal Communications Commission (FCC) is responsible for defining limitations and allowances for radio frequency communications—among other things—in the United States and its territories. This agency defines the regulations that are then implemented in IEEE and other standards.

fiber optic cable   A high-speed cabling technology that transmits light across glass fibers instead of electricity across copper wires.

fragmentation   The process of converting a single frame into multiple smaller frames in order to reduce retransmission overhead when occasional interference corrupts one part of the overall transmission in a WLAN. Also, the process of converting large TCP packets into multiple IP payloads for transmission in wired or wireless LANs.

frequency   The rate at which an RF wave, or any wave, repeats itself, commonly measured in hertz, MHz, or GHz.

gatekeeper   A device used on converged networks. Provides switching for VoIP communications.

H.323   A suite of protocols designed to support voice and multimedia communications across unreliable networks like TCP/IP.

IEEE   The Institute of Electrical and Electronics Engineers specifies standards based on regulations defined by regulatory bodies.

IEEE 802.11   The standard that defines the use of radio frequency signals to implement wireless LANs.

IEEE 802.1X   A standard, independent of the IEEE 802.11 standard, that defines port-based authentication. This standard is reference by IEEE 802.11 as being used to implement a robust security network (RSN).

interference   That which occurs when RF energy in the same frequency corrupts RF communications.

Internet Protocol (IP)   A protocol used for communicating data across a packet-switched internetwork. IP has the task of delivering packets from the source host to the destination host based on its address. IP defines addressing methods and structures for datagram encapsulation.

IVR (interactive voice response)   A term used to describe a system that can provide prompt and collect capabilities for callers in menu trees or self-service applications.

LNP   Local number portability (LNP) allows a subscriber to switch telephone service providers while retaining the same phone number. LNP can also be used when switching among VoIP service providers.

local exchange   Everything from the customer location to the central office.

MOS   The mean opinion score (MOS) is a subjective rating of the quality of a given multimedia or VoIP codec. Scores range from 1 (very bad) to 5 (excellent). A score of 4.0 or higher is usually considered very good.

network   A group of connected or interconnected people or things. In computer networks, it is a group of connected or interconnected computer systems.

Network layer   The layer of the OSI model (*see* OSI model) that is responsible for actual data transfer between logical devices. The IP protocol lives here. Also known as Layer 3.

network protocol   A collection of rules, recommendations, and options used to facilitate communications between computing devices.

octet   A term that describes eight bits of data. A byte is generally thought to be eight bits, but a byte can be fewer than eight bits and it can be more than eight bits. An octet is specifically an eight-bit byte.

OSI model   The Open Systems Interconnection (OSI) model provides a common basis for the purpose of system interconnection and includes a seven-layer approach. (*See* Application layer, Presentation layer, Session layer, Transport layer, Network layer, Data Link layer, and Physical layer.)

Quality of Service (QoS)   A term applied to a required level of quality. QoS is implemented through different technologies such as DiffServ, ToS, MPLS, and RSVP.

packet switching   A switching method used to segment a message into small parts and then send those parts across a shared network. Unlike circuit switching, a dedicated connection is not required. Also known as datagram switching.

Physical layer   The layer of the OSI model (*see* OSI model) that provides the actual transfer of bits on the network medium. Also known as Layer 1.

Power over Ethernet   A standard method for providing power to network devices over Ethernet cables. Also called PoE.

Presentation layer   The layer of the OSI model (*see* OSI model) that provides presentation services such as encryption and syntax management. Also known as Layer 6.

private branch exchange (PBX)   A phone switch located within and owned by a private organization. A PBX will usually connect to the PSTN, or it may connect through the Internet to other PBX systems.

protocol   *See* network protocol.

protocol analyzer   A tool used to decode packets on a network.

public switched telephone network (PSTN)   The PSTN is the public telephone network. Also known as plain old telephone system (POTS).

Quality of Service (QoS)   A variety of methods to provide different priority levels to various applications, users, or data flows. It can also be used to guarantee a certain level of performance to a data flow.

Real Time Protocol (RTP)   A standardized packet format for delivering audio and video over a network.

routing   The process of moving data packets from one network to another.

segmentation   The process of segmenting or separating the data into manageable or allowable sizes for transfer. Also known as fragmentation in the context of the lower levels of the OSI model.

Session layer   The layer of the OSI model (*see* OSI model) that provides for session initiation and management. This layer provides for connections between applications on a network. The functions of the Session layer are handled by TCP in the TCP/IP suite, or they may be handled by upper-layer protocols. Also known as Layer 5.

SIP   The session initiation protocol (SIP) is a call control protocol and is an alternative to H.323 for voice and multimedia communications.

throughput   The rate at which payload data can be transferred through a system.

traffic shaper   A device that allows network managers to control the data on their network more granularly. The manager may be able to block or slow particular types of traffic.

Transmission Control Protocol (TCP)   One of the core protocols of the Internet Protocol Suite. TCP operates at a higher level than IP, concerned only with the two end systems, for example an e-mail client and an e-mail server. TCP provides reliable, ordered delivery of a stream of bytes from one program on one computer to another program on another computer. TCP is capable of controlling message size, the rate at which messages are exchanged, and network traffic congestion.

Transport layer   The layer of the OSI model (*see* OSI model) that provides for data transport. This is the layer that most resembles the Transport layer of the TCP/IP model. Also known as Layer 4.

twisted pair cables   Networking cables that are implemented using multiple conductor cables. These cables are twisted in pairs. Both unshielded and shielded twisted pair cables exist.

User Datagram Protocol (UDP)   One of the core protocols of the Internet Protocol Suite. Using UDP, programs on networked computers can send each other short messages sometimes known as datagrams using Datagram Sockets. UDP does not guarantee reliability or packet ordering in the way that TCP does. Datagrams may arrive out of order, appear duplicated, or go missing without notice. Avoiding the overhead of checking whether every packet arrived makes UDP faster and more efficient for applications that do not need guaranteed delivery, such as voice.

virtual private network (VPN)   A session between two endpoints that encrypts all data transmitted across that session. Data is routed through the session as if it were a physical connection.

VLAN   A virtual local area network (VLAN) is a virtual network segment enabled through a Layer 2 switch that supports VLAN protocols. Nodes from many physical network segments are made to appear as if they were on the same segment by the VLAN switch.

VoIP   Voice over IP uses the traditional IP network to send voice data for communications.

VoWLAN   IP telephony over the wireless LAN is the use of a WLAN to transport IP voice communications.

Wi-Fi Alliance   An organization that certifies equipment to be interoperable with other equipment in the WLAN industry based on their certification standards.

INDEX

Note: Page numbers referencing figures are italicized and followed by an "*f*". Page numbers referencing tables are italicized and followed by a "*t*".

NUMERALS

C

D

F

T